Electronic Circuit Analysis

Fourth Edition

Prof. K. Lal Kishore

M.Tech, Ph.D, SMIEEE, FIETE, FIE, FAPAS, LMISTE, LMISHM

Vice-Chancellor

JNTU, Anantapur

Ananthapuramu – 515 002

BSP **BS Publications**

A unit of **BSP Books Pvt., Ltd.**

4-4-309/316, Giriraj Lane, Sultan Bazar,

Hyderabad - 500 095

Phone : 040 - 23445605, 23445688

Electronic Circuit Analysis, Fourth Edition ***by K. Lal Kishore***

Published by :

BSP BS Publications
A unit of **BSP Books Pvt., Ltd.**

4-4-309/316, Giriraj Lane, Sultan Bazar,
Hyderabad - 500 095
Phone : 040 - 23445605, 23445688
e-mail : info@bspbooks.net

ISBN : 978-93-85433-14-6 (HB)

DEDICATED TO

Sri ***Saraswati***

the goddess of knowledge

Preface to Fourth Edition

I am happy that 4th edition of the book "Electronic Circuit Analysis" is brought out. In this edition many changes are made, as per the requirements of the university syllabus. The topics cover the requirements of many university teachers and students. In some chapters, certain topics are repeated to have continuity, for convenience of readers, so that students can prepare chapter wise. New topics have been added as per the changes made in the syllabus of some universities.

For every chapter, the topics covered are listed and scope is given under the caption "In this chapter". At the end of every chapter, summary of the chapter is given briefly. Objective type questions with answers and essay type questions are also given for every chapter. This will enable students to prepare for examinations very well. This textbook will also be very useful for preparing for competitive examinations like Engineering services exam, GATE exam and other entrance examinations for the recruitment of various jobs. Answers to some GATE examination questions are also given.

This book is also useful for AMIETE, AMIE (Electronics), M.Sc (Electronics), Polytechnique and Diploma programs in electronics and B.Sc (Electronics) courses.

Though numbers of textbooks are available in electronics, there is no single textbook, covering all the topics mentioned here and dealing with topics in clear manner, so that it is useful to teachers as well as students. Suggestions to further improve the textbook are welcome.

The author thanks Prof. M.N. Giriprasad, HOD, ECE Department, JNTUCEA, Ms. J. Anusha, and Mr. C. Ravishankar Reddy, Academic Assistants of the Department for helping in the revision of the Textbook. The author also thanks the management of M/s. BS Publications Mr. Naresh Davergave, of BS Publications and all those who have helped in bringing out this book.

- Author

Preface to Third Edition

I am happy that the 3rd edition of the book "Electronic Circuit Analysis" is being broughtout. In this edition many changes are made, as per the changes made in the university syllabus. The topics meet the requirements of many university teachers and students. In some chapters, certain topics are repeated to have continuity, for convenience of readers and as per syllabus given, so that students can prepare chapter wise.

For every chapter, the topics covered are listed and scope is given under the title "In this chapter". At the end of every chapter, summary of the chapter is given briefly. Objective type questions with answers and essay type questions are also given for every chapter. This will enable students to prepare for examinations very well. This textbook will also be very useful for preparing for competitive examinations like Engineering services exam, GATE exam and other entrance examinations for the recruitment of various jobs.

This book is also useful for AMIETE, AMIE (Electronics), M.Sc (Electronics), Polytechnique and Diploma programs in electronics and B.Sc (Electronics) courses.

Number of textbooks are referred in writing this textbook. The author thanks publishers and authors of those textbooks.

Though number of textbooks are available in electronics, there is no single textbook available, covering all the topics mentioned here and dealing with the topics in clear manner so that it is useful to teachers as well as students. Suggestions to further improve the textbook are welcome.

The author thanks management of M/S. BS Publications Mr. Naresh Davergave, of BS Publications and all those who have helped in bringing out this book.

- Author

CONTENTS

CHAPTER 1

Single Stage Amplifiers

CHAPTER 2

Multistage Amplifiers

CHAPTER 3

BJT - Amplifiers, Frequency Response

CHAPTER 4

MOS Amplifiers

Chapter 5

Feedback Amplifiers

CHAPTER 6

Oscillators

CHAPTER 7

Large Signal Amplifiers

CHAPTER 8

Tuned Amplifiers

SYMBOLS

$g_{b'e}$	:	Input Conductance of BJT in C.E configuration between fictitious base terminal B' and emitter terminal E.
h_{ie}	:	Input impedance (resistance) of BJT in C.E configuration
h_{fe}	:	Forward short circuit current gain in C.E configuration
h_{re}	:	Reverse voltage gain in C.E. configuration
h_{oe}	:	Output admittance in C.E. configuration
$r_{bb'}$	:	Base spread resistance between base terminal B and fictitious base terminal B'.
C_e	:	Emitter junction capacitance
C_C	:	Collector junction capacitance
g_m	:	Transconductance or Mutual conductance
V_T	:	Volt equivalent of temperature $\frac{KT}{e} = \frac{T}{11,600}$
η	:	Diode constant $\eta = 1$ for G_e; $\eta = 2$ for S_i
$g_{b'c}$	:	Feedback conductance between B' and collector terminal C
g_{ce}	:	Output conductance between Collector and Emitter terminals.
C_D	:	Diffusion capacitance
Q	:	Charge
D_B	:	Diffusion constant for minority carriers in Base region
n	:	constant (= 1/2 for abrupt junctions)
W	:	Base width
ω	:	Angular frequency = $2\pi f$
f_T	:	Frequency at which C.E. short circuit current gain becomes unity
f_β	:	Frequency at which h_{fe} becomes 0.707 $h_{fe\ max}$. Frequency range upto f_β is referred as the B.W of the transistor circuit.

C_π : Incrimental capacitance in hybrid - π model

r_π : Incrimental resistance in hybrid - π model

A_{V_1} : Voltage gain of I stage amplifier circuit

A_{I_1} : Current gain of I stage amplifier circuit

B.W : Band width of the amplifier circuit.

f_C : Cut-off frequency

$f_L = f_1$: Lower cutoff frequency or Lower 3-db point or Lower half power frequency

$f_H = f_2$: Upper cutoff frequency or upper 3-db point or upper half power frequency

f_0 : Mid Band Frequency $f_O = \sqrt{f_1 f_2}$.

R_E : Emitter Resistor

C_E : Emitter Capacitor

$A_V(L.F)$: A_{V_L} = Voltage gain in the Low frequency range

$A_V(H.F)$: A_{V_H} = Voltage gain in the High frequency range

$A_V(M.F)$: A_{V_M} = Voltage gain in the Mid frequency range

P_O : Output Power

P_i : Input Power

A_P : Power Gain

ϕ : Phase angle

V_y : RMS value of voltage

I_y : RMS value of current

I_m : $(I_{Max} - I_{Min})$

V_m : $(V_{Max} - V_{Min})$

I_{P-P} : Peak to Peak value of current

P_{ac} : A.C. Output power

P_{DC} : DC Input power

η : Conversion Efficiency of the power amplifier circuit.

n : Transformer turns ratio (N_2/N_1)

N_2 : Number of turns of transformer Secondary winding

N_1 : Number of turns of transformer Primary winding

V_1 : Primary voltage of Transformer

V_2 : Secondary voltage of Transformer

R_t : Resistance of Tuned Circuit

R_P : Parallel resistance associated with the tuning coil (Inductor)

V_{be} : A.C voltage between base and emitter leads of transistor (BJT)

V_{BE} : D.C voltage between base and emitter leads of transistor (BJT)

Small subscripts are used for a.c. quantities.

Capital subscripts are used for a d.c. quantities.

Q_e	:	Effective Q factor of coil
δ	:	Fractional Frequency Variation
R_{tt}	:	Resistance of tapped tuned Circuit
Q_O	:	Quality factor of output circuit
M	:	Mutual Inductance
K_C	:	Critical value of the coefficient of coupling
M_C	:	Critical value of Mutual Inductance
R_S	:	Series Resistance in Voltage Regulators
S	:	Stability factor
S_T	:	Temperature coefficient in Voltage Regulator
R_O	:	Output Resistance
R_Z	:	Zener Diode Resistance
V_γ	:	Cut in voltage of junction diode
$V_O'{}_{(P-P)}$	:	Output ripple voltage
$V_i'{}_{(P-P)}$	:	Input ripple voltage

Brief History of Electronics

In science, we study about the laws of nature and verification and in technology, we study the applications of these laws to human needs.

Electronics is the science and technology of the passage of charged particles, in a gas or vacuum or semiconductor.

Before electronic engineering came into existence, electrical engineering flourished. Electrical engineering mainly deals with motion of electrons in metals only, where as Electronic engineering deals with motion of charged particles (electrons and holes) in metals, semiconductors and also in vacuum. Another difference is, in electrical engineering, the voltages and currents are very high KV, and Amperes, where as in electronic engineering one deals with few volts and mA. Yet another difference is, in electrical engineering; the frequencies of operation are 50 Hzs/60Hzs. In electronics it is KHzs, MHz, GHzs, (high frequency).

The beginning for Electronics was made in 1895 when H.A. Lorentz postulated the existence of discrete charges called *electrons*. Two years later, *J.J.Thomson* proved the same experimentally in 1897.

In the same year that is in 1897, Braun built the first tube based on the motion of electrons, the Cathode ray tube (CRT).

In 1904 Fleming invented the Vacuum diode called '*valve*'.

In 1906 a semiconductor diode was fabricated but they could not succeed, in making it work. So semiconductor technology met with premature death and vacuum tubes flourished.

In 1906 it self, De Forest put a third electrode into Fleming's diode and he called it Triode. A small change in grid voltage produces large change in plate voltage, in this device.

In 1912 Institute of Radio Engineering (IRE) was set up in USA to take care of the technical interests of electronic engineers. Before that in 1884 Institute of Electrical Engineers was formed and in 1963 both have merged into one association called IEEE (Institute of Electrical and Electronic Engineers).

The first radio broadcasting station was built in 1920 in USA.

In 1930 black and white television transmission started in USA.

In 1950 Colour television broadcasting was started.

The electronics Industry can be divided into 4 categories :

Components	:	Transistors, ICs, R, L, C components
Communications	:	Radio, TV, Telephones, wireless, landline communications
Control	:	Industrial electronics, control systems
Computation	:	Computers

Vacuum Tubes ruled the electronic field till the invention of transistors. The difficuty with vacuum tubes is with its excess generated heat. The filaments get heated to > 2000^0 K so that electronic emission takes place. The filaments get burnt and tubes occupy large space. So in 1945 Solid State Physics group was formed to invent semiconductor devices in Bell labs, USA.

1895 : H. A. Lorentz - Postulated existence of Electrons

1897 : J.J. Thomson - Proved the same

1904 : Fleming - Vacuum Diode

1906 : De. Forest - Triode

1920 : Radio Broadcasting in USA

1930 : Black and White TV USA

1947 : Shockley invented the junction transistor. (BJT)

1947 : Schokley BJT Invention

1950 : Colour Television

1959 : Integrated Circuit concept was announced by Kilby at an IRE convention.

1959 : KILBY etc. anounced ICs.

1969 : LSI, IC : Large Scale Integration, with more than 1000 but < 10,000 components per chip (integrated or joined together), device was announced.

1969 : SSI 10 - 100 comp/chip. LOGIC GATES, FFs.

1970 : Intel People, 9 months, chip with 1000 Transistors (4004μp)

1971 : μP - 4 bit INTEL

1971 : 4 bit Microprocessor was made by Intel group.

1975 : VLSI : Very large scale integration > 10,000 components per chip. ICs were made.

1975 : CHMOS - Complimentary High Metal Oxide Semiconductor ICs were announced by Intel.

1975 : MSI (Multiplenum, Address) 100 - 1000 comp/chip

1978 : LSI 8 bit μPs, ROM, RAM 1000 - 10,000 comp/chip

1980 : VLSI > 1,00,000 components/ser 16, 32 bit μPs

1981 : 16 bit μP > 1,00,000 components/ser 16, 32 bit μPs

1982 : 100,000 Transistors, 80286 Processor

1984 : CHMOS > 2,00,000 components/ser 16, 32 bit μPs

1985 : 32 bit μ p > 4,50,000 components/ser 16, 32 bit μPs

1986 : 64 bit μ p > 10,00,000 components/ser 16, 32 bit μPs

1987 : MMICS Monolithic Microwave Integrated Circuits

1989 : 1860 Intel's 64 bit CPU

1990 : ULSI > 500,000 Transistors ultra large scale

1992 : GSI > 10,00,000 Transistors Giant scale

100, 3 million Transistors, Pentium

1998 : 2 Million Gates/Die

2001 : 5 Million Gates / Die

2002 : 1000, 150 Million Transistors.

1 Gigabit Memory Chips

Nature is more SUPERIOR

2003 : 10 n.m. patterns, line width

2004 : Commercial Super Comp. 10TRILLION Flip Flops

2010 : Neoro - Computer Using Logic Structure Based on Human Brain

There are 10^7 cells/cm^3 in human brain

VLSI Technology Development :

3 μ Technology

↓

0.5 μ Technology

↓

0.12 μ Technology

ASICs (Application Specific Integrated Circuits)

HYBRID ICs

BI CMOS

MCMs (Multi Chip Modules)

3-D packages

Table showing predictions made by (International Technology Road map for Semiconductors) 1995 on VLSI Technology

	1995	1998	2001	2004	2007
Lithography (μ)	0.35	0.25	0.18	0.12	0.1
No. Gates/Die :	800K	2 M	5 M	10 M	20 M
No. Bits/Die					
Dram	64 M	256	1 G	4 G	16 G
Sram	16 M	64 M	256 M	1G	4 G
Wafer Dia (mm)	200	200-400	–400	–400	–400
Power (μW/Die)	15	30	40	40-120	40-200
Power Supply. V.	3.3	2.2	2.2	1.5	1.5
Frequency MHz	100	175	250	350	500

ITRS Predictions

Year	Tech. node (nm)	Number of transistors (M) Millions (M)	Number of wire levels	Freq. (MHz)	V_{dd} (V)	Power (W)
2001	130	89	7	1684	1.1	130
2002	115	112	7-8	2317	1.0	140
2003	100	142	8	3088	1.0	150
2004	90	178	8	3990	1.0	160
2005	80	225	8-9	5173	0.9	170
2006	70	283	9	5631	0.9	180
2007	65	357	9	6739	0.7	190
2010	45	714	9-10	11511	0.6	218
2013	32	1427	9-10	19348	0.5	251
2016	22	2854	10	28751	0.4	288

1 Single Stage Amplifiers

In this Unit,

- Single stage amplifiers in the three configurations of C.E, C.B, C.C, with design aspects are given.
- Using the design formulae for A_V, A_I, R_i, R_o etc., the design of single stage amplifier circuits is to be studied.
- Single stage JFET amplifiers in C.D, C.S and C.G configurations are also given.
- The Hybrid - π equivalent circuit of BJT, expressions for Transistor conductances and capacitances are derived.
- Miller's theorem, definitions for f_β and f_T are also given.
- Numerical examples, with design emphasis are given.

Introduction

An electronic amplifier circuit is one, which modifies the characteristics of the input signal, when delivered the output side. The modification in the characteristics of the input signal can be with respect to voltage, current, power or phase. Any one or all these characteristics power, or phase may be changed by the amplifier circuit.

1.1 CLASSIFICATION OF AMPLIFIERS

Amplifier circuits are classified in different ways as indicated below :

Types of Classification is Based on

(a) Frequency range

(b) Type of coupling

(c) Power delivered/conduction angle

(d) Signal handled.

(a) Frequency Range

AF (Audio Freq.)	:	40 Hzs – 15/20 kHz
RF (Radio Freq.)	:	> 20 kHz
Video Frequency	:	5 – 8 MHz
VLF (Very Low Freq.)	:	10 – 30 kHz
LF (Low Frequency)	:	30 – 300 kHz
Medium Frequency	:	300 – 3000 kHz
High Frequency	:	3 – 30 MHz
VHF (Very High Freq.)	:	30 – 300 MHz
UHF (Ultra High Freq.)	:	300 – 3000 MHz
SHF (Super High Freq.)	:	3000 – 30,000 MHz

(b) Types of Coupling

1. Direct coupled
2. RC coupled
3. Transformer coupled
4. LC Tuned Amplifiers
5. Series fed.

(c) Output power delivered/conduction angle

1. Low power (tens of mW or less).
2. Medium power (hundreds of mW).
3. High power (Watts).

Class A	360°
Class B	180°
Class AB	180 – 360°
Class C	< 180°
Class D	Switching type.
Class S	Switching type.

(d) Type of signal handled

1. Large signal   2. Small signal

In addition to voltage amplification A_V, current amplification A_I or power amplification A_P is expected from an amplifier circuit. The amplifier circuit must also have other characteristics like High input impedance (Z_i or R_i), Low output impedance (Z_o or R_o), Large Band Width (BW), High signal to Noise Ratio (S/N), and large *Figure of Merit (Gain BW product)*.

In order that the amplified signal is coupled to the load R_L or Z_L, for all frequencies of the input signal range, so that maximum power is transferred to the load, (the condition required for maximum power transfer is $|Z_0| = |Z_L|$ or $R_0 = R_L$) coupling the output of amplifier V_0 to load R_L or Z_L is important. When reactive elements are used in the amplifier circuit, and due to internal junction capacitances of the active device, the Z_i and Z_o of the amplifier circuit change with frequency. As the input signal frequency varies over a wide range, and for all these signals amplification and impedance matching have to be achieved, coupling of the output of the amplifier to the load is important.

Since the gain A_V, A_I or A_P that can be obtained from a single stage amplifier circuit where only one active device (BJT, JFET or MOSFET) is used, the amplifier circuits are cascaded to get large gain. Multistage amplifier circuits are discussed in the next chapter.

When the frequency of the input signal is high (greater than A.F. range) due to internal junction capacitances of the actual device, the equivalent circuit of the BJT used earlier is not valid. So another model of BJT valid for high frequencies, proposed by Giacoletto is studied in this chapter.

1.2 DISTORTION IN AMPLIFIERS

If the input signal is a sine wave the output should also be a true sine wave. But in all the cases it may not be so, which we characterize as distortion. Distortion can be due to the nonlinear characteristic of the device, due to operating point not being chosen properly, due to large signal swing of the input from the operating point or due to the reactive elements L and C in the circuit. Distortion is classified as :

(a) ***Amplitude distortion :*** This is also called non linear distortion or harmonic distortion. This type of distortion occurs in large signal amplifiers or power amplifiers. It is due to the nonlinearity of the characteristic of the device. This is due to the presence of new frequency signals which are not present in the input. If the input signal is of 10 KHz the output signal should also be 10 kHz signal. But some harmonic terms will also be present. Hence the amplitude of the signal (rms value) will be different $V_0 = A_V V_i$. But it will be V_0'.

(b) ***Frequency distortion :*** The amplification will not be the same for all frequencies. This is due to reactive component in the circuit.

(c) ***Phase-shift delay distortion :*** There will be phase shift between the input and the output and this phase shift will not be the same for all frequency signals. It also varies with the frequency of the input signal.

In the output signal, all these distortions may be present or any one may be present because of which the amplifier response will not be good.

Example : 1.1

The transistor parameters are given as,

$h_{ie} = 2$ k $h_{fe} = 50$ $h_{re} = 6 \times 10^{-4}$ $h_{oe} = 25$ μA/V

$h_{ic} = 2$ k $h_{fc} = -51$ $h_{rc} = 1$ $h_{oc} = 25$ μA/V

Find the individual as well as overall voltage gains and current gains. (Fig. 1.1(a)).

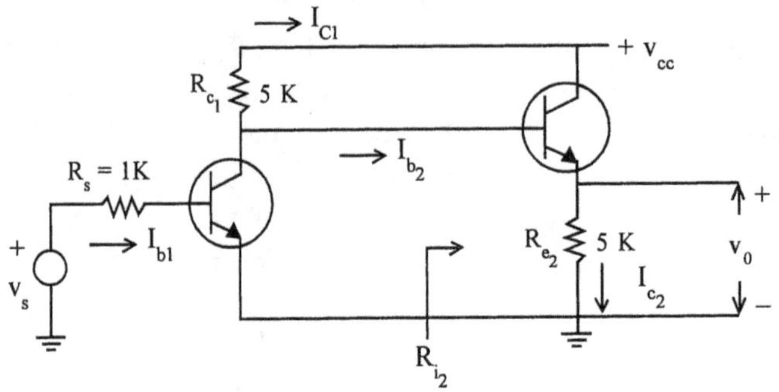

Fig. 1.1 (a) Circuit for Ex : 1.1

Solution :

It is advantageous to start the analysis with the last stage. Compute the current gain first, then the input impedance and voltage gain.

II stage : For the second stage $R_L = R_{e_2}$

$$A_{I_2} = \frac{-I_{e2}}{I_{b2}} = \frac{-h_{fc}}{1 + h_{oc}R_{e2}}$$

$$= \frac{51}{1 + 25 \times 10^{-6} \times 5 \times 10^3}$$

$$= \mathbf{45.3}$$

Input impedance $R_{i_2} = h_{ic} + h_{rc}\ A_{I_2}\ R_{e_2}$

$= 2 + 45.3 \times 5\text{k} = 228.5$ k.

∴ Input Z of the C.C. stage is very high.

Voltage gain of the second stage

$$A_{V_2} = \frac{V_0}{V_2} = A_{I_2} \frac{R_{e2}}{R_{i2}}$$

∵ $V_0 = I_{e_2} R_{e_2}$;

V_2 = Input voltage for the second stage = $R_{i_2} \cdot I_i$

∴
$$\frac{V_0}{V_2} = \frac{I_0 . R_{e2}}{I_i . R_{i2}}$$

$$= A_{I_2} \frac{R_{e2}}{R_{i_2}}$$

$$= \frac{45.3 \times 5}{228.5} = \mathbf{0.99}$$

I stage : For the first stage, the net load resistance in the parallel combination of R_{C_1} and R_{i_2} or

$$R_{i_2} = \frac{R_{c_1}.R_{i_2}}{R_{c_1} + R_{i_2}} = \frac{5 \times 228.5}{233.5} = 4.9 \text{ k}\Omega$$

Hence $$A_{I_1} = \frac{-I_{C_1}}{I_{b_1}} = \frac{-h_{fe}}{1 + h_{oe}R_{L_1}} = \frac{-50}{1 + 25 \times 10^{-6} \times 4.9 \times 10^3} = -44.5$$

The input impedance of the first stage will also be the input Z of the two stages since input Z of the second stage is also considered in determining the value of R_{L_1}. R_{i_1} depends on R_{L_1}.

$\therefore$ $R_{i_1} = h_{i_e} + h_{r_e} A_{I_1} R_{L_1}$ (from the standard formula)

$= 2 - 6 \times 10^{-4} \times 44.5 \times 4.9$

$R_{i_1} = 1.87 \text{ k}\Omega.$

Voltage gain of the first stage is,

$$A_{V_1} = \frac{V_2}{V_1} = \frac{A_{I_1}.R_{L_1}}{R_{i_1}}$$

$$= \frac{-44.5 \times 4.9}{1.87} = -116.6$$

$$A_I = \frac{-I_{e_2}}{I_{b_1}} = \frac{-I_{e_2}}{I_{b_2}} \cdot \frac{+I_{b_2}}{I_{c_1}} \cdot \frac{+I_{c_1}}{I_{b_1}}$$

$$= -A_{I_2} \cdot \frac{I_{b_2}}{I_{c_1}} . A_{I_1}$$

In the actual circuit, the current gets branched into I_{c1} and I_{b2} which depend upon the values of R_{C_1} and R_{i_2}. (Fig. 1.1(b)).

$\because$ $(-I_{b2} + I_{c1}) R_{c1} = I_{b2} \cdot R_{i2}$

$\therefore$ $+ I_{b2} (R_{i2} + R_{c1}) = I_{c1} \cdot R_{c1}$

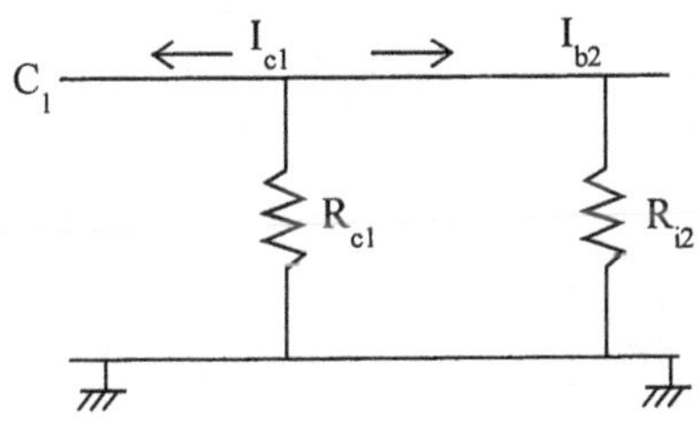

Fig. 1.1 (b) Current branching

$$\frac{I_{b2}}{I_{c1}} = \frac{R_{c1}}{R_{c1}+R_{i2}}$$

$$\therefore \quad A_I = A_{I2}\, A_{I1} \cdot \frac{R_{c1}}{R_{i2}+R_{c_1}}$$

$$= \frac{45.3\times(-44.5)\times 5}{228.5+5}$$

$$A_I = -43.2$$

$$A_V = \frac{V_0}{V_1} = \frac{V_0}{V_2} \cdot \frac{V_2}{V_1} = A_{V2} \cdot A_{V1}$$

$$A_V = 0.99 \times (-116.6) = -115.$$

$$\frac{I_{b2}}{I_{c1}} = \frac{R_{c1}}{R_{c1}+R_{i2}}$$

1.3 ANALYSIS OF CE, CC AND CB AMPLIFIERS

Small Signal Analysis means, we assume that the input AC signal peak to peak to amplitude is very small around the operating point Q as shown in Fig. 1.2. The swing of the signal always lies in the active region, and so the output is not distorted. In the *Large Signal Analysis,* the swing of the input signal is over a wide range around the operating point. The magnitude of the input signal is very large. Because of this the operating region will extend into the cutoff region and also saturation region.

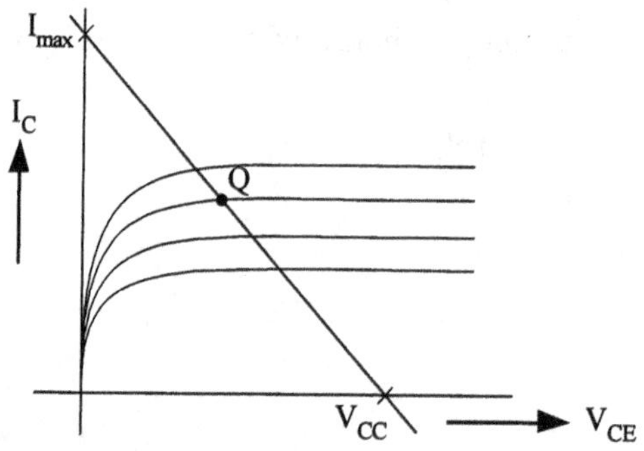

Fig. 1.2 Output characteristics of BJT

1.3.1 Common Emitter Amplifier

Common Emitter Circuit is as shown in the Fig. 1.3. The DC supply, biasing resistors and coupling capacitors are not shown since we are performing an *AC analysis*.

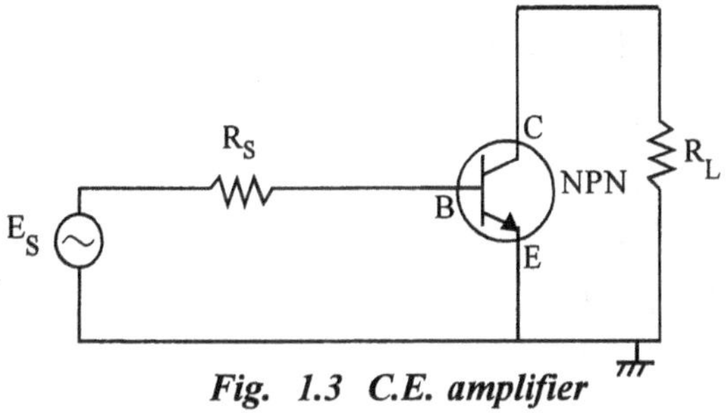

Fig. 1.3 C.E. amplifier

E_S is the input signal source and R_S is its resistance. The *h-parameter* equivalent for the above circuit is as shown in Fig. 1.4.

$$h_{ie} = \left.\frac{V_{be}}{I_b}\right|_{V_{ce}=0} \qquad h_{re} = \left.\frac{V_{be}}{V_{ce}}\right|_{I_b=0}$$

$$h_{oe} = \left.\frac{I_c}{V_{ce}}\right|_{I_b=0} \qquad h_{fe} = \left.\frac{I_c}{I_b}\right|_{V_{ce}=0}$$

The typical values of the *h-parameter* for a transistor in Common Emitter Configuration are,

$h_{ie} = 4\ k\Omega$,

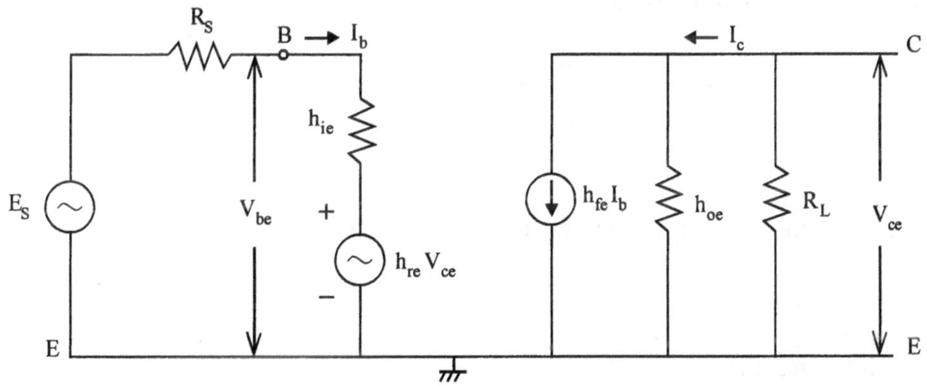

Fig. 1.4 h-parameter equivalent circuit

Since, $$h_{ie} = \frac{V_{be}}{I_b}$$

V_{be} is a fraction of volt 0.2V, I_b in μA, 100 μA and so on.

$$\therefore \qquad h_{ie} = \frac{0.2V}{50 \times 10^{-6}} = 4\ k\Omega$$

$$h_{fe} = I_c/I_b \simeq 100$$

I_C is in mA and I_B in μA.

$$\therefore \qquad h_{fe} >> 1 \simeq \beta$$

$h_{re} = 0.2 \times 10^{-3}$. Because, it is the *Reverse* Voltage Gain.

$$h_{re} = \frac{V_{be}}{V_{ce}}$$

and $$V_{ce} > V_{be};$$

$$h_{re} = \frac{\text{Input}}{\text{Output}}$$

Output is >> input, because amplification takes place. Therefore $h_{re} << 1$.

$$h_{oe} = 8\ \mu℧ \quad \text{and} \quad h_{oe} = \frac{I_c}{V_{ce}}$$

1.3.2 Input Resistance of the Amplifier Circuit (R_i)

The general expression for R_i in the case of Common Emitter Transistor Circuit is

$$R_i = h_{ie} - \frac{h_{fe}h_{re}}{h_{oe} + \dfrac{1}{R_L}} \qquad .....(1.1)$$

For Common Emitter Configuration,

$$R_i = h_{ie} - \frac{h_{fe}h_{re}}{h_{oe} + \dfrac{1}{R_L}} \qquad .....(1.2)$$

R_i depends on R_L. If R_L is very small, $\frac{1}{R_L}$ is large, therefore the denominator in the second term is large or it can be neglected.

$$\therefore \qquad R_i \cong h_{ie}$$

If R_L increases, the second term cannot be neglected.

$$R_i = h_{ie} - \text{(finite value)}$$

Therefore, R_i decreases as R_L increases. If R_L is very large, $\frac{1}{R_L}$ will be negligible compared to h_{oe}. Therefore, R_i remains constant. The graph showing R_i versus R_L is indicated in Fig. 1.5. R_i is not affected by R_L if $R_L < 1$ kΩ and $R_L > 1$ MΩ as shown in Fig. 1.5.

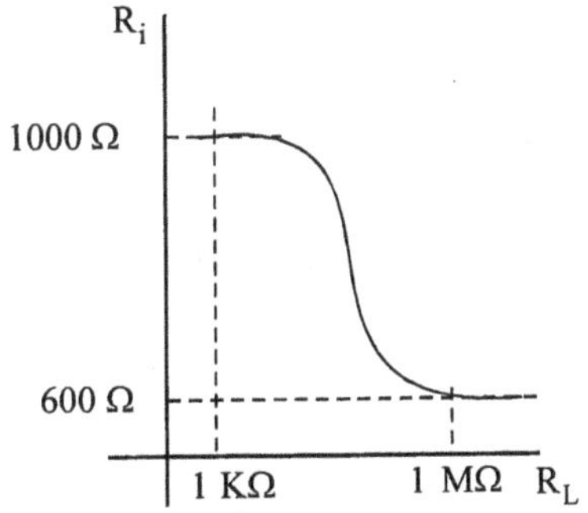

Fig. 1.5 Variation of R_i with R_L

R_i varies with frequency f because h-parameters will vary with frequency. h_{fe}, h_{re} will change with frequency f of the input signal.

1.3.3 Output Resistance of an Amplifier Circuit (R_0)

For Common Emitter Configuration,

$$R_o = \frac{1}{h_{oe} - \left(\frac{h_{re}h_{fe}}{h_{ie} + R_s}\right)} \qquad .....(1.3)$$

R_s is the resistance of the source. It is of the order of few hundred Ω.
R_o depends on R_s. If R_s is very small compared to h_{ie},

$$R_o = \frac{1}{h_{oe} - \frac{h_{re}h_{fe}}{h_{ie}}} \quad (\text{ independent of } R_s) \qquad .....(1.4)$$

Then, R_o will be large of the order of few hundred kΩ. If R_s is very large, then

$$R_o \simeq \frac{1}{h_{oe}} \simeq 150 \text{ k}\Omega.$$

The graph is as shown in Fig. 1.6.

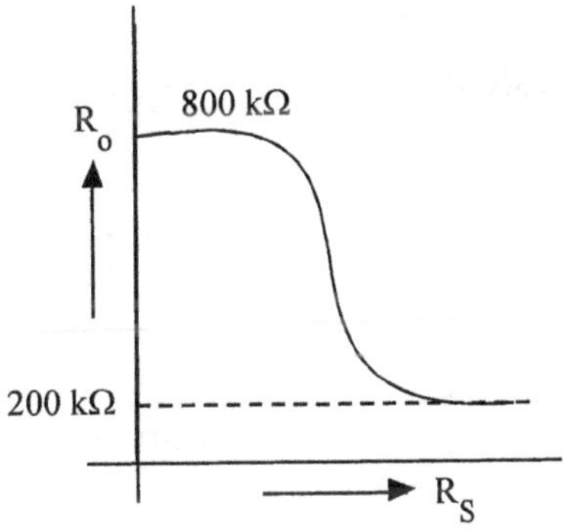

Fig. 1.6 Variation of R_o with R_S

1.3.4 Current Gain (A_i)

$$A_i = \frac{-h_{fe}}{1+h_{oe}R_L} \qquad .....(1.5)$$

If R_L is very small, $A_i \simeq h_{fe} \simeq 100$. So, Current Gain is large for Common Emitter Configuration. As R_L increases, A_i drops and when $R_L = \infty$, $A_i = 0$. Because, when $R_L = \infty$, output current I_0 or load current $I_L = 0$. Therefore, $A_i = 0$. Variation of A_i with R_L is shown in Fig. 1.7.

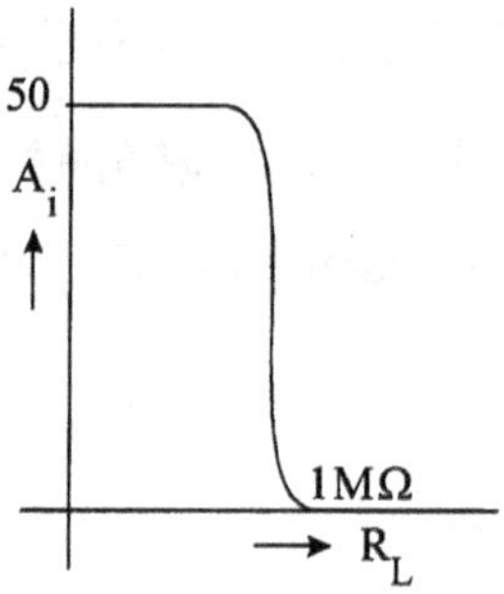

Fig 1.7 Variation of A_i with R_L

1.3.5 Voltage Gain (A_v)

$$A_V = \frac{-h_{fe}R_L}{h_{ie} + R_L(h_{ie}h_{oe} - h_{fe}h_{re})} \qquad .....(1.6)$$

If R_L is low, most of the output current flows through R_L. As R_L increases, output voltage increases and hence A_V increases. But if $R_L >> \frac{1}{h_{oe}}$, then the current from the current generator in the ***h-parameters*** equivalent circuit flows through h_{oe} and not R_L.

Then the, Output Voltage $= h_{fe} \cdot I_b \cdot \frac{1}{h_{oe}}$

(R_L is in parallel with h_{oe}. So voltage across h_{oe} = voltage across R_L). Therefore, V_0 remains constant as output voltage remains constant (Fig.1.8).

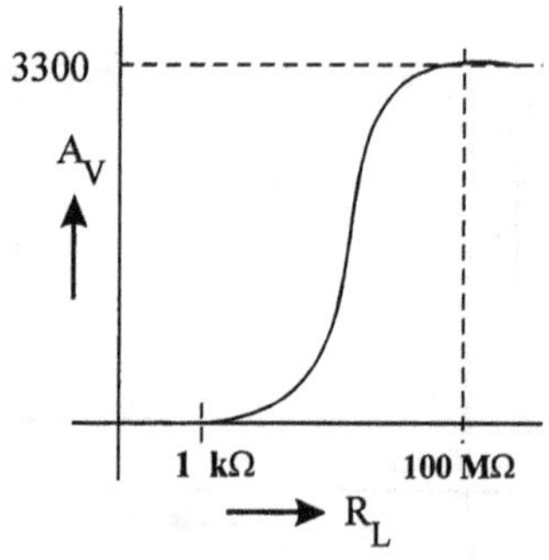

Fig. 1.8 Variation of A_v with R_L

1.3.6 Power Gain (A_P)

As R_L increases, A_I decreases. As R_L increases, A_V also increases.

Therefore, Power Gain which is the product of the two, A_V and A_I varies as shown in Fig. 1.9.

$$A_P = A_V A_I$$

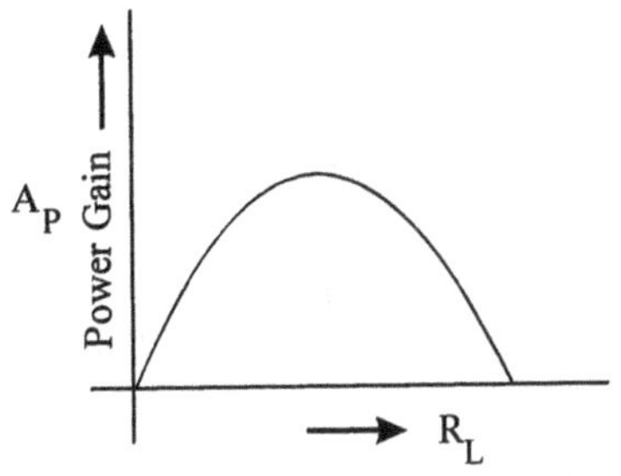

Fig. 1.9 Variation of A_P with R_L

Power Gain is maximum when R_L is in the range 100 kΩ – 1 MΩ i.e., when R_L is equal to the output resistance of the transistor. Maximum power will be delivered, under such conditions.

Therefore, it can be summarised as, Common Emitter Transistor Amplifier Circuit will have,

1. *Low to Moderate Input Resistance (300 Ω – 5 kΩ).*
2. *Moderately High Output Resistance (10 kΩ – 100 kΩ).*
3. *Large Current Amplification.*
4. *Large Voltage Amplification.*
5. *Large Power Gain.*
6. *180° phase-shift between input and output voltages.*

As the input current I_B, increases, I_C increases therefore drop across R_C increases and $V_0 = V_{CC} - V_I$ drop across R_C. Therefore, there is a phase shift of 180°.

The amplifier circuit is shown in Fig. 1.10.

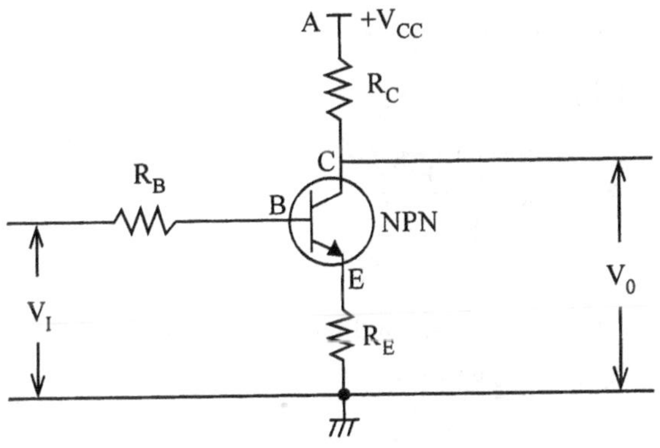

Fig. 1.10 CE amplifier circuit

1.3.7 Common Base Amplifier

The circuit diagram considering ***only AC*** is shown in Fig. 1.11.

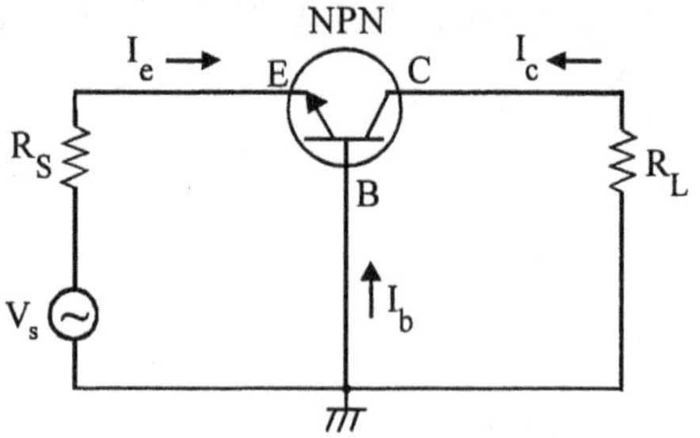

Fig. 1.11 CB amplifier

$$h_{ib} = \left.\frac{V_{eb}}{I_e}\right|_{V_{cb}=0}$$

V_{eb} is small fraction of a volt. I_e is in mA. So, h_{ib} is small.

$$h_{fb} = \left.\frac{I_c}{I_e}\right|_{V_{cb}=0} = -0.99 \text{ (Typical Value)}$$

$$I_c < I_e \qquad \therefore \qquad h_{fb} < 1$$

$$h_{ob} = \left.\frac{I_c}{V_{cb}}\right|_{I_e=0} = 7.7 \times 10^{-8} \text{ mhos (Typical Value)}$$

I_c will be very small because $I_e = 0$. This current flows in between base and collector loop.

$$h_{rb} = \left.\frac{V_{eb}}{V_{cb}}\right|_{I_e=0} = 37 \times 10^{-6} \text{ (Typical Value)}$$

h_{rb} is small, because V_{eb} will be very small and V_{cb} is large.

1.3.8 Input Resistance (R_i)

$$R_i = h_{ib} - \frac{h_{fb} \cdot h_{rb}}{h_{ob} + \frac{1}{R_L}}; \quad h_{fb} \text{ is } -ve \qquad .....(1.7)$$

when R_L is small < 100 kΩ, the second term can be neglected.

$$\therefore \qquad R_i = h_{ib} \simeq 30\ \Omega.$$

when R_L is very large, $\frac{1}{R_L}$ can be neglected.

$$R_i = h_{ib} - \frac{h_{fb}.h_{rb}}{h_{ob}}$$

So $R_i \simeq 500\ \Omega$ (Typical value) [$\because h_{fb}$ is negative]

$$\therefore \qquad R_i = h_{ib} + \frac{h_{fb}h_{rb}}{h_{ob}}$$

The variation of R_i with R_L is shown in Fig. 1.12. R_i varies from 20 Ω to 500 Ω.

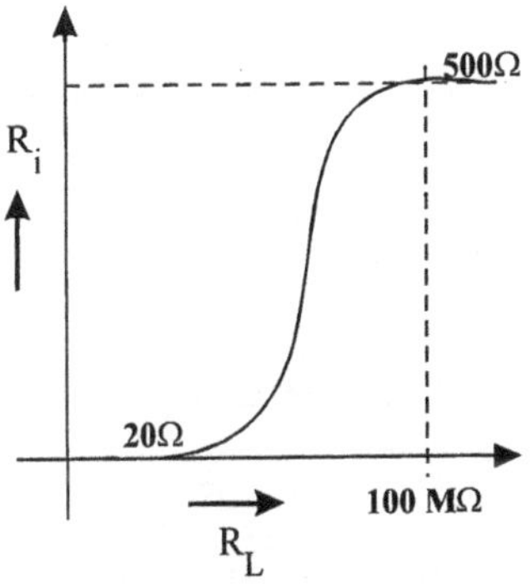

Fig. 1.12 Variation of R_i with R_L

1.3.9 Output Resistance (R_o)

$$R_o = \frac{1}{h_{ob} - \dfrac{h_{rb}h_{fb}}{h_{ib} + R_S}} \qquad .....(1.8)$$

If R_s is small, $R_o = 1\Big/\left(h_{ob} - \dfrac{h_{rb}h_{fb}}{h_{ib}}\right)$

But h_{fb} is negative.

$$\therefore \qquad R_o = \frac{1}{h_{ob} + \dfrac{h_{rb}h_{fb}}{h_{ib}}}$$

This will be sufficiently large, of the order of 300 kΩ. Therefore, value of h_{ob} is small. As R_s increases, $R_o = \dfrac{1}{h_{ob}}$ also increases. [This will be much larger because, in the previous case, in the denominator, some quantity is subtracted from h_{ob}.]

$$\therefore \qquad R_o = 12\ \text{M}\Omega$$

The variation of R_0 with R_s is shown in Fig. 1.13.

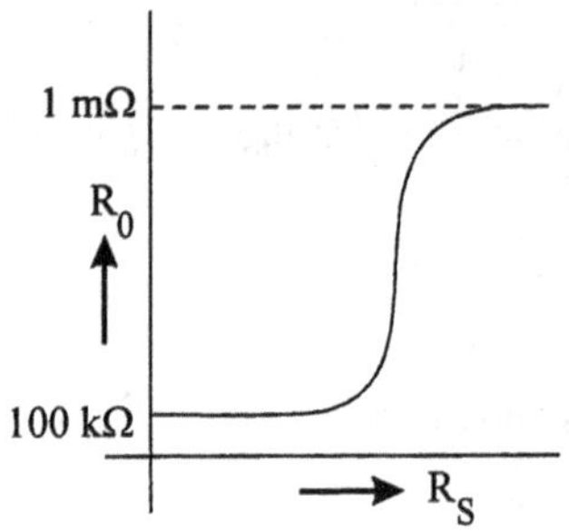

Fig. 1.13 Variation of R_0 with R_S

1.3.10 CURRENT GAIN (A_i)

$$A_i = \frac{-h_{fb}}{1+h_{ob}R_L} \quad .....(1.9)$$

A_i is < 1. Because $h_{fe} < 1$. As R_L increases, A_i decreases. A_i is negative due to h_{fb}. The variation of A_i with R_L is shown in Fig. 1.14.

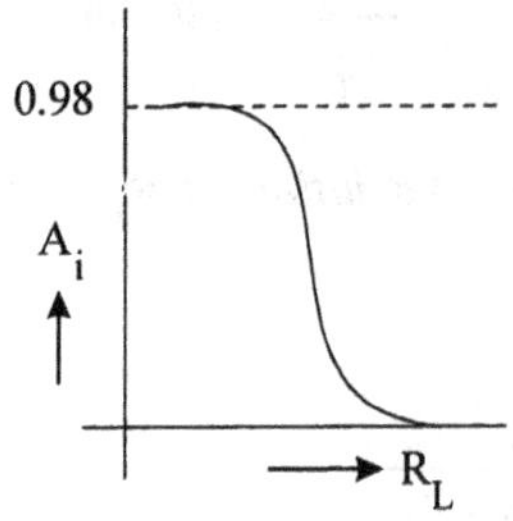

Fig. 1.14 Variation of A_i with R_L

1.3.11 VOLTAGE GAIN (A_v)

$$A_v = \frac{-h_{fb}R_L}{h_{ib}+R_L(h_{ib}h_{ob}-h_{fb}h_{rb})} \quad ....(1.10)$$

As R_L increases, A_V also increases. If R_L tends to zero, A_V also tends to zero. ($A_V \rightarrow 0$, as $R_L \rightarrow 0$). The variation of Voltage Gain A_V with R_L is shown in Fig. 1.15.

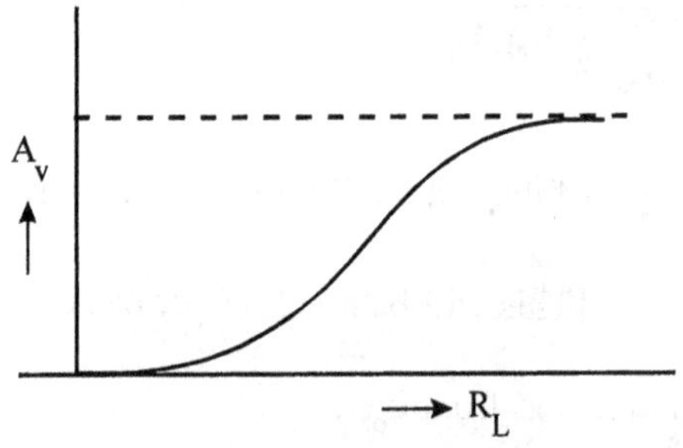

Fig. 1.15 Variation of A_v with R_L

1.3.12 Power Gain (A_P)

Power Gain $\qquad A_P = A_V \cdot A_I$

A_V increases as R_L increases. But A_I decreases as R_L increases. Therefore, Power Gain, which is product of both, varies with R_L as shown in Fig. 1.16.

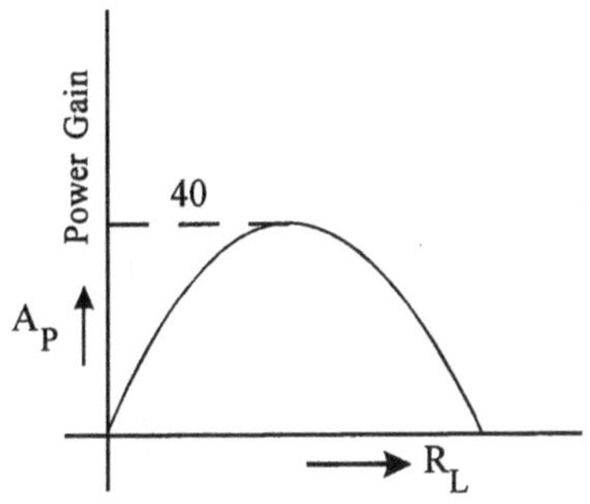

Fig 1.16 Variation of A_P with R_L

The characteristics of Common Base Amplifier with typical values are as given below.

1. *Low Input Resistance (few 100 Ω).*
2. *High Output Resistance (MΩ).*
3. *Current Amplification $A_i < 1$.*
4. *High Voltage Amplification and No Phase Inversion*
5. *Moderate Power Gain (30). $\because A_i < 1$.*

1.3.13 Common Collector Amplifier

The simplified circuit diagram for AC of a transistor (BJT) in Common Collector Configuration is as shown in Fig. 1.17 (without biasing resistors).

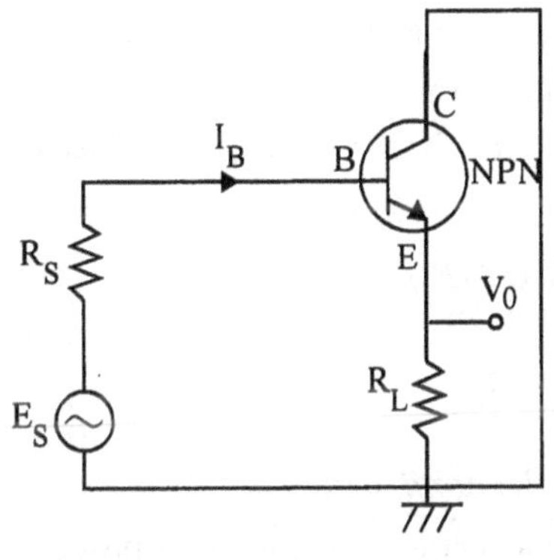

Fig. 1.17 CC Amplifier

The ***h-parameter*** equivalent circuit of transistor in Common Collector Configuration is shown in Fig. 1.18.

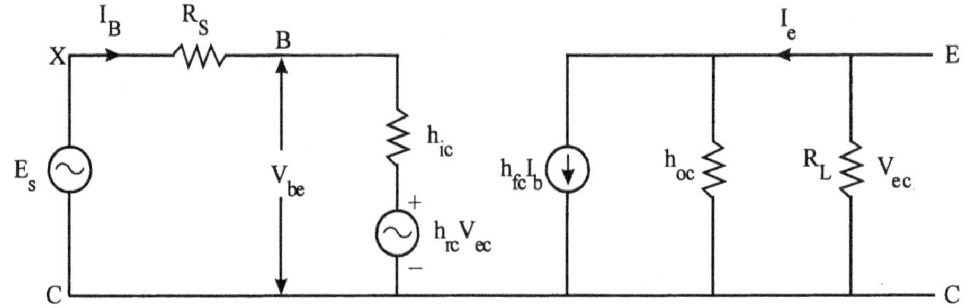

Fig. 1.18 h-parameter equivalent circuit

$$h_{ic} = \left.\frac{V_{bc}}{I_b}\right|_{V_{ec}=0} = 2{,}780\ \Omega \text{ (Typical Value)}$$

$$h_{oc} = \left.\frac{I_e}{V_{ec}}\right|_{I_b=0} = 7.7 \times 10^{-6} \text{ mhos (Typical Value)}$$

$$h_{fc} = -\left.\frac{I_e}{I_b}\right|_{V_{ce}=0} = 100 \text{ (Typical value)}$$

$\because I_e >> I_b$.

h_{fc} is negative because, I_E and I_B are in opposite direction.

$$h_{rc} = \left.\frac{V_{bc}}{V_{ec}}\right|_{I_b=0} ; \quad V_{bc} = V_{ec} \text{ (Typical Value)}$$

Because, $I_B = 0$, E - B junction is not forward baised.

$\therefore$ $V_{EB} = 0$.

For other circuit viz., Common Base and Common Emitter, h_r is much less than 1.

For Common Collector Configuration, $h_{rc} \simeq 1$.

The graphs (variation with R_C) are similar to Common Base Configuration.

Characteristics

1. *High Input Resistance* $\simeq 3\ k\Omega\ (R_i)$
2. *Low Output Resistance* $30\ \Omega.\ (R_0)$
3. *Good Current Amplification* $A_i >> 1$
4. $A_v \leq 1$
5. *Lowest Power Gain of all the configurations.*

Since, A_v is < 1, the output voltage (Emitter Voltage) follows the input signal variation. Hence it is also known as *Emitter Follower*. The graphs of variation with R_L and R_S are similar to Common Base amplifier.

Example : 1.2

For the circuit shown in Fig.1.19 estimate A_i, A_v, R_i and R_o using reasonable approximations. The ***h-parameters*** for the transistor are given as

$h_{fe} = 100 \quad h_{ie} = 2000\ \Omega \quad h_{re}$ is negligible and $h_{oe} = 10^{-5}$ mhos(℧).

$I_b = 100\ \mu A$.

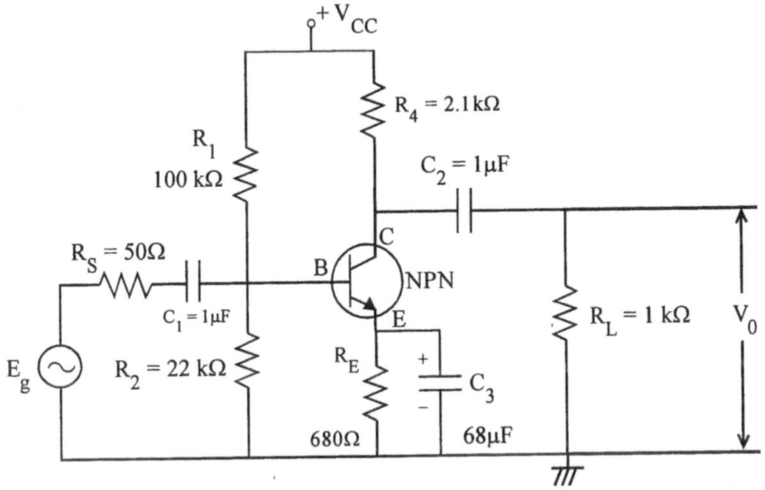

Fig. 1.19 CE amplifier circuit

Solution :

At the test frequency capacitive reactances can be neglected. V_{CC} point is at ground because the AC potential at $V_{CC} = 0$. So it is at ground. R_1 is connected between base and ground for AC. Therefore, $R_1 \| R_2$. R_4 is connected between collector and ground. So R_4 is in parallel with $1/h_{oe}$ in the output. The A.C. equivalent circuit in terms of ***h-parameters*** of the transistor is shown in Fig. 1.20.

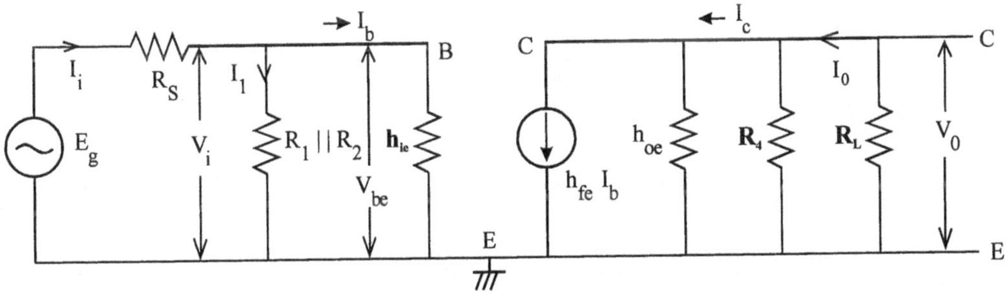

Fig. 1.20 Equivalent circuit

The voltage source $h_{re}\ V_{ce}$ is not shown since, h_{re} is negligible. At the test frequency of the input signal, the capacitors C_1 and C_2 can be regarded as short circuits. So they are not shown in the AC equivalent circuit. The emitter is at ground potential. Because X_{C_3} is also negligible, all the AC passes through C_3. Therefore, emitter is at ground potential and this circuit is in Common Emitter Configuration.

Input Resistance (R_i)

R_i input resistance looking into the base is h_{ie} only

The expression for R_i of the transistor alone $= h_{ie} - \left(\dfrac{h_{fe}\, h_{re}}{h_{oe} + \dfrac{1}{R_L}} \right)$.

R_L is very small and h_{re} is negligible. Therefore, the second term can be neglected. So R_i of the transistor alone is h_{ie}. Now R_i of the entire amplifier circuit, considering the bias resistors is,

$$R_i = h_{ie} \,||\, R_1 \,||\, R_2$$

$$\therefore \quad \frac{R_1 R_2}{R_1 + R_2} = \frac{100 \times 22}{100 + 22} = 18 \text{ k}\Omega$$

$$\therefore \quad R_i = \frac{18 \times 2}{18 + 2} = 1.8 \text{ k}\Omega$$

Output Resistance (R_0)

$$R_0 = \frac{1}{h_{oe} - \left[\dfrac{h_{re} h_{fe}}{h_{ie} + R_S} \right]} \quad \text{.....(1.11)}$$

Because, h_{re} is negligible, R_o of the transistor alone in terms of ***h-parameters*** of the transistor $= \dfrac{1}{h_{oe}}$. Now R_o of the entire amplifier circuit is,

$$\left(\frac{1}{h_{oe}} \,||\, R_4 \,||\, R_L \right) = (2.1 \times 10^{+3}) \;||\; (100 \text{ k}\Omega) \,||\, (1 \text{ k}\Omega)$$

$$= 2 \text{ k}\Omega. \,||\, (1 \text{ k}\Omega) = 0.67 \text{ k}\Omega$$

Current Gain (A_i)

To determine A_i the direct formula for A_i in transistor in Common Emitter Configuration is, $\dfrac{-h_{fe}}{1 + h_{oe} R_L}$.

But this cannot be used because the input current I_i gets divided into I_1 and I_b. There is some current flowing through the parallel configuration of R_1 and R_2. So the above formula cannot be used.

$$V_{be} = I_b \cdot h_{ie}$$

$$V_{be} = 10^{-4} \times (2000) = 0.2\text{V}. \text{ (This is AC Voltage not DC)}$$

Voltage across R_1 R_2 parallel configuration is also V_{be}.

$$\therefore \quad \text{Current } I_1 = \frac{V_{be}}{50 \times 10^3} = \frac{0.2}{50 \text{ k}\Omega} = 4 \ \mu\text{A}.$$

Therefore total input current,

$$I_i = I_1 + I_b = 4 + 100 = 104\ \mu A.$$

I_0 is the current through the 1 kΩ load.

$\frac{1}{h_{oe}}$ = 100 kΩ is very large compared with R_4 and R_L. Therefore, all the current on the output side, $h_{fe}\ I_b$ gets divided between R_4 and R_L only.

Therefore current through R_L is I_0,

$$I_0 = h_{fe}\ I_b \cdot \left[\frac{R_4}{R_4 + R_L}\right]$$

$$I_o = 100 \times 10^{-4}\ \frac{2.1\times10^3}{(2.1\times10^3 + 10^3)}$$

$$= 6.78\ \text{mA}.$$

Therefore current amplification,

$$A_i = \frac{I_0}{I_i}$$

$$= \frac{6.78\times10^{-3}}{104\times10^{-6}} = 65.$$

$$A_v = \frac{V_0}{V_i};\quad V_i = V_{be}$$

$$V_o = -I_0 \cdot R_L$$

$$= (-6.78 \times 10^3) \times (10^3)$$

$$= -6.78\text{V}$$

Because, the direction of I_0 is taken as entering into the circuit. But actually I_0 flows down, because V_0 is measured with respect to ground.

$$\therefore \qquad A_v = \frac{-6.78}{0.2}$$

$$= -33.9$$

Negative sign indicates that there is phase shift of 180^0 between input and output voltages, i.e. as base voltage goes more positive, (it is NPN transistor), the collector voltage goes more negative.

Example : 1.3

For the circuit shown, in Fig. (1.21), estimate A_v and R_i. $\frac{1}{h_{oe}}$ is large compared with the load seen by the transistor. All capacitors have negligible reactance at the test frequency.

$$h_{ie} = 1\ k\Omega,\ h_{fe} = 99\ \ h_{re} \text{ is negligible.}$$

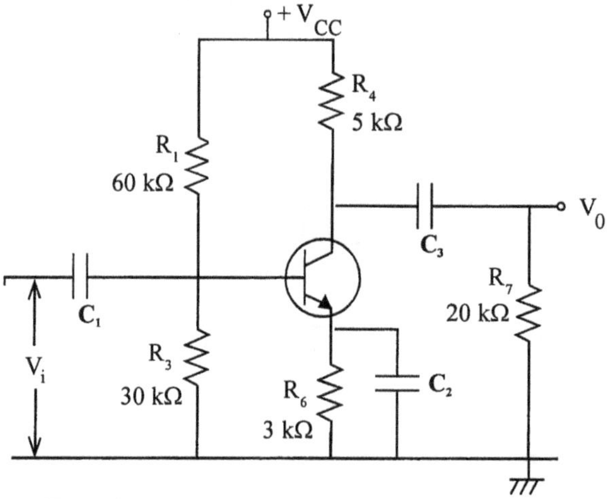

Fig. 1.21 AC amplifier circuit (Ex : 1.2)

Solution :

The same circuit can be redrawn as,

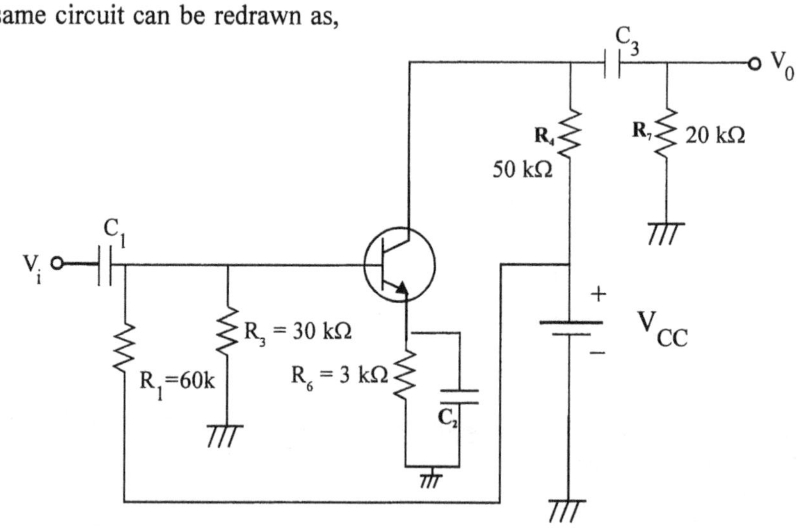

Fig. 1.22 Redrawn circuit of AC amplifier

In the second circuit also, R_4 is between collector and positive of V_{CC}. R_1 is between $+V_{CC}$ and base. Hence both the circuits are identical. Circuit in Fig. 1.21 is same as circuit in Fig. 1.22. In the AC equivalent circuit, the direct current source should be shorted to ground. Therefore, R_4 is between collector and ground and R_1 is between base and ground. Therefore, R_4 is in parallel with R_7 and R_1 is in parallel with R_3 (Fig. 1.23).

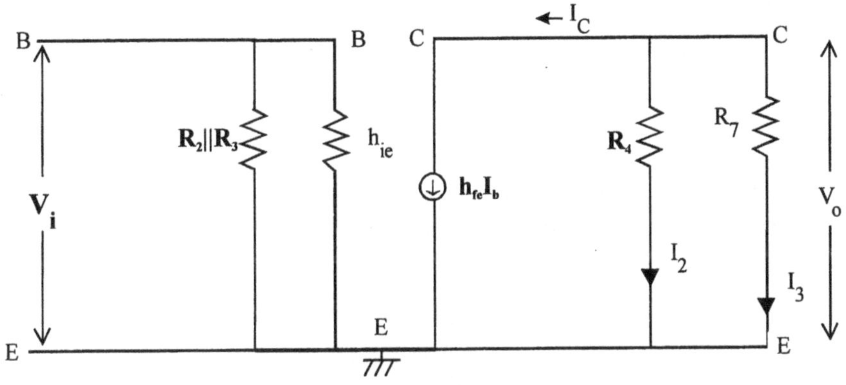

Fig. 1.23 h - Parameter equivalent circuit

$$R_2 \parallel R_3 = \frac{60 \times 30}{60+30} = \frac{1800}{90} = 20 \text{ k}\Omega.$$

$$R_4 \parallel R_7 = R_L = \frac{5 \times 20}{5+20} = 4 \text{ k}\Omega.$$

Therefore, the circuit reduces to, (as shown in Fig. 1.24).

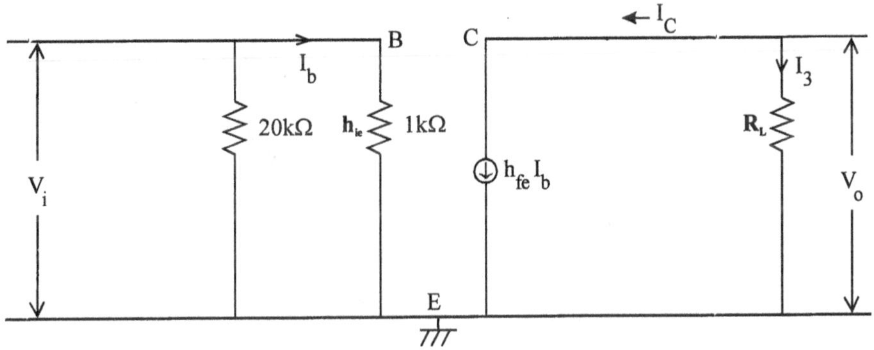

Fig. 1.24 Simplified circuit of Fig. 1.23

$$I_b = \frac{V_i}{h_{ie}} \qquad (\because h_{re} \text{ is negligible})$$

$$I_c = h_{fe}\, I_b = \frac{h_{fe}.V_i}{h_{ie}}$$

$$\therefore \quad V_0 = -I_c \,.\, R_L = -\frac{h_{fe}V_i.R_L}{h_{ie}}$$

$$A_v = \frac{V_0}{V_i} = -\frac{h_{fe}R_L}{h_{ie}} = \frac{-99\,(14.28 \times 10^3)}{10^3}$$

$$A_v = -400$$

R_i is the parallel combination of 20 kΩ and h_{ie}.

$$\frac{20 \times 1\ \text{k}\Omega}{20+1} = 950\ \Omega$$

Example : 1.4

Given a single stage transistor amplifier with *h - parameter* as h_{ic} = 1.1 kΩ, h_{rc} = 1, h_{fc} = –51, h_{oc} = 25 μA/v. Calculate A_I, A_V, A_{Vs}, R_i, and R_o for the Common Collector Configuration, with $R_s = R_L$ = 10 k.

Solution :

$$A_I = \frac{-h_{fc}}{(1+h_{oc} R_L)} = \frac{51}{1+25\times10^{-6}\times10^4} = 40.8$$

$$R_i = h_{ic} + h_{rc} A_I R_L = 1.1 \times 10^3 + 1 \times 40\ .\ 8 \times 10^4 = 409.1\ \text{k}\Omega$$

$$A_v = \frac{A_I . R_L}{R_i} = \frac{40.8\times10^4}{409.1\times10^3} = 0.998$$

$$A_{vs} = \frac{A_V . R_i}{R_i + R_s} = \frac{0.998\times409.1}{419.1} = 0.974$$

$$R_0 = \frac{1}{h_{oc} - \frac{h_{fc}.h_{rc}}{h_{ic}+R_s}} = \frac{1}{25\times10^{-6} + \frac{51\times1}{(1.1+10)10^3}} = \frac{1}{4.625\times10^{-3}}$$

$$R_0 = 217\Omega$$

Example : 1.5

For any transistor amplifier prove that

$$R_i = \frac{h_i}{1-h_r A_V}$$

Solution :

$$R_i = h_i - \frac{h_f.h_r}{h_o + \frac{1}{R_L}}$$

But $$A_I = \frac{-h_f}{1+h_o.R_L}$$

$$\therefore \quad R_i = h_i + h_r A_I R_L \qquad .....(1.12)$$

$$\therefore \qquad A_v = \frac{A_I . R_L}{R_i}$$

$$R_L = \frac{A_V . R_i}{A_I}$$

Substituting this value of R_L in equation (1.12)

$$R_i = h_i + \frac{h_r . A_I . A_V . R_i}{A_I} = h_i + h_r . A_V . R_i$$

$$R_i [1 - h_r A_v] = h_i \qquad \therefore \boxed{R_i = \frac{h_i}{1 - h_r A_v}}$$

Example : 1.6

For a Common Emitter Configuration, what is the maximum value of R_L for which R_1 differs by not more than 10% of its value at $R_2 = 0$?

$h_{ie} = 1100\ \Omega$; $\qquad h_{fe} = 50$

$h_{re} = 2.50 \times 10^{-4}$; $\qquad h_{oe} = 25\ \mu$ A/v

Solution :

Expression for R_i is,

$$R_i = h_{ie} - \frac{h_{fe} . h_{re}}{h_{oe} + \frac{1}{R_L}} .$$

If $R_2 = 0$, $R_i = h_{ie}$. The value of R_L for which $R_i = 0.9\ h_{ie}$ is found from the expression,

$$0.9\ h_{ie} = h_{ie} - \frac{h_{fe} . h_{re}}{h_{oe} + \frac{1}{R_L}}$$

or

$$\frac{h_{fe} . h_{re}}{h_{oe} + \frac{1}{R_L}} = h_{ie} - 0.9\ h_{ie} = 0.1\ h_{ie}$$

$$\frac{h_{fe} . h_{re}}{0.1 h_{ie}} = h_{oe} + \frac{1}{R_L}$$

$$\frac{1}{R_L} = \frac{h_{fe} . h_{re}}{0.1 h_{ie}} - h_{oe} = \frac{h_{fe}\ h_{re} - 0.1\ h_{oe}\ h_{re}}{0.1 h_{ie}}$$

or

$$R_L = \frac{0.1 h_{ie}}{h_{fe} h_{re} - 0.1 h_{oe}\ h_{ie}} = \frac{0.1 \times 1100}{50 \times 2.5 \times 10^{-4} - 0.1 \times 1100 \times 25 \times 10^{-6}}$$

$$R_L = 11.3\ k\Omega$$

1.4 MILLER'S THEOREM

Fig. 1.25(a) shows an amplifier with a capacitor between input and output terminals. It is called as feedback capacitor. When the gain K is large, the feedback will change the input Z and output Z of the circuit.

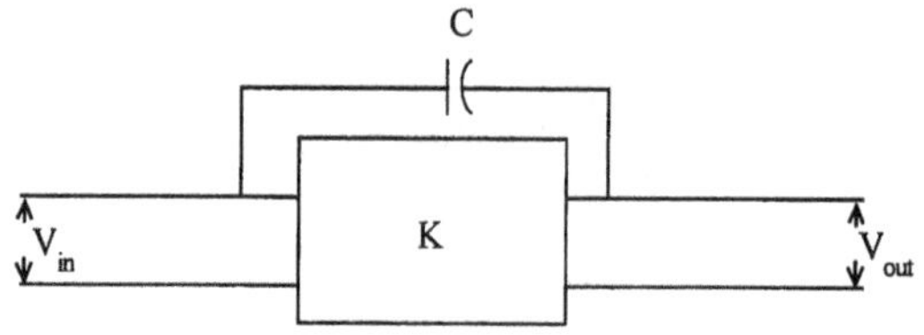

Fig. 1.25 (a) Feedback capacitor

A circuit as shown above is difficult to analyze, because of capacitor. So according to the Miller's theorem, the feedback capacitor can be split into two values, one as connected in the input side and the other on the output side, as shown in Fig. 1.25 (b).

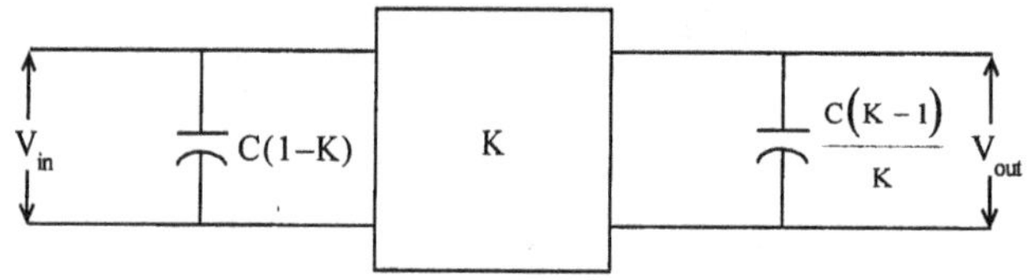

Fig. 1.25 (b) Splitting of feedback capacitor using Miller's theorem.

1.4.1 MATHEMATICAL PROOF OF MILLER'S THEOREM

The AC current passing through capacitor (C) in Fig. 1.25 (a) is

$$I_C = \frac{V_{in} - V_{out}}{\left(\frac{1}{j\omega c}\right)} = \frac{(V_{in} - V_{out})}{-J\,X_C}$$

$$V_{out} = K\,V_{in}$$

$$\therefore \quad I_C = \frac{(V_{in} - KV_{in})}{-jX_C} = \frac{V_{in}(1-K)}{-jX_C}$$

$$\frac{V_{in}}{I_C} = Z_{in} = \frac{V_{in}}{\frac{V_{in}(1-K)}{-jX_C}} = \frac{-jX_C}{(1-K)}$$

$$= \frac{-j}{2\pi f\,C(1-K)}$$

Since $$X_C = \frac{1}{2\pi f C}$$

$\frac{V_{in}}{I_C}$ is the input Z as seen from the input terminals.

$$\therefore \quad Z_{in} = \frac{-j}{2\pi f[C(1-K)]}$$

$$\therefore \quad C_{in} = C\,(1-K)$$

Similarly output capacitance can be derived as follows :

Current in the capacitor,

$$I_C = \frac{V_{out} - V_{in}}{-jX_C} = \frac{V_{out}\left(1 - \frac{V_{in}}{V_{out}}\right)}{-j\,X_C}$$

$$I_C = \frac{V_{out}\left(1 - \frac{1}{K}\right)}{-jX_C}$$

$$= \frac{V_{out}}{I_C} = Z_{out}$$

$$= \frac{V_{out}}{\frac{V_{out}\left(1 - \frac{1}{K}\right)}{-jX_C}}$$

$$Z_{out} = \frac{-jX_C}{\left(\frac{A-1}{A}\right)} = \frac{-j}{2\pi f C\left(\frac{K-1}{K}\right)}$$

$$\therefore \quad C_{out}\,(\text{Miller}) = C\left(\frac{K-1}{K}\right)$$

FREQUENCY EFFECTS

LEAD NETWORK (FIG. 1.26)

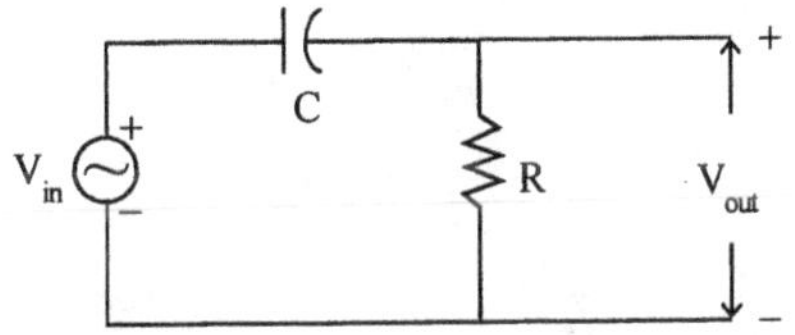

Fig. 1.26 Lead network

$$I = \frac{V_{in}}{R + \frac{1}{j\omega C}}$$

$$V_0 = I.\,R = \frac{V_{in}.R}{\left(R + \frac{1}{j\omega C}\right)}$$

V_0 leads with respect to V_{in}. So it is called as *lead network* for the above circuit (Fig. 1.25), at low frequencies, $X_C = \infty$.

$\therefore$ V_0 is low since (I is low). As f increases X_c decreases. Hence I flows, and V_o increases.

$\therefore$ Gain increases. Hence the frequency response is as shown.

Cut-off Frequency

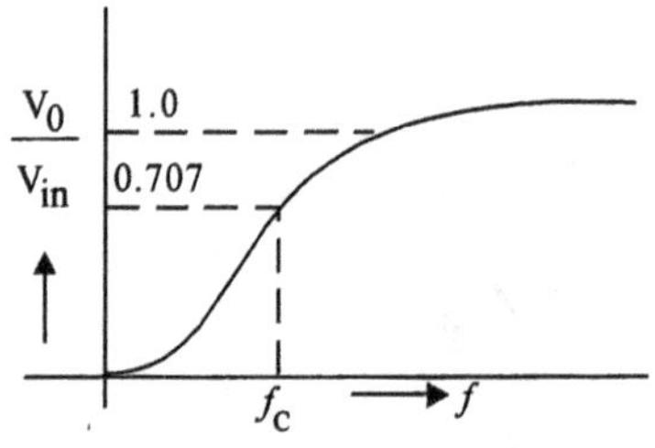

Fig. 1.27 Frequency response

$$V_{out} = \frac{R}{\sqrt{R^2 + X_C^2}}.V_{in}$$

$$\frac{V_{out}}{V_{in}} = \frac{R}{\sqrt{R^2 + X_C^2}}$$

Cut-off frequency is the frequency at which $\frac{V_{out}}{V_{in}} = \frac{1}{\sqrt{2}}$.

This happens when, $X_c = R$

$$\frac{1}{2\pi f_c C} = R \quad \text{or} \quad \boxed{f_C = \frac{1}{2\pi RC}}$$

Lower cut-off frequency.

Half Power Point

At $\quad f = f_C, \quad P = \frac{V_{out}^2}{2R}$

Power is half of maximum power when V_{out} is maximum. Therefore it is also called as half power point.

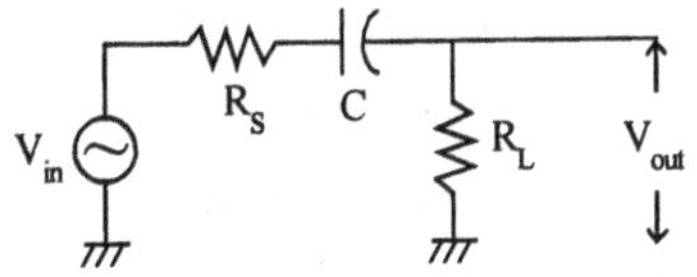

Fig. 1.28 Equivalent circuit

Considering Source Resistance :

$$\frac{V_{out}}{V_{in}} = \frac{R}{\sqrt{(R_S + R_L) + X_c^2}}$$

$$\text{At } f_C , \quad R_S + R_L = X_C$$

$$\therefore \qquad f_C = \frac{1}{2\pi(R_S + R_L).C}$$

In the midband frequency region $X_C \cong 0$.

$$\therefore \qquad \frac{V_{out}}{V_{in}} = \frac{R_L}{R_S + R_L}$$

$$A_{mid} \text{ (midband voltage gain)} = \frac{R_L}{R_S + R_L}$$

Stiff Coupling

If $X_c = 0.1\ (R_S + R_L)$ it is called *Stiff Coupling.*

The coupling capacitor must have (or bypass capacitor).

$$X_c = \frac{1}{10} R_E$$

This is known as *Stiff Coupling.*

Amplifier Analysis

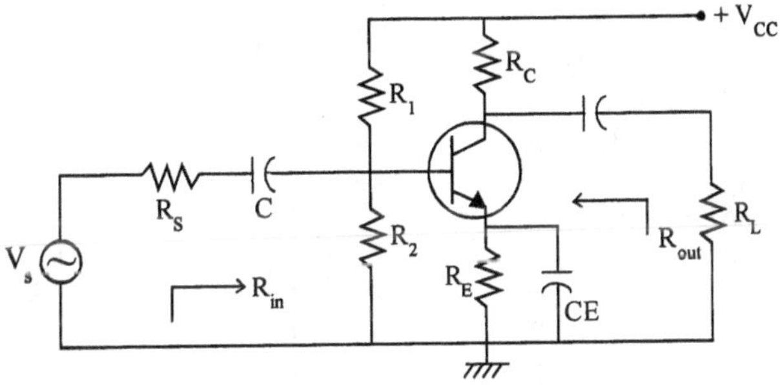

Fig. 1.29 Amplifier circuit

The equivalent circuit from input side, not considering R_S. (Fig. 1.30).
The equivalent circuit from output side, not considering R_L.

$$R_{in} = R_1 \parallel R_2 \parallel \beta r_e' \quad \text{(Not cosidering } R_S\text{)}$$
$$R_{out} \cong R_C \quad \text{(Not cosidering } R_L\text{)}$$

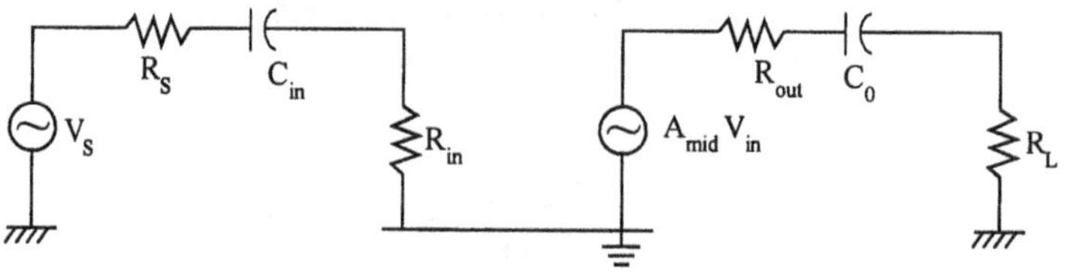

Fig. 1.30 Equivalent circuit

Considering the input side as a load network,

$$f_{in} = \frac{1}{2\pi(R_S + R_{in})C_{in}}$$

Similarly output lead network has cut-off frequency

$$f_{out} = \frac{1}{2\pi(R_{out} + R_L)C_{out}}$$

Lag Networks

$$V_0 = I.\, X_c$$
$$= \frac{V_i \,.\, X_C}{\left(R + \dfrac{1}{j\omega C}\right)}$$

$$V_0 = \frac{\dfrac{1}{j\omega C}.V_{in}}{\sqrt{R^2 + X_C{}^2}}$$

$$V_0 = \frac{-j(V_i / \omega C)}{\sqrt{R^2 + X_C^2}} \quad \text{therefore it is lag network}$$

V_0 lags with respect to V_i.

$$\frac{V_{out}}{V_{in}} = \frac{X_c}{\sqrt{R^2 + X_c{}^2}}$$

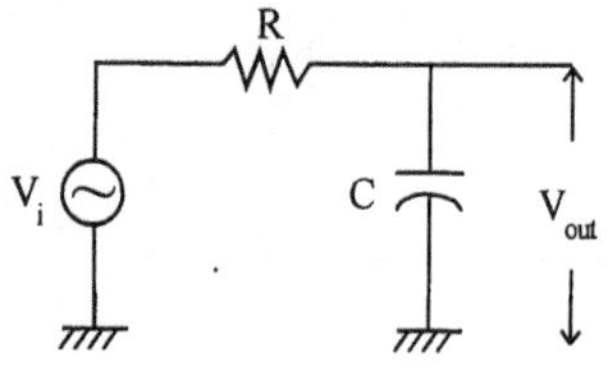

Fig. 1.31 Lag network

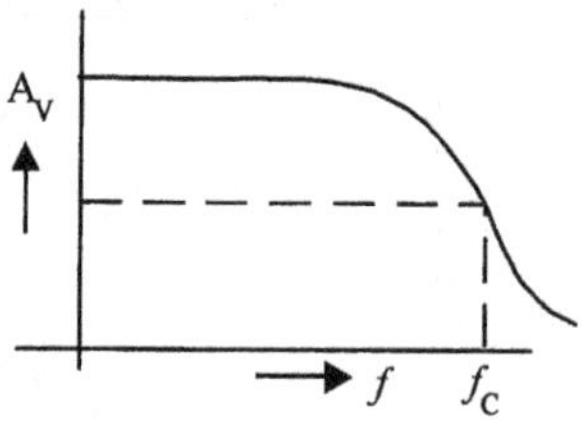

Fig. 1.32

at f_C, $R = X_C$

$\therefore$ $$\boxed{f_C = \frac{1}{2\pi RC}}$$

Midband gain $A_{mid} = \frac{R_L}{R_S + R_L}$

Decibel

Power gain $= G = \frac{P_2}{P_1}$

P_2 = Output power.

P_1 = Input power.

Decibel power gain $= G' = 10 \log_{10} G$.

If $G = 100$, $G' = 10 \log 100 =$ **20 db**.

If $G = 2$, $G' = 10 \log 2 = 3.01$db.

Usually, it is rounded off to 3.

Negative Decibel

If a < 1, a' will be negative.

If $G = \frac{1}{2}$,

$G' = 10 \log \frac{1}{2}$ = **– 3.01 db**

If

G	G'
1	0 db
10	10 db
100	20 db
1000	30 db
10, 000	40 db

Ordinary Gains Multiply :

If two stages are there,

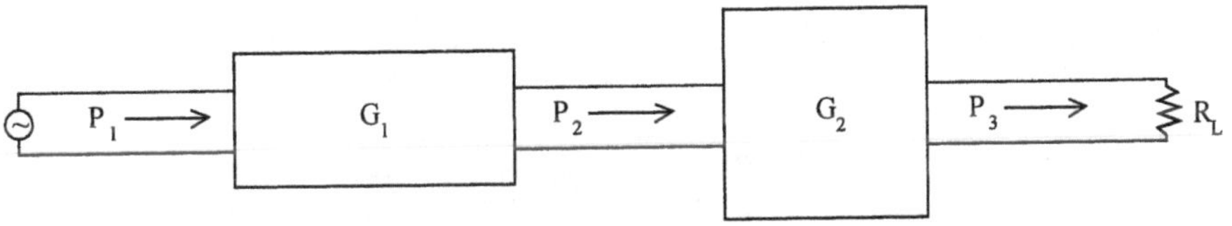

Fig. 1.33 Cascaded amplifier stages

$$G_1 = \frac{P_2}{P_1};\ G_2 = \frac{P_3}{P_2}$$

Overall gain $$G = \frac{P_3}{P_1}$$

$$= \frac{P_2}{P_1} \cdot \frac{P_3}{P_2}$$

$$\mathbf{G = G_1\, G_2}$$

Decibel gains add up.

Decibel Voltage Gain

If $A =$ Normal Voltage Gain $\frac{V_2}{V}$, decibel voltage gain $A' = 20 \log A$. $A = \frac{V_2}{V_1}$

If R_1 is output resistance,

Input power $$P_1 = \frac{V_1^2}{R_1}$$

If R_2 is output resistance, $P_2 = \frac{V_2^{\ 2}}{R_2}$

$\therefore$ Power gain $$G = \frac{P_2}{P_1} = \frac{V_2^{\ 2}}{V_1^{\ 2}} \cdot \frac{R_1}{R_2}$$

If the impedances are matched, i.e. input resistance $R_1 =$ output resistance R_2,

Power gain $$G = \frac{V_2^{\ 2}}{V_1^{\ 2}} = A^2$$

In decibels, $10 \log G = G' = 10 \log A^2$.

In decibels, $10 \log G = G' = 20 \log A$.

Cascaded Stages (Fig. 1.34)

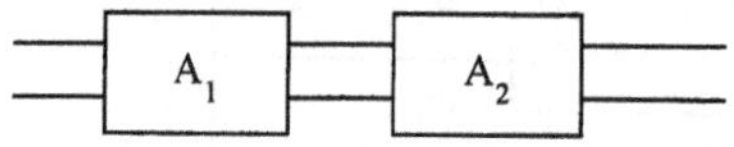

Fig. 1.34 Cascaded stages

$$A = A_1 \times A_2$$
$$A_1 = A_1' + A_2' \text{ (in decibels)}$$

Stiff Coupling

When the capacitor is chosen such that, $X_c = \frac{-R_E}{10}$, it is called as stiff coupling. Because, for the circuit shown.

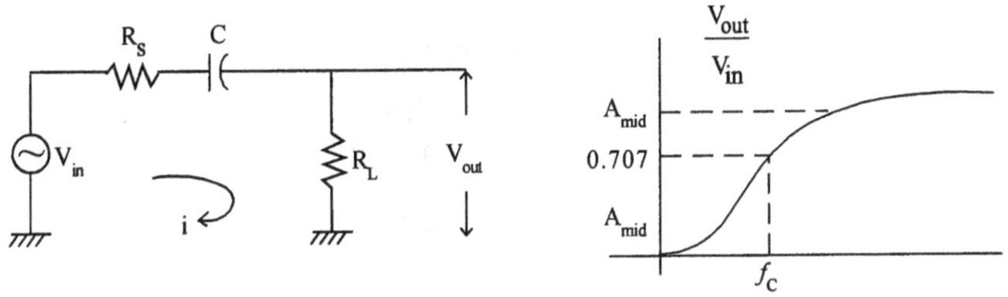

Fig. 1.35 Stiff coupling network

For the circuit, shown above, in the mid frequency range, X_c is negligible.

$$\therefore \quad i = \frac{V_{in}}{(R_S + R_L)}$$

$$V_{out} = i.\ R_L = \left(\frac{V_{in}}{R_S + R_L}\right) R_L.$$

$\therefore$ $A_{(mid)}$ = Voltage gain in the mid frequency range is,

$$A_{mid} = \frac{V_{out}}{V_{in}} = \frac{R_L}{R_S + R_L}$$

For the complete circuit, $\frac{V_{out}}{V_{in}} = \frac{R_L}{\sqrt{(R_S + R_L)^2 + X_c^2}}$

When $\quad (R_S + R_L) = X_c, = \frac{1}{2\pi f_C C}$

$$\frac{V_{out}}{V_{in}} = \frac{R_L}{\sqrt{2}(R_S + R_L)}$$
$$= 0.707\ A_{mid}$$

$$\therefore \quad f_C = \frac{1}{2\pi(R_S + R_L).C}$$

When $\quad X_c = 0.1\ (R_S + R_L),$

$$\frac{V_{out}}{V_{in}} = \frac{R_L}{\sqrt{(R_S+R_L)^2+[0.1\,(R_S+R_L)]^2}}$$

$$\frac{V_{out}}{V_{in}} = 0.995\ A_{mid}.$$

X_c is made = 0.1 (R_S + R_L), at the lowest frequency. (f_C lower)

∴ At the frequency, A = 0.995 A_{mid}. So it is called as *Stiff Coupling*.

$$A = \frac{V_{out}}{V_{in}} = \frac{X_c}{\sqrt{R^2+X_c^{\ 2}}} = \frac{1}{\sqrt{1+\left(\frac{R}{X_c}\right)^2}}$$

$$\because \quad \frac{R}{X_c} = \frac{R}{\frac{1}{2\pi f_C}} = 2\pi fRC = \frac{f}{f_C}$$

$$\because \quad f_c = \frac{1}{2\pi RC}$$

$$\therefore \quad A = \frac{1}{\sqrt{1+\left(\frac{f}{f_C}\right)^2}}$$

Decibel voltage gain $A' = 20 \log \frac{1}{\sqrt{1+\left(\frac{f}{f_c}\right)^2}}$

f_C = cutoff frequency.

When $\frac{f}{f_C} = 0.1,$

$$A' = 20 \log \frac{1}{\sqrt{1+(0.1)^2}} \cong 0\text{db}$$

When $\frac{f}{f_C} = 1,$

$$A' = 20 \log \frac{1}{\sqrt{1+1^2}} = -3.01\text{ db} \cong 3\text{ db}$$

When $\frac{f}{f_C} = 10,$

$$A' = 20 \log \frac{1}{\sqrt{1+10^2}} = -20 \text{ db}$$

When $\frac{f}{f_C} = 100,$

$$A' = 20 \log \frac{1}{\sqrt{1+100^2}} = -40 \text{ db}$$

When $\frac{f}{f_C} = 1000,$

$$A' = 20 \log \frac{1}{\sqrt{1+1000^2}} = 60 \text{ db}$$

$\therefore$ when $f = \frac{f_C}{10},$ $A' = 0$

$f = f_C,$ $A' = -3$ db

$f = 10 f_C,$ $A' = -20$ db

and so on.

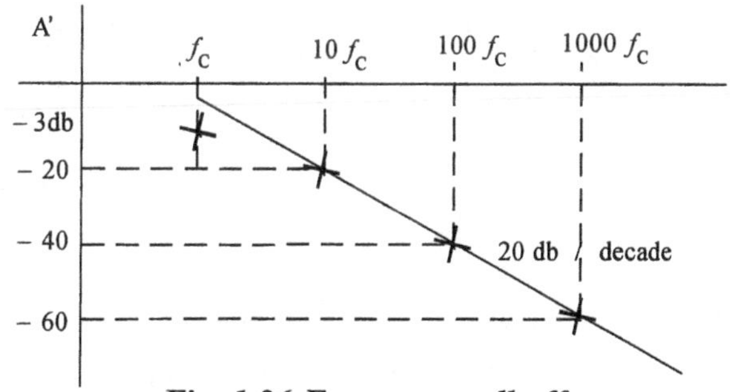

Fig. 1.36 Frequency roll-off

$\therefore$ A change of **20db** per decade *change* in frequency.

An Octave is a factor of 2 in frequency change

When *f changes from 100 to 200 Hzs, it has changed by one octave.*

When, *f* changes from 100 to 400 Hzs, it is two octaves for lead network, Bode Plot is

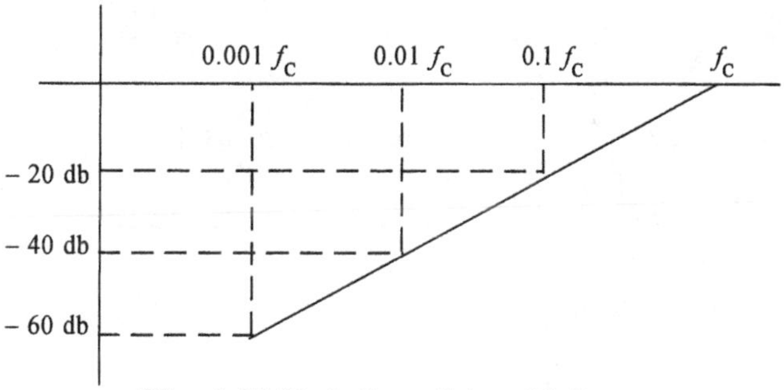

Fig. 1.37 Variation of A_V with f

1.5 DESIGN OF SINGLE STAGE AMPLLIFIERS

To form transistor amplifier configuration, we connect a load impedance Z_L and a signal source as shown in Fig. 1.38.

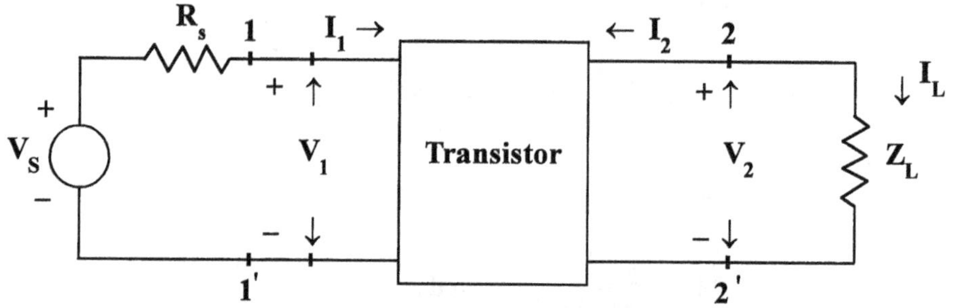

Fig 1.38 Amplifier circuit

V_s is the signal source, R_s is the resistance of the signal source, and Z_L is the load impedance. Transistor can be connected in C.E, C.B. and C.C. Configuration. To analyse these circuits i.e. to determine the current gain A_I, Voltage gain A_V, input impedance, output impedance etc, we can use the h parameters. So the equivalent circuit for the above transistor amplifier circuit in general form without indicating C.E, C.B, or C.C. Configuration, can be denoted as in the Fig. 1.39.

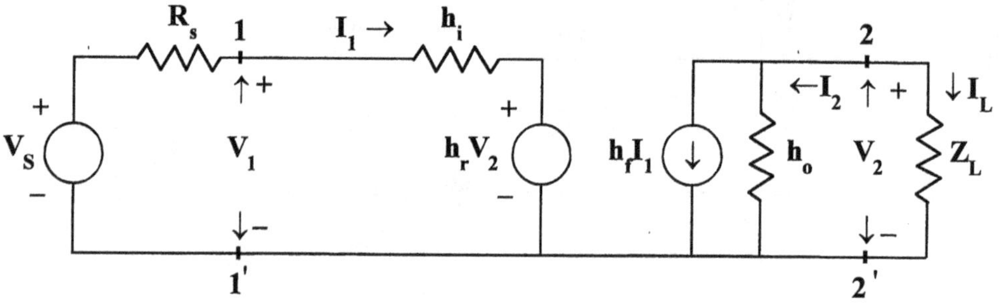

Fig 1.39 h-parameter equipvalent circuit (general repersentation)

Current Amplification,

$$A_I = \frac{I_L}{I_1};$$

But $\quad I_L = -I_2$

$$\therefore \quad A_I = -\frac{I_2}{I_1};$$

Negative sign is because I_2 is always represented as flowing into the current source. For a transistor PNP or NPN type, if it is flowing out of the transistor, I_2 is represented as $-I_2$. Now voltage across Z_2 is taken with 2 as +ve and 2^1 as -ve The directions of I_L are as shown and $I_L = -I_2$.

From the above circuit we have,

$$I_2 = h_f I_1 + h_o V_2$$

But $\quad V_2 = I_L Z_L = -I_2 . Z_L \quad (\text{Since } I_L = -I_2)$

$\therefore \quad I_2 = h_f I_1 - h_o . Z_L . I_2$

or $\quad I_2 + h_o Z_L I_2 = h_f I_1 .$

$I_2 (1 + h_o Z_L) = h_f I_1$

$$\therefore \quad \boxed{A_I = -\frac{I_2}{I_1} = \frac{-h_f}{1 + h_o Z_L}} \qquad \text{.......... (1.13)}$$

Negative sign indicates that I_1 and I_2 are out of phase by 180^0.

1.5.1 Input Impedance (Z_i)

Input impedance of any circuit is the impedance we measure looking back into the amplifier circuit. Now amplifier terminals are 1and 1^1. R_s is the resistance of the signal source. So to determine the output z of the amplification alone, it need not be considered.

$$Z_i = \frac{V_1}{I_1}$$

But $\quad V_1 = h_i I_1 + h_r V_2 .$

$$\text{Hence} \quad Z_i = \frac{V_1}{I_1} = h_i + h_r . \frac{V_2}{I_1} . \qquad \text{.......... (1.14)}$$

$$\text{But} \quad V_2 = -I_2 . Z_L , \quad \therefore \frac{V_2}{I_1} = \left(\frac{-I_2}{I_1}\right) Z_L = A_I Z_L \qquad \text{.......... (1.15)}$$

Substituting the values of V_2 in Equation (1.15),

$$\boxed{Z_i = h_i + h_r . A_I . Z_L}$$

But $\quad A_I = \dfrac{-h_f}{1 + h_o z_L}$

$$\therefore \quad Z_i = h_i - \frac{h_f . h_r z_L}{1 + h_o z_L}$$

$$= h_i - \frac{h_f . h_r}{\frac{1}{z_L} + h_o}$$

$$\boxed{Z_i = h_i - \frac{h_f . h_r}{Y_L + h_o}} \qquad \text{.......... (1.16)}$$

Y_L is load admittance. Therefore, Input impedance is a function of load impedance. If $Z_L = 0$, $Y_L = \infty$,

$$\therefore \quad Z_I = h_i .$$

1.5.2 Voltage Gain (A_v)

$$A_v = \frac{V_2}{V_1}$$

But $\quad V_2 = +I_L . Z_L = -I_2 . Z_L \qquad$ Since, $I_2 = -I_L$

But $\quad I_2 = -A_I . I_1$

$\therefore \quad V_2 = +A_I . I_1 Z_L .$

$$\frac{V_2}{V_1} = \frac{A_I I_1 Z_L}{V_1}$$

But $\quad \dfrac{V_1}{I_1} = Z_i$

$$z_i = h_i - \frac{h_f . h_r}{\dfrac{1}{R_L} + h_o}$$

$$\therefore \quad A_V = \frac{A_I Z_L}{Z_i}$$

$$\boxed{A_V = -\frac{-h_f . z_L}{h_i + z_L (h_i h_o - h_f h_r)}} \qquad \text{.......... (1.17)}$$

1.5.3 Output Admittance (Y_0)

The output impedance can be determined by using two assumptions $Z_L = \infty$, and $V_s = 0$

Y_o is defined as $\dfrac{I_2}{V_2}$ with $Z_L = \infty$

But $\quad I_2 = h_f I_1 + h_o v_2 .$

Dividing by V_2,

$$\frac{I_2}{V_2} = Y_O = \frac{h_f . I_1}{V_2} + h_o . \qquad \text{.......... (1.18)}$$

From the equivalent circuit with $V_s = 0$,

$$R_s . I_1 + h_i I_1 + h_r V_2 = 0.$$

Dividing by V_2 Through out, we get

$$\frac{R_S . I_1}{V_2} + \frac{h_i I_1}{V_2} + h_r = 0$$

or
$$\frac{I_1}{V_2} = \frac{-h_r}{h_i + R_S} \qquad \text{.......... (1.19)}$$

Substitute this value in the equation for Y_o

$$Y_o = h_f \left(\frac{-h_r}{h_i + R_s} \right) + h_o$$

$$\boxed{Y_o = h_o - \frac{h_f . h_r}{h_i + R_s}}$$

Therefore, $Z_o = 1/Y_o$

Therefore, output admittance is a function of Source Resistance R_s where as Z_i is a function of Y_L. If $R_s = 0$, $Y_o \cong h_o$. Since ($h_f h_r / h_i$) is very small.

1.5.4 Voltage Gain (A_{VS}) considering Source Resistance (R_S)

$$\therefore \quad A_{vs} = \frac{V_2}{V_S} = \frac{V_2}{V_1} \times \frac{V_1}{V_S} = A_V . \frac{V_1}{V_S}.$$

This is the equivalent circuit of the input side of the amplifier circuit. (Fig. 1.40).

$$\therefore \quad V_1 = \frac{V_S . Z_i}{Z_i + R_S}$$

$$\therefore \quad A_{vs} = \frac{A_V . Z_i}{Z_i + R_S}$$

$$A_v = \frac{V_2}{V_1}.$$

$$\therefore \quad A_V Z_i = \frac{V_2}{V_1} . Z_i$$

But $\quad V_2 = I_L . Z_L \; ; \; V_1 = I_1 Z_i$

$$\therefore \quad A_V . Z_i = \frac{I_L . Z_L . Z_i}{I_1 . Z_L} = A_I . Z_L$$

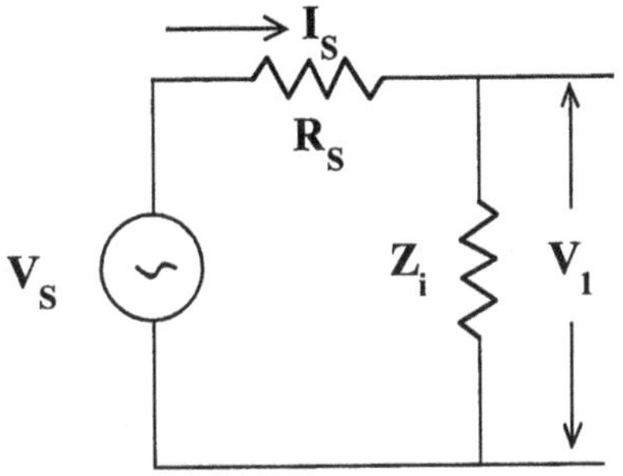

Fig 1.40 Voltage source.

$$\therefore \quad \boxed{A_{vs} = \frac{A_I Z_L}{R_S + Z_i}.} \qquad \text{If } R_s = 0,\ A_{vs} = A_v . \qquad \text{.......... (1.20)}$$

Hence A_V is the voltage gain of an ideal voltage source with zero terminal resistance.

1.5.5 Current Gain (A_{IS}) considering Source Resistance (R_S)

The input source can also be represented as a current source I_S in parallel with resistance R_s or voltage source V_s in series with resistance R_s. Now let us consider the input source as a current source in parallel with R_s. The equivalent circuit is (Fig. 1.41)

$$A_{IS} = \frac{I_L}{I_S} \; ; \; I_L = -I_2$$

$$\therefore \quad A_{IS} = \frac{I_L}{I_S} = -\frac{I_2}{I_S}$$

$$A_{IS} = -\frac{I_2}{I_S} = -\frac{I_2}{I_1} . \frac{I_1}{I_S}$$

$$= A_I . \frac{I_1}{I_S} \qquad \text{.......... (1.21)}$$

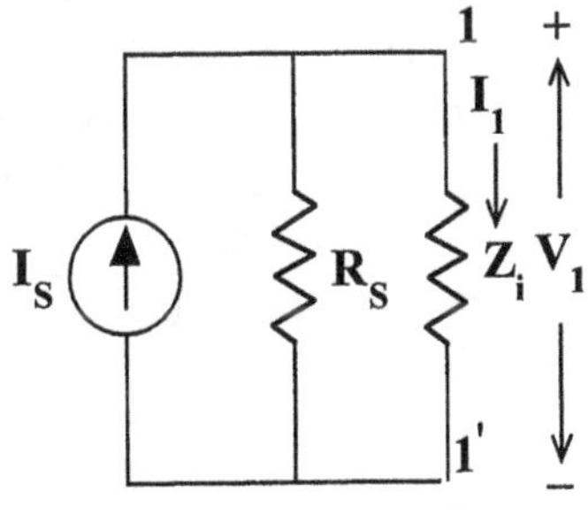

Fig 1.41 Current source

$$\therefore \quad A_I = \frac{-I_2}{I_1}.$$

From the Fig 6.27, $I_1 = \dfrac{I_S.R_S}{R_S + Z_i}$

$$\therefore \quad A_{IS} = \frac{A_I \times R_S}{(R_S + Z_i)} = \frac{A_I.R_S}{(R_S + Z_i)}$$

If $R_S = \infty$, $A_I = A_{IS}$. Therefore, A_I is the current gain for an ideal source

Now $$A_{VS} = \frac{A_I.Z_L}{Z_i + R_s} \qquad \text{.......... (1.22)}$$

$$A_{IS} = \frac{A_I.R_S}{Z_i + R_S} \qquad \text{.......... (1.23)}$$

Dividing (1.10) and (1.11), $\dfrac{A_{VS}}{A_{IS}} = \dfrac{Z_L}{R_S}$.

or $$\boxed{A_{VS} = A_{IS}.\frac{Z_L}{R_S}} \qquad \text{.......... (1.24)}$$

This equation is independent of the transistor parameters This is valid if the equivalent current and voltage sources have the same resistance.

1.5.6 Power Gain (A_P)

$$A_P = \frac{P_2}{P_1} = \frac{V_2 I_L}{V_1 I_1}$$

$$= A_V . A_I$$

$$\because \quad A_V = \frac{A_I.Z_L}{Z_i},$$

$$\therefore \quad \boxed{A_P = \frac{A_I^2 Z_L}{Z_i}} \qquad \text{........ (1.25)}$$

COMPARISON OF THE CE, CB, CC CONFIGURATIONS

C.E : Of the three, it is the most versatile. Its voltage and current gain are > 1. Input and output resistance vary least with R_s and R_L. R_i and R_o values lie between maximum and minimum for all the three configurations. Phase shift of 180^0 between V_i and V_o. Power Gain is Maximum.

C.B : $A_I < 1$ $A_V > 1$. R_i is the lowest and R_o is the highest . It has few applications. Sometimes it is used to match low impedance source V_i and V_o. No phase shift.

C.C : $A_V < 1$. A_I is very high. It has very ***high input Z and low output Z. So it is used as a buffer*** between high Z source and low impedance load. It is also called as ***emitter follower.***

To analyse circuit consisting of a number of Transistors, each Transistor should be replaced by its equivalent circuit in h-parameters. The emitter base and collector points are indicated and other circuit elements are connected without altering the circuit Configuration. This way the circuit analysis becomes easy.

Example 1.7

Find the Common Emitter Hybrid parameters in terms of the Common Collector Hybrid parameters for a given transistor.

Solution

We have to find h_{ie}, h_{re}, h_{fe} and h_{oe} in terms of h_{ic}, h_{rc}, h_{fc} and h_{oc}.

The transistor circuit in Common Collector Configuration is shown in Fig. 1.42. The h-parameter equivalent circuit is shown in Fig. 1.43.

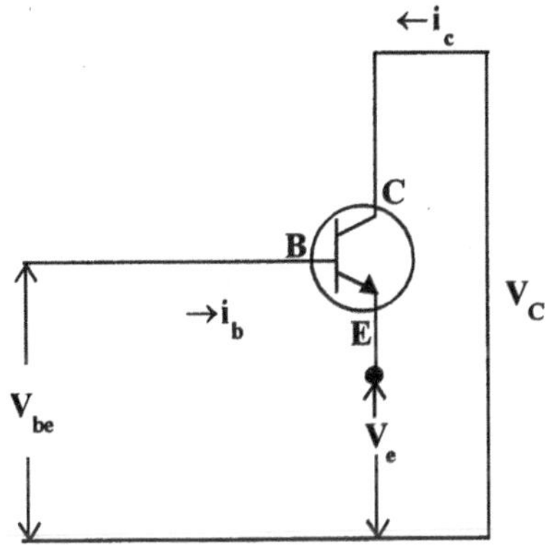

Fig 1.42 For example 1.7

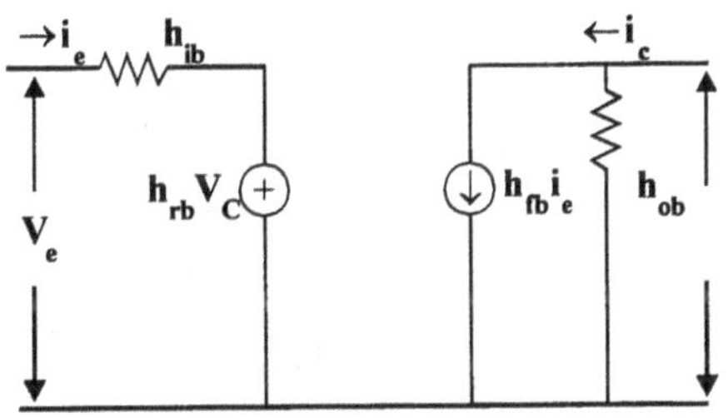

Fig 1.43 Equivalent circuit

C.C :

$$v_b = h_{ic} \cdot i_b + h_{rC}\, v_e$$

$$i_e = h_{fc}\, i_b + h_{oC}\, v_e.$$

$$i_e = h_{fc}\, i_b - i_1$$

$$= (h_{fe}\, i_b - V_{ce}\, h_{oc})$$

$$v_{be} = i_b \cdot h_{ic} - h_{rc}\, v_{ce} + v_{ce} = i_b \cdot h_{ic} \qquad \text{Since } v_{ce} = 0$$

But $$h_{ie} = \left. \frac{V_{be}}{I_b} \right|_{V_{ce}=0}$$

$$\therefore \quad \boxed{h_{ie} = h_{ic}.} \qquad \text{........ (1.26)}$$

$$h_{fe} = \left. \frac{i_c}{i_b} \right| V_{ce} = 0$$

$$i_c = -i_b - i_e = -i_b - h_{fc} \cdot i_b . + h_{oc}\, v_{ce}$$

But $v_{ce} = 0$

$$= -i_b - (h_{fc}\, i_b - h_{oc}\, v_{ce})$$

$$= -i_b - h_{fc}\, i_b + h_{oc}\, v_{ce} \qquad v_{ce} = 0$$

$$h_{fe} = \frac{-i_b - h_{fc} i_b}{i_b} = -(1 + h_{fc})$$

$$\boxed{h_{fe} = -(1 + h_{fc})} \qquad \text{........ (1.27)}$$

$$h_{re} = \left.\frac{V_{be}}{V_{ce}}\right| i_{b=0}$$

$$V_{be} = i_b\, h_{ie} - h_{rc}\, V_{ce} + V_{ce}.$$

$$i_b = 0$$

∴ $$h_{re} = -\frac{h_{rc} V_{ce} + V_{ce}}{V_{ce}} = 1 - h_{rc}$$

$$\boxed{h_{re} = 1 - h_{rc}} \qquad \text{........ (1.28)}$$

$$h_{oe} = \left.\frac{i_c}{V_{ce}}\right| i_b = 0$$

$$i_c = -i_e - i_b$$

But $i_b = 0$.

$$i_c| = i_e$$

$$i_e = V_{ce}\, hoc$$

∴ $$\frac{i_e}{V_{ce}} = \frac{i_e}{V_{ce}} = h_{oc}$$

∴ $$i_c = -i_e = -h_{fc}\, i_b + V_{ce}\, h_{oc}$$

∴ $$\boxed{h_{oe} = h_{oc}} \qquad \text{........ (1.29)}$$

Since, $i_b = 0$

∴ $$i_c = V_{ce} \times h_{oc}$$

∴ $h_{oe} = h_{oc}$ is admittance in mhos (℧).

1.6 HIGH INPUT RESISTANCE TRANSISTOR CIRCUITS

In some applications the amplifier circuit will have to have very high input impedance. Common Collector Amplifier circuit has high input impedance and low output impedance. But its $A_V < 1$. If the input impedance of the amplifier circuit is to be only 500 kΩ or less the Common Collector Configuration can be used. This circuit is known as the ***Darlington Connection*** (named after Darlington) or ***Darlington Pair Circuit***. (Fig. 1.44).

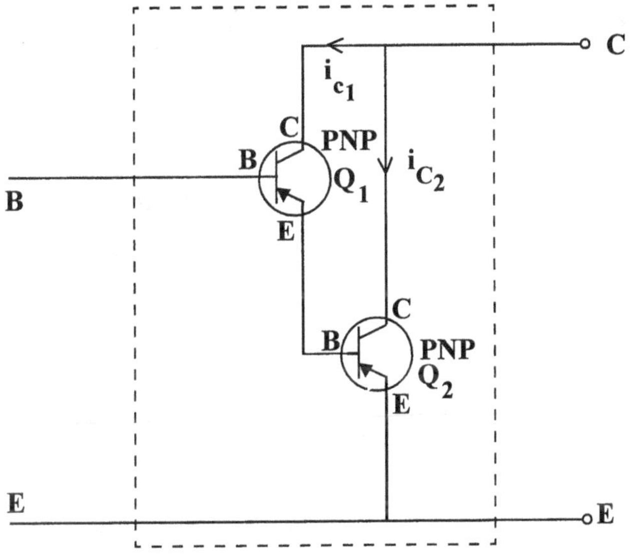

Fig 1.44 Darlington pair circuit.

In this circuit, the two transistors are in Common Collector Configuration. The output of the first transistor Q_1 (taken from the emitter of the Q_1) is the input to the second transistor Q_2 at the base. The input resistance of the second transistor constitutes the emitter load of the first transistor. So, Darlington Circuit is nothing but two transistors in Common Collector Configuration connected in series. The same circuit can be redrawn as AC equivalent circuit. So, DC is taken as ground shown in Fig.1.45. Hence, 'C' at ground potential. Collectors of transistors Q_1 and Q_2 are at ground potential. The AC equivalent Circuit is shown in Fig. 1.45.

There is no resistor connected between the emitter of Q_1 and ground i.e., Collector Point. So, we can assume that infinite resistance is connected between emitter and collector. For the analysis of the circuit, consider the equivalent circuit shown in Fig. 1.45 and we use Common Emitter ***h-parameters***, h_{ie}, h_{re}, h_{oe} and h_{fe}.

For PNP transistor, I_c leaves the transistor, I_e enters the transistor and I_b leaves the transistor.

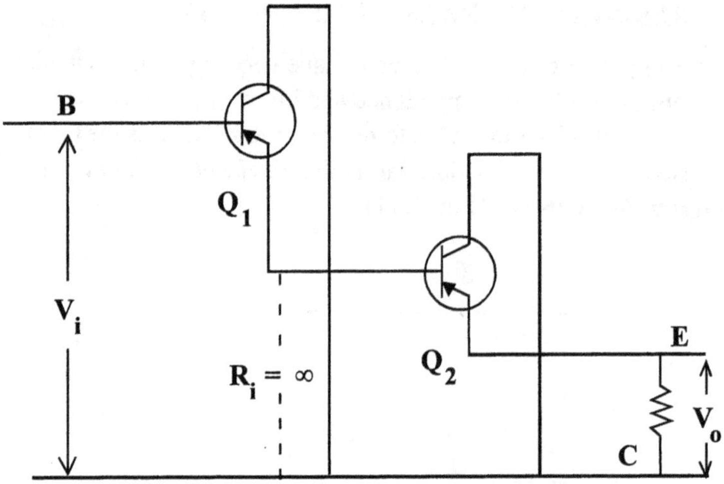

Fig 1.45 Equivalent circuit.

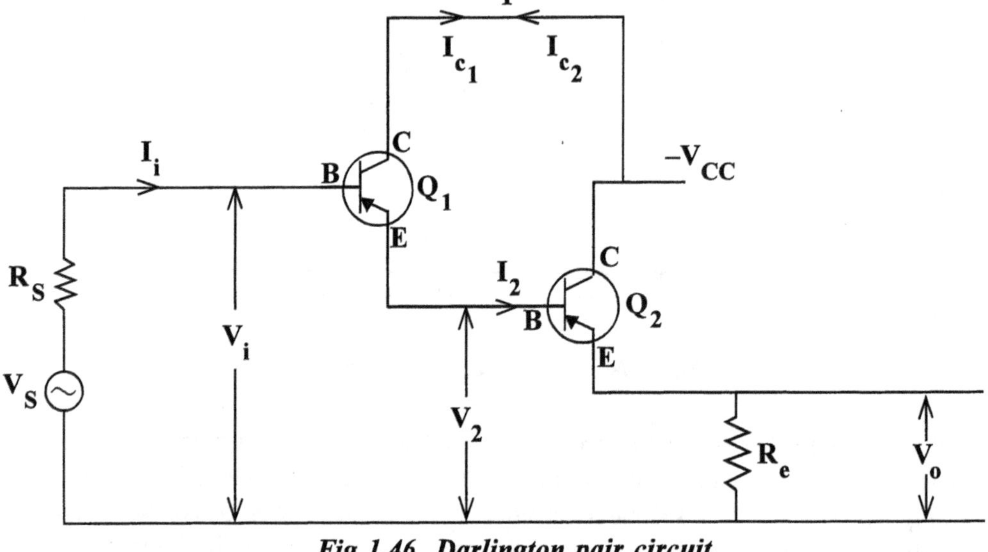

Fig 1.46 Darlington pair circuit.

1.6.1 CURRENT AMPLIFICATION FOR DARLINGTON PAIR

$$I_c = I_{c1} + I_{c2}.$$

$$I_{c1} = I_{b1}\, h_{fe};$$

$$I_{c_2} = I_{b_2}\, h_{fe}. \text{ (Assuming identical transistor and } h_{fe} \text{ is same)}$$

But $\quad I_{b_2} = I_{e_1}$ (the emitter of Q_1 is connected to the base of Q_2)

$$\therefore \quad I_c = I_{b1}\, h_{fe} + I_{e1}\, h_{fe} \qquad \text{.....(1.30)}$$

$$I_{e_1} = I_{b_1} + I_{c_1} = I_{b_1}\,(1 + h_{fe}) \qquad \text{.....(1.31)}$$

$$\therefore \quad I_{c_1} = h_{fe}\, I_{b_1}$$

Substituting equation (1.31) in (1.30),

$$I_c = I_{b_1} h_{fe} + I_{b_1}(1 + h_{fe})h_{fe} = I_{b_1}(2h_{fe} + h_{fe}^2)$$

But $h^2_{fe} >> 2h_{fe}$

Since, h_{fe} is of the order of 100.

$$\therefore \quad I_c = I_{b_1} h_{fe}^2$$

It means that we get very large current amplification $\left(A_I = \frac{I_c}{I_{b_1}}\right)$' in the case of Darlington Pair Circuit, it is of the order h_{fe}^2 i.e. $100^2 = 10{,}000$.

$$\therefore \quad \boxed{A_I = \frac{I_c}{I_{b_1}} \cong (h_{fe})^2}$$

1.6.2 Input Resistance (R_i)

Input resistance R_{i_2} of the transistor Q_2 (which is in Common Collector Configuration) in terms of ***h- parameters*** in Common Emitter Configuration is,

$$R_{i_2} = h_{ie} + (1+h_{fe}) R_L$$

But $\quad h_{ie} << h_{fe}R_L$, and $h_{fe} R_E >> h_{ie}$ $A_{I_2} = \frac{I_0}{I_2} = (1 + h_{fe})$

Here R_L is R_e, since, output is taken across emitter resistance.

$$\therefore \quad Ri_2 \cong (1+ h_{fe}) R_e$$

The input resistance Ri_1 of the transistor Q_1 is, since it is in Common Collector Configuration,

$$R_i = h_{ic} + h_{rc} A_I.R_L.$$

Expressing this in term of Common Emitter ***h-parameters***,

$$hic \cong h_{ie}; h_{rc} \cong 1.$$

(For Common Collector Reverse Voltage Gain = 1) and R_L for transistor Q_1 is the input resistance of transistor Q_2.

$$\therefore \quad R_{i_1} = h_{ie} + A_{I_1} R_{i_2}. \; R_{i_2} \text{ is large,}$$

Therefore, $h_{oe}.R_{i_2} \leq 0.1$. and A_I # $1+ h_{fe}$

$$R_{i_1} \cong A_{I_1} R_{i_2}$$

$$R_{i_2} = (1 + h_{fe}) R_e$$

But the expression for Common Collector Configuration in terms of Common Emitter ***h-parameters*** is

$$A_I = \frac{1+h_{fe}}{1+h_{oe}.R_L}$$

Here, $R_L = R_{i2}$ and $R_{i2} = (1 + h_{fe}) R_e$.

$$\therefore \quad A_{I1} = \frac{1+h_{fe}}{1+h_{oe}(1+h_{fe})R_e}$$

$h_{oe} R_e$ will be < 0.1 and can be neglected.

$\therefore \quad h_{0e}$ value is of the order of μ mhos(micro mhos)

$$\therefore \quad A_{I1} = \frac{1+h_{fe}}{1+h_{oe}h_{fe}R_e}$$

$$\therefore \quad R_{i1} \simeq A_{I1} \cdot R_{i2}$$

$$\boxed{R_i \simeq \frac{(1+h_{fe})^2 R_e}{1+h_{oe}h_{fe}R_e}}$$

This is a very high value. If we take typical values, of $R_e = 4\ k\Omega$, using ***h-parameters***,

$R_{i_2} = 205\ k\Omega$.

$R_i = 1.73\ M\Omega$.

$A_I = 427$.

Therefore, Darlington Circuit has ***very high input impedance*** and very ***large current gain*** compared to Common Collector Configuration Circuit .

1.6.3 Voltage Gain (A_v)

General expression for A_v for Common Collector in term of ***h-parameters*** is

$$A_v = 1 - \frac{h_{ie}}{R_i};\ h_{ie} \cong h_{ic} \quad \text{or } A_{V1} = \frac{V_2}{V_i} = \left[\frac{1-h_{ie}}{R_{i1}}\right]$$

But $\quad R_i \cong A_{I1}.R_2. \quad \therefore A_{V1} = \left(-\frac{h_{ie}}{A_{I_1} R_{i_2}}\right)$

$$\therefore \quad A_{V2} = \frac{V_0}{V_2} = \left(1 - \frac{h_{ie}}{R_{i_2}}\right)$$

Therefore, over all Voltage Gain $A_v = A_{v1} \times A_{v2}$

$$= \left(1 - \frac{h_{ie}}{A_{I_1}.R_{i_2}}\right)\left(1 - \frac{h_{ie}}{R_{i_2}}\right)$$

$$A_v \cong \left(1 - \frac{h_{ie}}{Ri_2}\right) \quad \because A_{I_1} R_{i_2} >> h_{ie} \quad \because A_{i_1} \text{ is} >> 1$$

Therefore, A_v is always < 1.

1.6.4 Output Resistance (R_0)

The general expression for R_o of a transistor in Common Collector Configuration in terms of Common Emitter ***h-parameters*** is,

$$R_o = \frac{R_s + h_{ie}}{1+h_{fe}}$$

$$\therefore \quad R_{o1} = \frac{R_s + h_{ie}}{1+h_{fe}}$$

Now for the transistor Q_2, R_s is R_{o1}.

$$\therefore \qquad R_{o2} = \frac{\frac{R_S + h_{ie}}{1+h_{fe}} + h_{ie}}{1+h_{fe}}$$

Therefore, R_{o2} is the output resistance of the Darlington Circuit.

$$\therefore \qquad R_{o2} = \frac{R_s + h_{ie}}{(1+h_{fe})^2} + \frac{h_{ie}}{1+h_{fe}}$$

This is a small value, since, $1+h_{fe}$ is >> 1.

Therefore, the characteristic of Darlington Circuit are

1. *Very High Input Resistance (of the order of $M\Omega$).*
2. *Very Large Current Gain (of the order of 10, 000).*
3. *Very Low Output Resistance (of the order of few Ω).*
4. *Voltage Gain, $A_v < 1$.*

Darlington Pairs are available in a single package with just three leads, like one transistor in integrated form.

Disadvantages :

We have assumed that the ***h-parameters*** of both the transistor are identical. But in practice it is difficult to make out. ***h-parameters*** depend upon the operating point of Q_1 and Q_2. Since the emitter current of transistor Q_1 is the base current for transistor Q_2, the value of $I_{c2} >> I_{c1}$

1. *The quiescent or operating conditions of both the transistor will be different. h_{fe} value will be small for the transistor Q_1. $\because h_{fe} = (I_c/I_b).I_{b2}$ is less*
   CDIL make CIL997 is a transistor of Darlington Pair Configuration with $h_{fe} = 1000$.
2. *The second drawback is leakage current of the 1*[st] *transistor Q_1 which is amplified by the second transistor Q_2 ($\because \; I_{e1} = I_{b2}$).*
   Hence overall leakage current is more. Leakage Current is the current that flows in the circuit with no external bias voltages applied
   - (a) *The **h-parameters** for both the transistors will not be the same.*
   - (b) *Leakage Current is more.*
   - (b) *Darlington transistor pairs in single package are available with h_{fe} as high as 30,000*

1.6.5 R-C Coupled Amplifier Circuit (Single Stage)

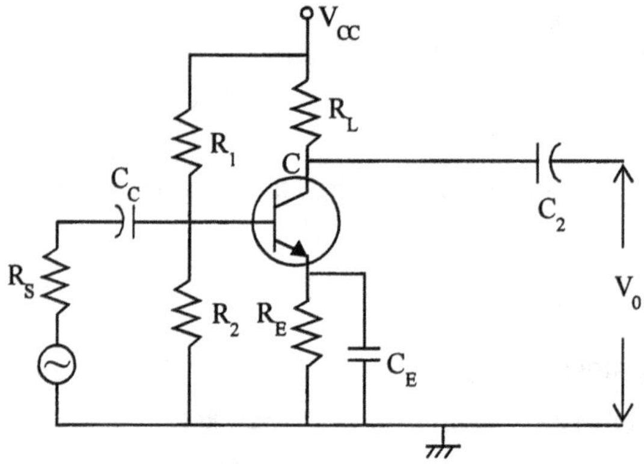

Fig. 1.47 RC Coupled amplifier circuit

The circuit is shown in Fig. 1.47.

R_L is the load resistor that develops the output voltage from the transistor. C_2 is used to couple the AC component of the output to R_L. (Fig. 1.47).

In the self bias circuit, R_E the emitter resistors is connected between emitter and ground. But through R_E, a negative feedback path is there. So to prevent A.C. negative feedback, if a capacitor is connected in parallel with R_E, and it is chosen such that X_E provides least resistance path compared to R_E, AC signal passes through C_E and not through R_E. Therefore, there will not be any negative feedback for AC signals.

$$X_E \text{ is chosen such that } X_E \leq \frac{R_E}{10}$$

and X_E is chosen at the frequency f_1. For frequencies higher than f_1, X_E any way will be less.

The equivalent circuit for the above transistor configuration is shown in Fig. 1.48.

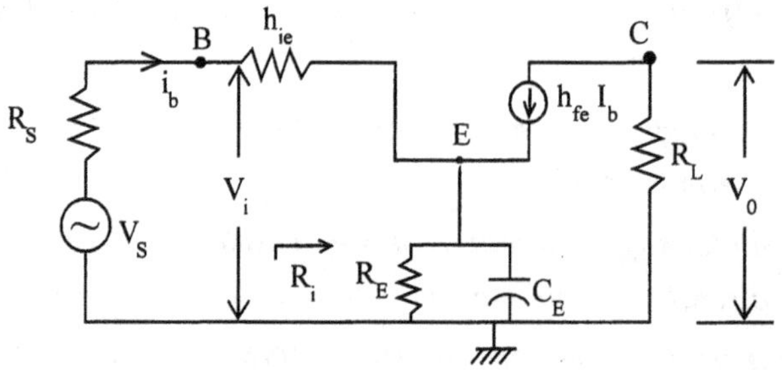

Fig. 1.48 Equivalent circuit

Suppose the value of C_C is very large. Then at the low and medium frequencies, the impedance is negligible. So the *equivalent* circuit is as shown. We shall consider the effect of C_E later.

From the *equivalent* circuit,

$$V_0 = -h_{fe} \cdot i_b \cdot R_L$$

negative sign is used since it is NPN transistor. The collector current is flowing into the transistor.

Input current, $I_b = V_s / R_s + R_i$

For a transistor amplifier circuit in common emitter configuration,

$$R_i = h_{ie} + (1 + h_{fe}) Z_E.$$

Where $Z_E = R_E$ in parallel with $C_E = \dfrac{R_E}{1 + j\omega C_E R_E}$

$$\therefore \quad V_0 = -h_{fe} \cdot R_L \times \frac{V_S}{R_S + h_{ie} + \dfrac{(1+h_{fe})R_E}{1 + j\omega C_E R_E}}$$

$$\therefore \quad A_V = \frac{V_0}{V_S}$$

A_V at low frequencies, (LF)

$$A_V(LF) = \frac{V_0}{V_S}$$

$$\boxed{A_V(LF) = \frac{-h_{fe} \cdot R_L}{R_S + h_{ie} + \dfrac{(1+h_{fe})R_E}{1 + j\omega C_E R_E}}}$$

When ω is large, $\dfrac{(1+h_{fe})R_E}{1 + j\omega C_E R_E}$ can be neglected. So in the Mid Frequency range, (M.F.),

$$\therefore \quad \boxed{A_V(M.F) = \frac{-h_{fe} R_L}{R_S + h_{ie}}}$$

In this expression, there is no f or ω term. Hence in the mid frequency range, A_V is independent of f or the gain remains constant, irrespective of change in frequency.

1.6.6 Effect of Coupling Capacitor on Low Frequency Response

Consider the circuit shown in Fig. 1.49

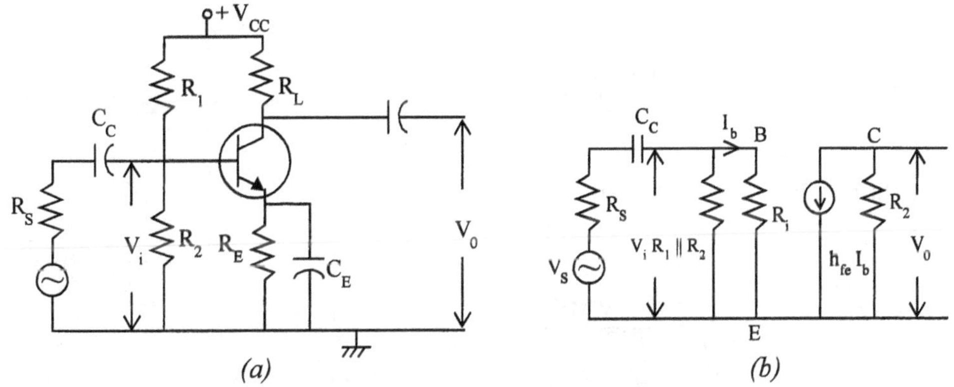

Fig. 1.49 Effect of coupling capacitors

Suppose that the value of C_E is such that its effect on the frequency response can be neglected and the value of X_C at low frequencies is such that it is not a simple short circuit for A.C. signals, so that its effect has to be considered.

The effect of C_E is, voltage drop across C_C will reduce V_i with a corresponding drop in V_0.

The frequency at which the gain drops by a factor of $\frac{1}{\sqrt{2}}$, is the lower 3 db frequency.

$$f_1 = \frac{1}{2\pi(R_S + R_i')C_C}$$

where $R_i' = R_1 \parallel R_2 \parallel R_i$

and $R_i = h_{ie}$, for an ideal capacitor C_E.

These expressions are valid when emitter is bypassed. In AC *equivalent circuit, emitter is assumed to be at GND potential. Therefore C_E has no effect.*

Large values of capacitors are required (10 μF, 5 μF etc.) in transistor amplifiers for coupling and bypass purposes for good low frequency response. Since these values are large, only electrolytic capacitors are used. Capacitors of such large values are not available in other types of capacitors. These capacitors are bulky. So audio amplifier circuits, Integrated Circuits (I.C) often have large capacitors connected externally to the (I.C) itself to provide the required low frequency response.

Example : 1.8

In the Resistance Capacitance (RC) coupled amplifier, $A_{v_m} = 50$, $f_1 = 50$ Hz and $f_2 = 100$ kHz. Find the values of frequencies at which the gain reduces to 40 on either side of midband region.

Solution :

$$A_{VH} = \frac{A_{V\,MF}}{1 + j\left(\frac{f}{f_2}\right)}$$

Phase shift angle $\phi = 180^o - \tan^{-1}\left(\frac{f}{f_2}\right)$

$$A_{V\,LF} = \frac{A_{V\,MF}}{\sqrt{1 + \left(\frac{f_1}{f}\right)^2}}$$

$$A_{VL} = \frac{A_{VM}}{1 + j\left(\frac{f_1}{f}\right)}$$

$$\phi = 180^o + \tan^{-1}\left(\frac{f_1}{f}\right)$$

At the frequency f, the gain is $A_{V\,L.F} = 40$.

$A_{VL.F} = 40$, $A_{VM..F} = 50$, $f_1 = 50$ Hzs, $f = ?$

$$\frac{40}{50} = \frac{1}{\sqrt{1+\left(\frac{50}{f}\right)^2}} \qquad \therefore\ f = 66.66 \text{ Hz.}$$

$$A_{V\,H.F} = \frac{A_{V\,M.F}}{\sqrt{1+\left(\frac{f}{f_2}\right)^2}}$$

At frequency f, $A_{V\,H.F} = 40$. $\therefore$ $f = ?$ $f_2 = 100$ kHz.

$$\frac{40}{50} = \frac{1}{\sqrt{1+\left(\frac{f}{100\times10^3}\right)^2}}; \qquad f = 75 \text{ kHz.}$$

Example : 1.9

An amplifier of 40db gain has both its input and output load resistances equal to 600Ω. If the amplifier input power is – 30db, what is the output power and voltage ? Assume that reference power level used is 1mW in a 600 Ω resistance.

Solution :

Since it is logarithmic scale, for dbs,

P_0 (db) = $(P_i)_{db}$ + (gain $_{db}$);

$P_0 = P_i \times A_P$ Output power (P_0) = Input Power (P_i) × Power Gain (A_P)

$P_0 = ?$ $P_i = -30$db, gain = 40db

$\therefore$ $P_0 = -30 + 40 =$ **10db.**

Reference power level = 1 mW.

$$\therefore \quad 10 \log = \left(\frac{P_0}{1\text{mW}}\right) = 10\text{db}$$

$$10 \log \frac{P_0}{0.001} = 10$$

$$\log \frac{P_0}{0.001} = 1$$

$$\frac{P_0}{0.001} = 10$$

$$\because \quad \log 10 = 1$$

$$\therefore \quad P_0 = 10 \times 0.001$$

$$= \mathbf{10mW}$$

(With respect to, 1mW, the input and output powers are compared and expressed in db.

$$\therefore \quad 10 \log \frac{P_0}{1\text{mW}} = 10\text{db})$$

RMS output voltage is,

$$P_0 = \frac{V_0^2}{R}$$

or

$$V_0 = \sqrt{P_0.R} = \sqrt{10 \times 10^{-3} \times 600}$$

$$= \sqrt{6} = \mathbf{2.45\ V}$$

Example : 1.10

An amplifier has $R_i = 0.5\ k\Omega$ and $R_0 = 0.05\ k\Omega$. The amplifier gives an output voltage of 1V peak for an input voltage of 1mV peak. Find A_V, A_i, A_P in db.

Solution :

$$V_0\text{ (peak)} = 1\text{V}.$$

$$V_{0\text{ (rms)}} = \frac{V_0\text{(peak)}}{\sqrt{2}}; \quad V_{0\text{ (rms)}} = \frac{1\text{V}}{\sqrt{2}}$$

$$A_V = 20 \log\left(\frac{V_0}{V_i}\right)$$

$$= 20 \log \left(\frac{1}{1\text{mV}} \right) = 60 \text{ db}$$

$$I_{i\,(peak)} = V_{i\,(Peak)} / R_i$$
$$= 1\text{mV}/500$$
$$= 2 \times 10^{-6} \text{ A.}$$

$$I_{0\,(peak)} = V_{0\,(Peak)} / R_0$$
$$= 1/50$$
$$I_{0\,(Peak)} = 0.02 \text{ A.}$$

$$\therefore \text{ Current Gain } (A_i) = 20 \log \left(\frac{0.02}{2 \times 10^{-6}} \right) = \mathbf{80\ db}$$

$$P_i = \frac{V_i^2}{R_i} = 10^{-9} \text{ W;}$$

$$P_0 = \frac{V_0^2}{R_0} = 10^{-2} \text{ W.}$$

$$\text{Power gain,} \quad A_P = 10 \log \left(\frac{P_0}{P_i} \right) = \mathbf{70\ db}$$

SUMMARY

- Electronic amplifiers are classified based on (a) Frequency range (b) Type of coupling (c) Output power (d) Type of signal handled.
- Power amplifiers are classfied as (a) Class A (b) Class B (c) Class AB (d) Class C (e) Class D and (f) Class S amplifiers.
- The type of distortion that can occur are classified as (a) amplitude (b) frequency (c) Phase distrotions
- h-parameter analysis enables to derive expressions for A_I, A_v, R_i, R_o, A_p of amplifiers in terms of transistor parameters and circuit components. These equations are derived.
- Gain is expressed in decibles.
- Comparison between the three types, C.E, CB, CC amplifiers is given and expressions are derived for A_I, A_V, Z_i, Z_0, A_p etc.
- Darlington pair amplifier circuit gives high R_i (MΩ) large current gain (10K), low voltage gain (<1) and low R_0 (few Ω). The circuit is explained and equations are derived.
- Single stage R-C coupled amplifier circuit is given and expressions for lower cut-off frequency and upper cut-off frequency are given.

OBJECTIVE TYPE QUESTIONS

1. The units of h-parameters are ....
2. h-parameters are named as hybrid parameters because ....
3. The general equations governing h-parameters are

   V_1 = ...

   I_2 = ...
4. The parameter h_{re} is defined as h_{re} = ...
5. h-parameters are valid in the .... frequency range.
6. Typical values of h-parameters in Common Emitter Configuration are ....
7. The units of the parameter h_{rc} are ....
8. Conversion Efficiency of an amplifier circuit is ....
9. Expression for current gain A_I in terms of h_{fe} and h_{re} are A_I = ...
10. In Common Collector Configuration, the values of $h_{rc} \simeq$ ....
11. In the case of transistor in Common Emitter Configuration, as R_L increases, R_i ....
12. Current Gain A_I of BJT in Common Emitter Configuration is high when R_L is ....
13. Power Gain of Common Emitter Transistor amplifier is ....
14. Current Gain A_I in Common Base Configuration is ....
15. Among the three transistor amplifier configurations, large output resistance is in .... configuration.
16. Highest current gain, under identical conditions is obtained in .... transistor amplifier configuraiton.
17. C.C Configuration is also known as .... circuit.

ESSAY TYPE QUESTIONS

1. Write the general equations in terms of h-parameters for a BJT in Common Base Amplifiers configuration and define the h-parameters.
2. Convert the h-parameters in Common Base Configuration to Common Emitter Configuration, deriving the necessary equations.
3. Compare the transistor (BJT) amplifiers circuits in the three configurations with the help of h-parameters values.
4. Draw the h-parameter equivalent circuits for Transistor amplifiers in the three configurations.
5. With the help of necessary equations, discuss the variations of A_V, A_I, R_i R_O, A_P with R_S and R_L in Common Emitter Configuration.
6. Discuss the Transistor Amplifier characteristics in Common Base Configuration and their variation with R_S and R_L with the help of equations.
7. Compare the characteristics of Transistor Amplifiers in the three configurations.

ANSWERS OF OBJECTIVE TYPE QUESTIONS

1. Ω, mhos and constants
2. the units of different parameters are not the same
3. $h_{11}I_1 + h_{12}V_2$
   $h_{21}I_1 + h_{22}V_2$
4. $\left.\frac{\partial V_B}{\partial V_C}\right|_{I_B=K}$
5. Audio
6. $h_{ie} = 1\ k\Omega$, $h_{re} = 1.5 \times 10^{-4}$,
   $h_{oe} = 6\mu mhos$, $h_{fe} = 200$
7. No units (constant)
8. $\frac{\text{AC Signal Power Delivered to the Load}}{\text{DC Input Power}} \times 100$
9. $\frac{h_{fe}}{1+h_{oe}R_L}$
10. 1
11. Decreases
12. Low
13. Large
14. < 1
15. Common Base configuration
16. Common Collector Configuration
17. Voltatge follower/buffer

2 Multistage Amplifiers

In this Unit,

- Cascading of single stage amplifiers is discussed.
- Expressions for overall voltage gain are derived.
- Lower cut-off frequency, upper cut-off frequency, when n-stages are cascaded are given.
- Phase response of an amplifier, decibel voltage gain, stiff coupling terms are explained.
- Other types of transistor circuits, Darlington pair circuits, Boot strapped sweep circuit, Cascode amplifiers are also discussed.
- Numerical examples are also given.

2.1 MULTISTAGE AMPLIFIERS METHODS OF INTER STAGE COUPLING

If the amplification obtained from a single stage amplifiers is not sufficient, two or more such amplifiers are connected in Cascade or Series i.e., the output of the first stage will be the input to the second stage. This voltage is further amplified by the second stage and so we get large amplification or large output voltage compared to the input. In the multistage amplifiers, the output of the first stage should be coupled to the input of the second stage and so on : Depending upon the type of coupling, the multistage amplifiers are classified as:

1. Resistance and Capacitance Coupled Amplifiers (RC Coupled)
2. Transformer Coupled Amplifiers
3. Direct Coupled DC Amplifiers
4. Tuned Circuit Amplifiers (R-L coupled)

2.1.1 Resistance and Capacitance Coupled Amplifiers (RC Coupled)

This type of amplifier is very widely used. It is least expensive and has good frequency response. In the multistage resistive capacitor coupled amplifiers, the output of the first stage is coupled to the next through coupling capacitor and R_L. In two stage Resistor Capacitor coupled amplifiers, there is no separate R_L between collector and ground, but R_C, the resistance between collector and V_{CC} (R_C) itself acts as R_L in the AC equivalent circuit.

2.1.2 Transformer Coupled Amplifiers

Here the output of the amplifier is coupled to the next stage or to the load through a transformer. With this overall circuit gain will be increased $\left(\because \frac{N_2}{N_1} = \frac{V_2}{V_1}\right)$ and also impedance matching can be achieved.

But such transformer coupled amplifiers *will not have broad frequency response* i.e., $(f_2 - f_1)$ is small since inductance of the transformer windings will be large. So Transformer coupling is done for power amplifier circuits, where impedance matching is critical criterion for maximum power to be delivered to the load.

2.1.3 Direct Coupled (DC) Amplifiers

Here DC stands for direct coupled and not (direct current). In this type, there is no reactive element. L or C used to couple the output of one stage to the other. The AC output from the collector of one stage is directly given to the base of the second stage transistor directly. So type of amplifiers are used for large amplification of DC and using low frequency signals. Resistor Capacitor coupled amplifiers can not be used for amplifications of DC or low frequency signals since X_C the capacitive reactance of the coupling capacitor will be very large or open circuit for DC ($X_C = 1/2\,\pi f_C$. If $f = 0$ or low, then $X_C \rightarrow \infty$.)

2.1.4 Tuned Circuit Amplifiers

In this type there will be one RC or LC tuned circuit between collector and V_{CC}, in the place of R_C. These amplifiers will amplify signals of only fixed frequency f_0 which is equal to the resonance frequency of the tuned circuit LC. These are also used to amplify signals of a narrow band of frequencies centered around the tuned frequency f_0.

2.1.5 BANDWIDTH OF AMPLIFIERS

The gain provided by an amplifier circuit is not the same for all frequencies because the reactance of the elements connected in the circuit and the device reactance value depend upon the frequency. *Bandwidth of an amplifier is the frequency range over which the amplifier stage gain is reasonably constant within* $\pm$ *3 db, or 0.707 of* A_V *Max Value.*

Based upon the B.W. of the amplifiers, they can be classified as :

1. ***Narrow band amplifiers :*** Amplification is restricted to a narrow band of frequencies around a centre frequency.

   There are essentially tuned amplifiers.

2. ***Untuned amplifiers :*** These will have large bandwidth. Amplification is desired over a considerable range of frequency spectrum.

Untuned amplifiers are further classified w.r.t bandwidth.

1. DC amplifiers (Direct Coupled)	:	DC to few kHz
2. Audio frequency amplifiers (AF)	:	20 Hz to 20 kHz
3. Broad band amplifier	:	DC to few MHz
4. Video amplifier	:	100 Hz to few MHz

2.1.5.1 DISTORTION IN AMPLIFIERS

If the input signal is a sine wave the output should also be a true sine wave. But in all the cases it may not be so, which we characterize as distortion. Distortion can be due to the nonlinear characteristic of the device, due to operating point not being chosen properly, due to large signal swing of the input from the operating point or due to the reactive elements L and C in the circuit. Distortion is classified as :

(a) ***Amplitude distortion :*** This is also called non linear distortion or harmonic distortion. This type of distortion occurs in large signal amplifiers or power amplifiers. It is due to the nonlinearity of the characteristic of the device. This is due to the presence of new frequency signals which are not present in the input. If the input signal is of 10 KHz the output signal should also be 10 KHz signal. But some harmonic terms will also be present. Hence the amplitude of the signal (rms value) will be different $V_0 = A_V V_i$. But it will be V_0'.

(b) ***Frequency distortion :*** The amplification will not be the same for all frequencies. This is due to reactive component in the circuit.

(c) ***Phase-shift delay distortion :*** There will be phase shift between the input and the output and this phase shift will not be the same for all frequency signals. It also varies with the frequency of the input signal.

In the output signal, all these distortions may be present or any one may be present because of which the amplifier response will not be good.

Example : 2.1

The transistor parameters are given as,

$h_{ie} = 2$ k	$h_{fe} = 50$	$h_{re} = 6 \times 10^{-4}$	$h_{oe} = 25\ \mu A/V$
$h_{ic} = 2$ k	$h_{fc} = -51$	$h_{rc} = 1$	$h_{oc} = 25\ \mu A/V$

Find the individual as well as overall voltage gains and current gains, for the circuit shown in Fig. 2.1(a).

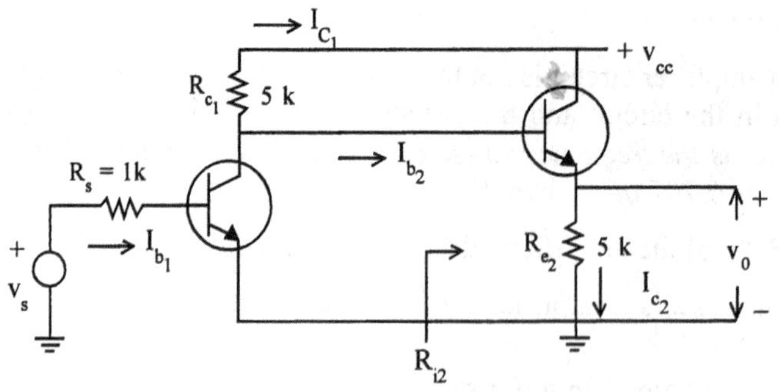

Fig. 2.1 (a) Circuit for Ex : 2.1

Solution :

It is advantageous to start the analysis with the last stage. Compute the current gain first, then the input impedance and voltage gain.

II stage : For the second stage $R_L = R_{e_2}$

$$A_{I_2} = \frac{-I_{e2}}{I_{b2}} = \frac{-h_{fc}}{1+h_{oc}R_{e2}}$$

(negative (–) sign is there because for NPN transistor, I_e is negative (–) since it is leaving the transistor)

$$= \frac{51}{1+25\times10^{-6}\times5\times10^{3}}$$

$$= \mathbf{45.3}$$

Input impedance $R_{i_2} = h_{ic} + h_{rc}\ A_{I_2}\ R_{e_2}$

$= 2 + 45.3 \times 5\text{ k} = 228.5\text{ k}.$

∴ Input Z of the C.C. stage is very high.

Voltage gain of the second stage

$$A_{V_2} = \frac{V_0}{V_2} = A_{I_2}\frac{R_{e_2}}{R_{i_2}}$$

∵ $V_0 = I_{e_2} R_{e_2}$;

V_2 = Input voltage for the second stage = $R_{i_2} . I_i$

∴ $$\frac{V_0}{V_2} = \frac{I_0.R_{e2}}{I_i.R_{i2}}$$

$$= A_{I_2}\frac{R_{e2}}{R_{i_2}}$$

$$= \frac{45.3\times5}{228.5} = \mathbf{0.99}$$

I stage : For the first stage, the net load resistance in the parallel combination of R_{C_1} and R_{i_2} or

$$R_{i_2} = \frac{R_{c_1}.R_{i_2}}{R_{c_1}+R_{i_2}} = \frac{5\times 228.5}{233.5} = 4.9\ \text{k}\Omega$$

Hence
$$A_{I_1} = \frac{-I_{c_1}}{I_{b_1}} = \frac{-h_{fe}}{1+h_{oe}R_{L_1}} = \frac{-50}{1+25\times10^{-6}\times4.9\times10^{3}} = -44.5$$

The input impedance of the first stage will also be the input Z of the two stages since input Z of the second stage is also considered in determining the value of R_{L_1}. R_{i_1} depends on R_{L_1}.

$\therefore$ $R_{i_1} = h_{i_e} + h_{r_e} A_{I_1} R_{L_1}$ (from the standard formula)

$= 2 - 6 \times 10^{-4} \times 44.5 \times 4.9$

$R_{i_1} = 1.87\ \text{k}\Omega.$

Voltage gain of the first stage is,

$$A_{V_1} = \frac{V_2}{V_1} = \frac{A_{I_1}.R_{L_1}}{R_{i_1}}$$

$$= \frac{-44.5\times4.9}{1.87} = -116.6$$

$$A_I = \frac{-I_{e_2}}{I_{b_1}} = \frac{-I_{e_2}}{I_{b_2}}.\frac{+I_{b_2}}{I_{c_1}}.\frac{+I_{c_1}}{I_{b_1}}$$

$$= -A_{I_2}.\frac{I_{b_2}}{I_{c_1}}.A_{I_1}$$

In the actual circuit, the current gets branched into I_{c_1} and I_{b_2} which depend upon the values of R_{C_1} and R_{i_2} [Fig. 2.1(b)].

$\because$ $(-I_{b_2} + I_{c_1}) R_{c_1} = I_{b_2} . R_{i_2}$

$\therefore$ $+ I_{b_2} (R_{i_2} + R_{c_1}) = I_{c_1} . R_{c_1}$

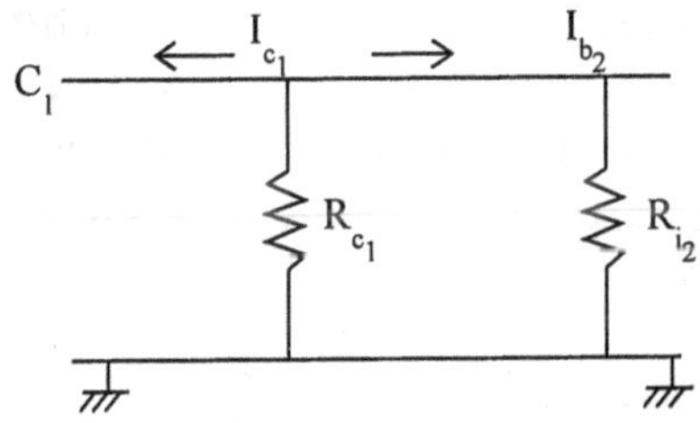

Fig. 2.1 (b) Current branching

$$\frac{I_{b_2}}{I_{c_1}} = \frac{R_{c_1}}{R_{c_1} + R_{i_2}}$$

$$\therefore \quad A_I = A_{I_2} A_{I_1} \cdot \frac{R_{c_1}}{R_{i_2} + R_{c_1}}$$

$$= \frac{45.3 \times (-44.5) \times 5}{228.5 + 5}$$

$$A_I = -43.2$$

$$A_V = \frac{V_0}{V_1} = \frac{V_0}{V_2} \cdot \frac{V_2}{V_1} = A_{V_2} \cdot A_{V_1}$$

$$A_V = 0.99 \times (-116.6) = -115.$$

2.1.6 R-C Coupled Amplifier Circuit (Single Stage)

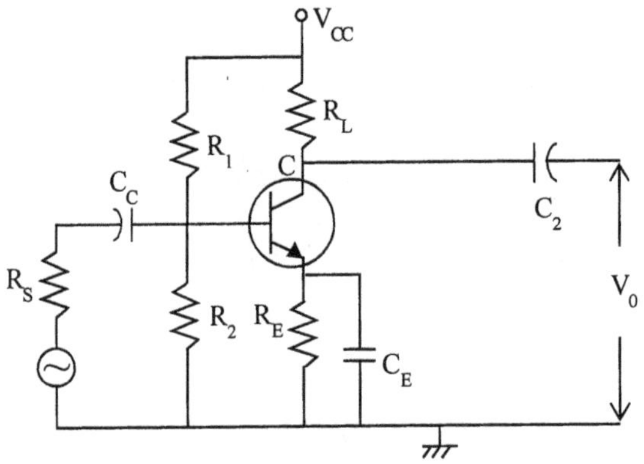

Fig. 2.2 RC coupled amplifier circuit

The circuit is shown in Fig. 2.2.

R_L is the load resistor that develops the output voltage from the transistor. C_2 is used to couple the AC component of the output to R_L.

In the self bias circuit, R_E the emitter resistors is connected between emitter and ground. But through R_E, a negative feedback path is there. So to prevent A.C. negative feedback, if a capacitor is connected in parallel with R_E, and it is chosen such that X_E provides least resistance path compared to R_E, AC signal passes through C_E and not through R_E. Therefore, there will not be any negative feedback for AC signals.

$$X_E \text{ is chosen such that } X_E \leq \frac{R_E}{10}$$

and X_E is chosen at the frequency f_1. For frequencies higher than f_1, X_E any way will be less.

The equivalent circuit for the above transistor configuration is,

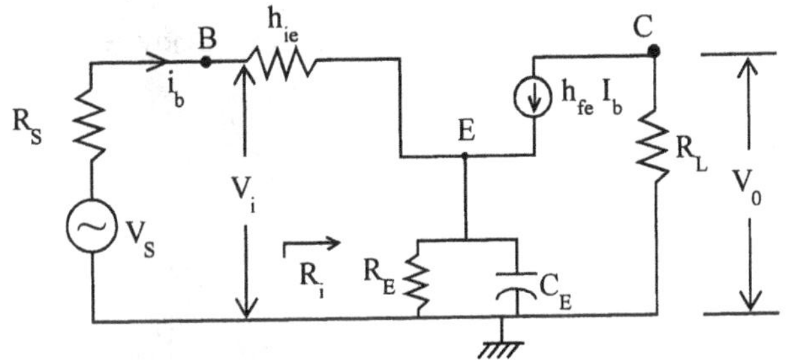

Fig. 2.3 Equivalent circuit

Suppose the value of C_C is very large. Then at the low and medium frequencies, the impedance is negligible. So the *equivalent* circuit is as shown. We shall consider the effect of C_E later.

From the *equivalent* circuit,

$$V_0 = -h_{fe} \cdot i_b \cdot R_L$$

negative sign is used since it is NPN transistor. The collector current is flowing into the transistor.

Input current, $I_b = V_s / R_s + R_i$

For a transistor amplifier circuit in common emitter configuration,

$$R_i = h_{ie} + (1 + h_{fe}) Z_E.$$

Where $Z_E = R_E$ in parallel with $C_E = \dfrac{R_E}{1 + j\omega C_E R_E}$

$$\therefore \quad V_0 = -h_{fe} \cdot R_L \times \frac{V_S}{R_S + h_{ie} + \dfrac{(1+h_{fe})R_E}{1+j\omega C_E R_E}}$$

$$\therefore \quad A_V = \frac{V_0}{V_S}$$

A_V at low frequencies, (LF)

$$A_V(LF) = \frac{V_0}{V_S}$$

$$\boxed{A_V(LF) = \frac{-h_{fe} \cdot R_L}{R_S + h_{ie} + \dfrac{(1+h_{fe})R_E}{1+j\omega C_E R_E}}}$$

When ω is large, $\dfrac{(1+h_{fe})R_E}{1+j\omega C_E R_E}$ can be neglected. So in the Mid Frequency range, (M.F.),

$$\therefore \quad \boxed{A_V(M.F) = \frac{-h_{fe} R_L}{R_S + h_{ie}}}$$

In this expression, there is no f or ω term. Hence in the mid frequency range, A_V is independent of f or the gain remains constant, irrespective of change in frequency.

2.1.7 Effect of Coupling Capacitor on Low Frequency Response

Suppose that the value of C_E is such that its effect on the frequency response can be neglected and

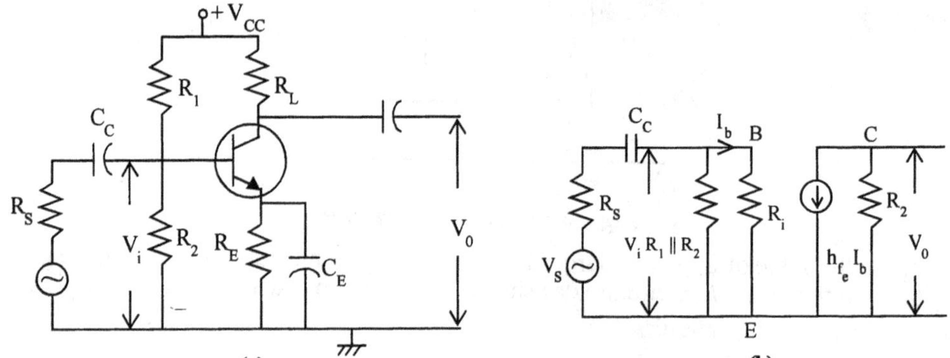

(a) (b)

Fig. 2.4 Effect of coupling capacitors

the value of X_C at low frequencies is such that it is not a simple short circuit for A.C. signals, so that its effect has to be considered [Fig. 2.4 (a) & (b)].

The effect of C_E is, voltage drop across C_C will reduce V_i with a corresponding drop in V_0.

The frequency at which the gain drops by a factor of $\frac{1}{\sqrt{2}}$, is the lower 3 db frequency.

$$f_1 = \frac{1}{2\pi(R_S + R_i')C_C}$$

where $R_i' = R_1 \parallel R_2 \parallel R_i$

and $R_i = h_{ie}$, for an ideal capacitor C_E.

These expressions are valid when emitter is bypassed. In AC *equivalent circuit, emitter is assumed to be at GND potential. Therefore C_E has no effect.*

Large values of capacitors are required (10μF, 5μF etc.,) in transistor amplifiers for coupling and bypass purposes for good low frequency response. Since these values are large, only electrolytic capacitors are used. Capacitors of such large values are not available in other types of capacitors. These capacitors are bulky. So audio amplifier circuits, Integrated Circuits (I.C) often have large capacitors connected externally to the (I.C) itself to provide the required low frequency response.

Example : 2.2

In the Resistance Capacitance (RC) coupled amplifier, $A_{v_m} = 50$, $f_1 = 50$ Hz and $f_2 = 100$ k Hz. Find the values of frequencies at which the gain reduces to 40 on either side of midband region.

Solution :

$$A_{VH} = \frac{A_{V\,MF}}{1 + j\left(\dfrac{f}{f_2}\right)}$$

Phase shift angle $\quad \phi = 180^o - \tan^{-1}\left(\dfrac{f}{f_2}\right)$

$$A_{V\,LF} = \frac{A_{V\,MF}}{\sqrt{1 + \left(\dfrac{f_1}{f}\right)^2}}$$

$$A_{VL} = \frac{A_{VM}}{1 + j\left(\dfrac{f_1}{f}\right)}$$

$$\phi = 180^o + \tan^{-1}\left(\frac{f_1}{f}\right)$$

At the frequency f, the gain is $A_{V\,L.F} = 40$.

$A_{VL.F} = 40, \qquad A_{VM..F} = 50, \qquad f_1 = 50$ Hzs, $f = ?$

$$\frac{40}{50} = \frac{1}{\sqrt{1 + \left(\dfrac{50}{f}\right)^2}} \qquad \therefore\ f = 66.66 \text{ Hz.}$$

$$A_{V\,H.F} = \frac{A_{V\,M.F}}{\sqrt{1 + \left(\dfrac{f}{f_2}\right)^2}}$$

At frequency f, $\quad A_{V\,H.F} = 40. \quad \therefore \quad f = ? \qquad f_2 = 100$ kHz.

$$\frac{40}{50} = \frac{1}{\sqrt{1 + \left(\dfrac{f}{100 \times 10^3}\right)^2}}; \qquad f = 75 \text{ kHz.}$$

Example : 2.3

An amplifier of 40db gain has both its input and output load resistances equal to 600Ω. If the amplifier input power is – 30db, what is the output power and voltage ? Assume that reference power level used is 1mW in a 600 Ω resistance.

Solution :

Since it is logarithmic scale, for dbs,

$$P_o\,(db) = (P_i)_{db} + (gain_{\,db});$$

$$P_o = P_i \times A_P \quad \text{Output power } (P_o) = \text{Input Power } (P_i) \times \text{Power Gain } (A_P)$$

$$P_o = ?\, P_i = -30db, \text{ gain} = 40db$$

$$\therefore \quad P_o = -30 + 40 = \mathbf{10db.}$$

Reference power level = 1mW.

$$\therefore \quad 10 \log = \left(\frac{P_o}{1mW}\right) = 10db$$

$$10 \log \frac{P_o}{0.001} = 10$$

$$\log \frac{P_o}{0.001} = 1$$

$$\frac{P_o}{0.001} = 10$$

$$\because \quad \log 10 = 1$$

$$\therefore \quad P_o = 10 \times 0.001$$

$$= \mathbf{10mW}$$

(With respect to, 1mW, the input and output powers are compared and expressed in db.

$$\therefore \quad 10 \log \frac{P_o}{1mW} = 10db$$

RMS output voltage is,

$$P_o = \frac{V_o^2}{R}$$

or

$$V_o = \sqrt{P_o.R} = \sqrt{10 \times 10^{-3} \times 600}$$

$$= \sqrt{6} = \mathbf{2.45\ V}$$

Example : 2.4

An amplifier has $R_i = 0.5K\Omega$ and $R_0 = 0.05K\Omega$. The amplifier gives an output voltage of 1V peak for an input voltage of 1mV peak. Find A_V, A_i, A_P in db.

Solution :

$$V_o \text{ (peak)} = 1V.$$

$$V_{o\,(rms)} = \frac{V_o\text{(peak)}}{\sqrt{2}}; \qquad V_{o\,(rms)} = \frac{1V}{\sqrt{2}}$$

$$A_V = 20 \log\left(\frac{V_o}{V_i}\right)$$

$$= 20 \log\left(\frac{1}{1mV}\right) = 60 \text{ db}$$

$$I_{i\,(peak)} = V_{i\,(Peak)} / R_i$$

$$= 1mV/500$$

$$= 2 \times 10^{-6} A.$$

$$I_{o\,(peak)} = V_{o\,(Peak)} / R_o$$

$$= 1/50$$

$$I_{o\,(Peak)} = 0.02 \text{ A}.$$

$$\therefore \text{ Current Gain } (A_i) = 20 \log\left(\frac{0.02}{2\times10^{-6}}\right) = \mathbf{80\ db}$$

$$P_i = \frac{V_i^2}{R_i} = 10^{-9} \text{ W};$$

$$P_o = \frac{V_o^2}{R_o} = 10^{-2} \text{ W}.$$

$$\text{Power gain,} \quad A_P = 10 \log\left(\frac{P_o}{P_i}\right) = \mathbf{70\ db}$$

2.2 ANALYSIS OF CASCADED RC COUPLED BJT AMPLIFIER

A RC coupled amplifier using N-P-N transistors in CE configuration shown in Fig. (2.5). Here one power supply is employed for two transistors used in RC coupled BJT amplifier. Here from Fig. (2.5) we notice the resistors R_1, R_2 and R_E which provides biasing and stabilization network.

In the Fig. (2.5) the network connection is arranged as the signal developed across collector resistor R_C of the first stage is coupled to the base of the second stage through the coupling capacitor

C_C. Here the coupling from one stage (output of first stage, CE amplifier is fed to as input for second CE amplifier) to the next is established by a coupling capacitor C_C followed by a connection to a shunt resistor, so these amplifiers are caused RC coupled amplifiers.

Here the purpose of input capacitor C_{in} is to couple ac signal voltage to the base of transistor Q_1. Without capacitor C_{in} the signal source will be in parallel with resistor R_2 and the bias voltage of the base will be affected.

Thus the purpose of C_{in} is to allow only the alternating current from signal source to flow into the input circuit.

Here the purpose of emitter by pass capacitor C_E, offers low reactance path to the signal. If it is not present, then the voltage drop across R_E will reduce the effective voltage across base-emitter terminals and thus gain is reduced.

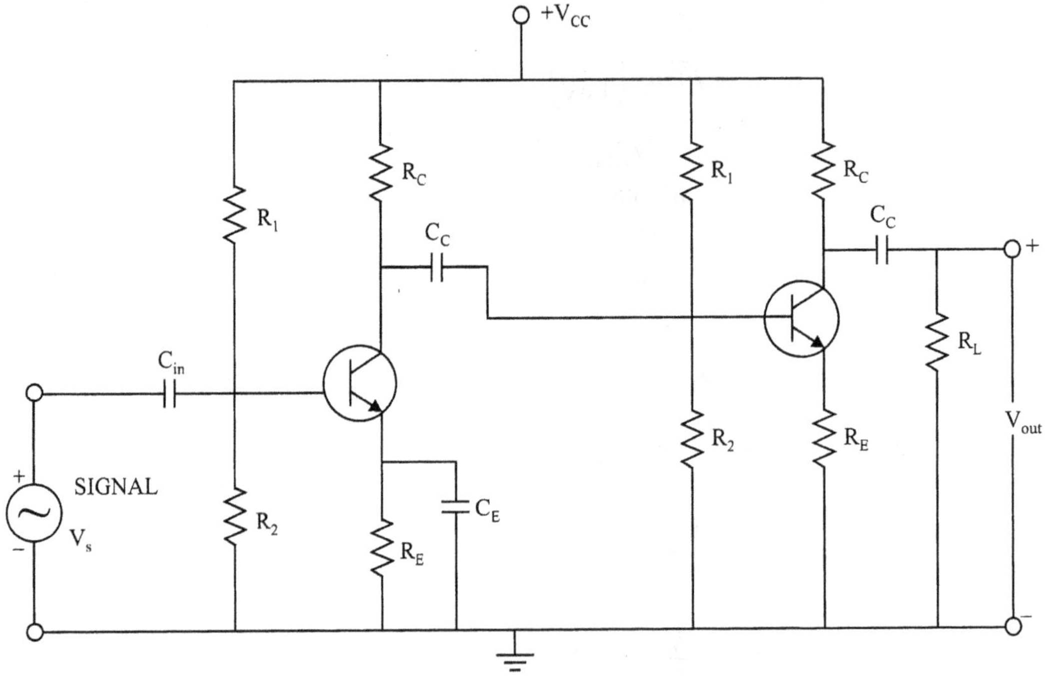

Fig. 2.5 Two stage R-C coupled transistor amplifier

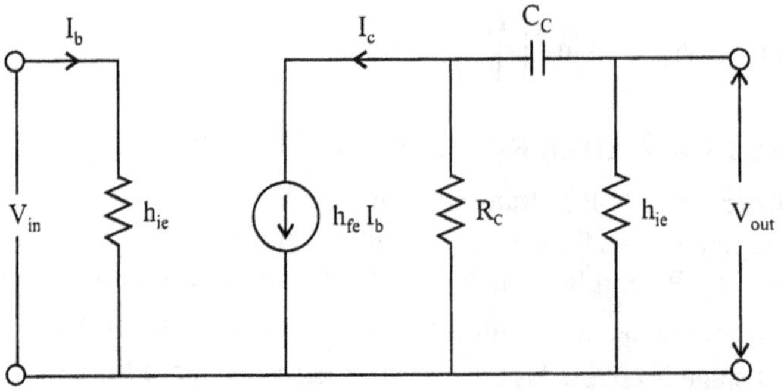

Fig. 2.6 Approximate model of an R-C coupled transistor amplifier

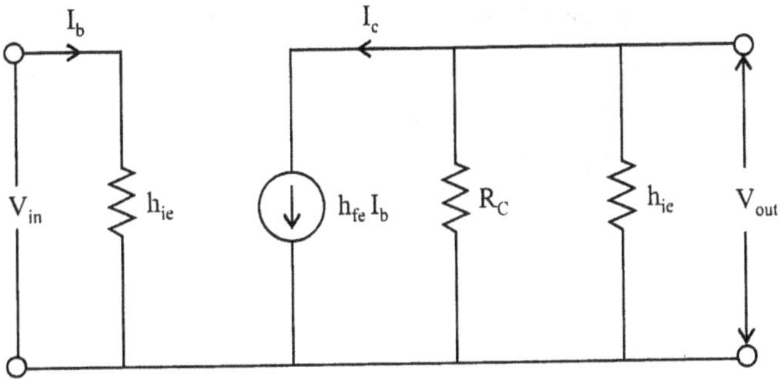

Fig. 2.7 Equivalent circuit for mid frequency range

The coupling capacitor C_C transmits ac signal but blocks the dc voltage of the first stage from reaching the base of the second stage coupling capacitor is also called blocking capacitor.

OPERATION

When ac signal is applied to the base of first amplifier, it appears in the amplified form across R_C.

- The signal across R_E is transmitted to the base of next stage of the amplifier through coupling capacitor C_C.
- As we go on increasing stages the overall gain increased.
- The phase is same at the output because two transistors involved here that is output phase of single stage is 180° and double stage is 360° that is same phase.
- Here the overall gain is less than the product of the gains of individual stages.
- The reason behind is when a second stage is made to follow the first one, the effective load resistance of the first stage is reduced because of the shunting effect of the input resistance of the second stage. This reduces the gain of the stage which is loaded by the next stage.

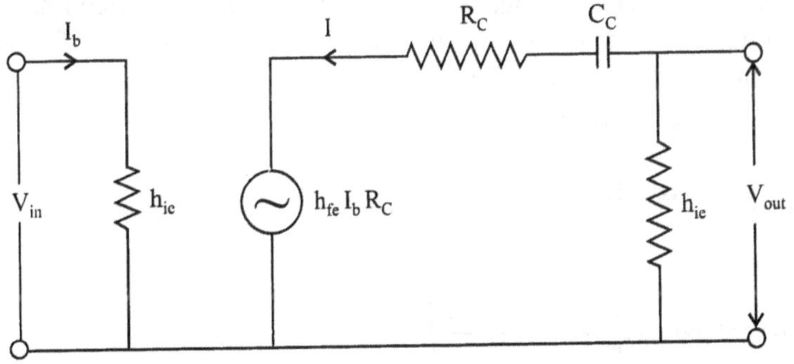

Fig. 2.8 Thevenin's equivalent circuit for low-frequency range

ANALYSIS

As to draw approximate model of the circuit shown in Fig. (2.5), following assumptions are made.

1. h_{re} is so small so that voltage source $h_{re}V_{out}$ can be neglected.
2. $\frac{1}{h_{oe}}$ is so large that it can acts as open circuit.

3. The resistors R_1 and R_2 are very large when compare to h_{ie}.
4. The reactance of emitter-bypass capacitor C_E for any given input frequency so small parallel with combination of R_E and C_E consider as a short circuit.

Here the approximate model of the circuit is Fig. (2.5) drawn as Fig. (2.6).

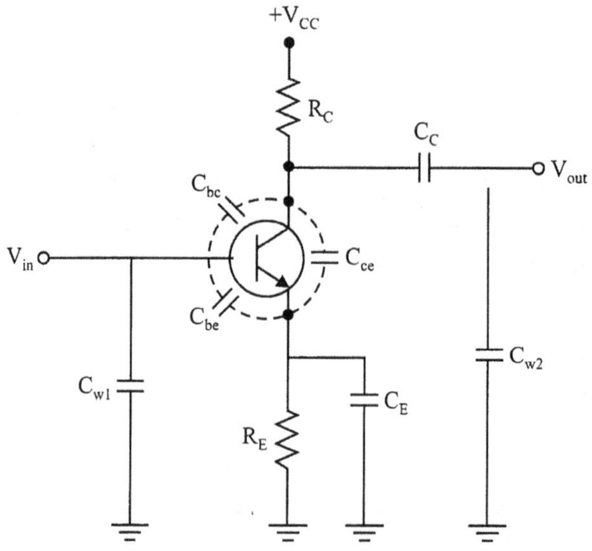

Fig. 2.9 RC Coupled amplifier at high frequencies.

For the purpose of analysis the entire frequency range may be divided into following three cases.

(a) ***Mid-Frequency Range***: At mid frequencies, the impedance offered by the coupling capacitor C_C is so small that it can be effectively considered as a short circuit and its effect can be neglected. Equivalent circuit for mid-frequency range is given in Fig. 2.7 and its Thevenin's equivalent circuit is given in Fig. 2.8.

From circuit shown in Fig. 2.8

$$\text{Current, } I = \frac{h_{fe} I_b R_C}{R_C + h_{ie}} \qquad \text{.....(2.1)}$$

$$\text{So current gain, } A_{in} = \frac{I}{I_b} = \frac{h_{fe} R_C}{R_C + h_{ie}} \qquad \text{.....(2.2)}$$

$$\text{output voltage, } V_{out} = h_{ie} I = \frac{h_{ie} h_{fe} I_b R_C}{R_C + h_{ie}} \qquad \text{.....(2.3)}$$

$$\text{Input voltage, } V_{in} = h_{ie} I_b \qquad \text{.....(2.4)}$$

$$\text{So voltage gain } A_{vm} = \frac{V_{out}}{V_{in}} = \frac{h_{ie}h_{fe}I_b R_C/(R_C + h_{ie})}{h_{ie}I_b}$$

$$= \frac{h_{fe}R_C}{R_C + h_{ie}} \qquad .....(2.5)$$

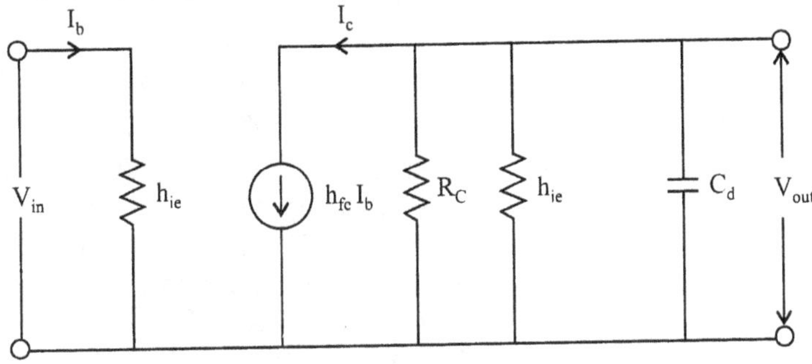

Fig. 2.10 Equivalent circuit for high frequency range

From equations (2.2) and (2.5) it is obvious that current and voltage gains are equal.

(b) ***Low-frequency Range***: In low-frequency range, the impedance offered by coupling capacitor C_C is comparable to the collector resistance R_C. It largely affects the current amplification. Therefore, it becomes necessary to include it in the equivalent circuit, as shown in Fig. 2.6.

Thevenin's equivalent circuit is shown in Fig. 2.10. From circuit shown in Fig. 2.10.

$$\text{Current, } I = \frac{h_{fe}I_b R_C}{h_{ie} + R_C - j/\omega C_C} \qquad ....(2.6)$$

$$\text{So current gain, } A_{il} = \frac{I}{I_b} = \frac{h_{fe}R_C}{h_{ie} + R_C - j/\omega C_C} \qquad .....(2.7)$$

$$\text{Output voltage, } V_{out} = h_{ie}\, I = \frac{h_{fe}h_{ie}I_b R_C}{h_{ie} + R_C - j/\omega C_C} \qquad .....(2.8)$$

$$\text{Input voltage, } V_{in} = h_{ie}\, I_b$$

$$\text{So voltage gain, } A_{vl} = \frac{V_{out}}{V_{in}} = \frac{h_{ie}h_{fe}I_b R_C}{h_{ie} + R_C - j/\omega C_C} = \frac{h_{fe}R_C}{h_{ie} + R_C - j/\omega} \qquad .....(2.9)$$

From eq. (2.9) it is obvious that in the low-frequency range voltage gain decreases with the decrease in frequency.

(c) ***High-Frequency Range*:** In high frequency range, the reactance offered by coupling capacitor C_C is very small and it may be consider as a short circuit.

The important factor that comes into picture at high frequencies is the inter-electrode capacitances.

These capacitances are due to formation of depletion layers at the junctions. The inter-electrode capacitances are shown by dotted lines in Fig. 2.9. The capacitance C_{bc} between base and collector connects the output with input. Because of this, negative feedback takes place in the circuit and gain is reduced.

The feedback effect increases with the increase in frequency because with the increase in frequency reactive impedance of the capacitor decreases.

The capacitance C_{be}, between the base and emitter, offers a low impedance path at the input side at high frequencies. This reduces the input impedance of the device and consequently effective input signal is reduced. Thus gain falls. Similarly, the capacitance C_{ce} produces a shunting effect at high frequencies on the output side. It is important to note the point that C_{bc} is the most important capacitance because feedback takes place from output circuit to input circuit through this capacitance. This is known as Miller effect. Besides these capacitances, there are wiring capacitances C_{w1} and C_{w2} as shown in Fig. 2.9. These are the capacitances between the connecting wires of the circuit and ground.

It can be shown that C_{be} and C_{bc} may be replaced with a single capacitor C_d across the input resistance h_{ie} of the transistor. The value of shunt capacitance c_d in the input circuit of the first stage is small because it depends on the output impedance of the first transistor, which is small. But in the output circuit of the first stage c_d is increased by stray capacitance of the wiring. The reactance $1/\omega C_d$ will have appreciable shunting effect on R_2 and h_{ie}. The equivalent circuits are given in Fig. 2.10 and Fig. 2.11.

From Thevenin's equivalent circuit shown in Fig. 2.11.

$$\text{Current, } I = \frac{h_{fe}I_b \dfrac{R_C h_{ie}}{R_C + h_{ie}}}{\dfrac{R_c h_{ie}}{R_C + h_{ie}} + \dfrac{1}{j\omega C_d}}$$

$$= \frac{h_{fe}I_b R_C h_{ie}}{R_C h_{ie} + \dfrac{1}{j\omega c_d}(R_C + h_{ie})} \qquad \text{.....(2.10)}$$

$$\text{So current gain, } A_{in} = \frac{I}{I_b} = \frac{h_{fe}h_{ie}R_C}{h_{ie}R_C + \dfrac{1}{j\omega C_d}(h_{ie} + R_C)}$$

$$\text{Output voltage, } V_{out} = I \times \frac{1}{j\omega c_d} = \frac{h_{fe}h_{ie}I_bR_C/(h_{ie}+R_C)}{\frac{h_{ie}R_C}{h_{ie}+R_C}+\frac{1}{j\omega c_d}} \times \frac{1}{j\omega C_d}$$

$$= \frac{h_{fe}h_{ie}I_bR_C}{h_{ie}R_cj\omega C_d + h_{ie} + R_C} \qquad .....(2.11)$$

input voltage, $V_{in} = h_{ie}\, I_b$

$$\therefore \quad \text{Voltage gain, } A_{vh} = \frac{V_{out}}{V_{in}} = \frac{h_{fe}R_C}{h_{ie}R_Cj\omega C_d + h_{ie} + R_C} \qquad .....(2.12)$$

From eq. (2.12) it is obvious that with the increase in input frequency the magnitude of voltage gain falls off

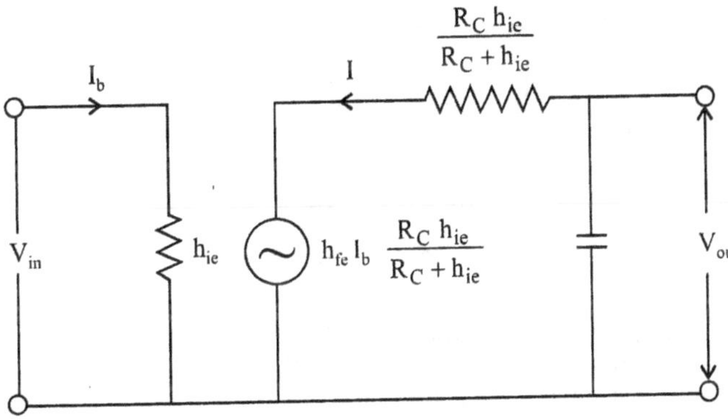

Fig. 2.11 Thevenin's equivalent circuit

2.3 n-STAGE CASCADED AMPLIFIER

Consider the schematic shown in Fig. (2.12).

2.3.1 Overall Voltage Gain

The resultant voltage gain is given by the product of the individual voltage gains of each stage.

$$A_{V1} = \frac{V_2}{V_1} = A_1 \angle \theta_1$$

A_1 = Magnitude of the voltage gain of the first stage and θ_1 is the phase angle between output and input voltage of this stage

$$\frac{V_0}{V_1} = \frac{V_2}{V_1} \cdot \frac{V_3}{V_2} \cdot \frac{V_4}{V_3} \cdots \frac{V_n}{V_{n-1}} \cdot \frac{V_o}{V_n}$$

$$\begin{aligned}A_V &= A_{V_1} A_{V_2} \ldots A_{V_n}\\ &= A_{V_1} \angle\theta_1 \,.\, A_{V_2} \angle\theta_2 \,.\, A_{V3} \angle\theta_3 \ldots\ldots A_{V_n} \angle\theta_n\\ &= A_{V_1} A_{V_2} \ldots A_{V_n} \angle\theta_1 + \angle\theta_2 + \angle\theta_3 \ldots\ldots \angle\theta_n\\ &= A_V \angle\theta. \quad [\text{Since } A_V = A_{V_1} A_{V_2} A_{V_3} \ldots A_{V_n} \text{ and } \theta = \theta_1 + \theta_2 + \ldots + \theta_n]\end{aligned}$$

2.3.2 Current Gain

$$A_I = \frac{I_O}{I_{b_1}}$$

It is the ratio of the output current I_0 of the last stage to the input current or base current I_{b_1} of the first stage.

$$A_I = \frac{I_O}{I_{b_1}} = \frac{-I_{c_n}}{I_{b_1}}$$

Where I_{cn} is equal to the collector current of the nth stage transistor.

$$\frac{I_n}{I_{b_1}} = \frac{I_1}{I_{b_1}} \cdot \frac{I_2}{I_1} \ldots\ldots \frac{I_n}{I_{n-1}}$$

$$= A'_{I_1} \, A'_{I_2} \ldots\ldots A'_{I_n}$$

A_{I_1} = The base to collector current gain of the first stage

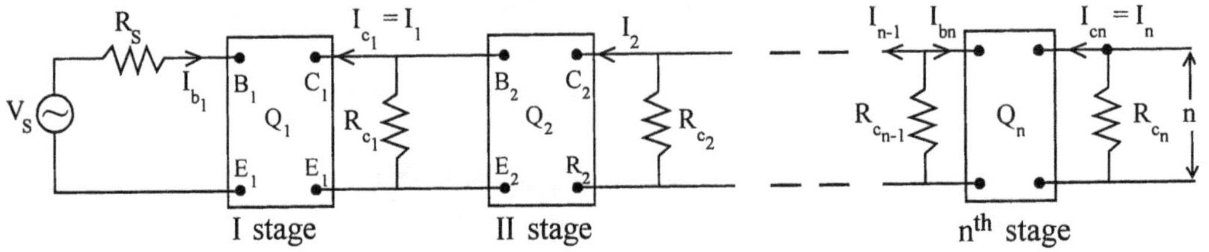

Fig. 2.12 Multistage amplifiers

A'_{I_n} = Base collector current gain of the nth stage

2.3.3 Power Gain

$$A_P = \frac{\text{Output power}}{\text{Input power}} = \frac{-V_0 I_n}{V_1 I_{b1}} = A_V \,.\, A_I$$

But
$$A_V = \frac{I_n \cdot R_{c_n} (R_0)}{I_{b_1} \cdot R_{i_1}}$$

$$A_V = \frac{\text{Output Voltage}}{\text{Input Voltage}}$$

But $I_n/I_{b_1} = A_I$

$$\therefore \quad A_V = A_I = \frac{R_{c_n}\,(R_0)}{R_{i_1}}$$

R_{cn} = Collector Resistance of nth stage transistor
R_{i1} = Input resistance of I stage amplifier

$$\therefore \quad A_P = (A_I)^2 \cdot \frac{R_{c_n}}{R_{i_1}}$$

2.3.4 Choice of Transistor in a Cascaded Amplifier Configuration

By connecting transistor in cascade, voltage gain gets multiplied. But what type of configuration should be used ? Common Collector (CC) or Common Base (CB) or Common Emitter (CE)? To get voltage amplification and current amplification, only Common Emitter (CE) configuration is used. Because for Common Collector, the voltage gain is less than 1 for each stage. So the overall amplification is less than 1.

Common Base Configuration is also not used since A_I is less than 1.

$$\therefore \quad A_V = A_I \times \frac{R_L}{R_i}$$

Effective load resistance R_L is parallel combination of R_C and R_i of the following stage, (next stage) (since in multi stage connection, the output of one stage is the input to the other stage). This parallel combination is less than R_i. Therefore, $\frac{R_L}{R_i} < 1$.

The current gain A_I in common base configuration is $h_{fb} < 1$ or $\cong 1$.

Therefore overall voltage gain $\cong$ 1. Therefore Common Base configuration is not used for cascading.

So only Common Emitter configuration is used. ($h_{fe} >> 1$)

Therefore overall voltage gain and current gains are >1 in Common Emitter configuration.

For a single stage Resistor Capacitor coupled amplifier, bandwidth is $(f_2 - f_1) \cong f_2$. f_1 is very small compared to f_2. f_2 is usually of the order of kHz or MHz. f_1 is of the order of few HZ.

$$A_H = \frac{A_M}{\sqrt{1 + \left(\frac{f}{f_2}\right)^2}}$$

$$\left(\frac{A_H}{A_M}\right)^n = \left\{\frac{1}{\sqrt{1+\left(\frac{f}{f_2}\right)^2}}\right\}^n$$

$$\therefore \qquad \frac{1}{\sqrt{2}} = \left\{\frac{1}{\sqrt{1+\left(\frac{f_{2n}}{f_2}\right)^2}}\right\}^n$$

If f_{2n} is the upper 3 db frequency

$$\left(\frac{A_H}{A_M}\right)^n = \left(\frac{1}{\sqrt{2}}\right) \quad \text{for } f = f_{2n}$$

f_2 is the frequency at which

$$A_H = \frac{A_M}{\sqrt{2}},$$

When n stages of RC coupled amplifiers are connected in cascade, the upper 3db frequency $(f_2)^n$ for which the overall voltage gain falls to $\frac{1}{\sqrt{2}}$ of its midband value is,

$$\left\{\frac{1}{\sqrt{1+\left(\frac{f_{2n}}{f_2}\right)^2}}\right\}^n = \frac{1}{\sqrt{2}}$$

or $$\left\{\sqrt{1+\left(\frac{f_{2n}}{f_2}\right)^2}\right\}^n = \sqrt{2}$$

$$\left\{1+\left(\frac{f_{2n}}{f_2}\right)^2\right\}^n = 2$$

$$1+\left(\frac{f_{2n}}{f_2}\right)^2 = (2)^{\frac{1}{n}}$$

$$\left(\frac{f_{2n}}{f_2}\right)^2 = \left[(2)^{\frac{1}{n}} - 1\right]$$

$$\therefore \quad \boxed{\frac{f_{2n}}{f_2} = \sqrt{2^{1/n} - 1}}$$

$$A_L = \frac{A_M}{\sqrt{1+\left(\frac{f_1}{f}\right)^2}};$$

$$\left(\frac{A_L}{A_M}\right)^n = \left\{\frac{1}{\sqrt{1+\left(\frac{f_1}{f}\right)^2}}\right\}^n$$

If $f = f_{1n}$,

then $$\left(\frac{A_L}{A_M}\right)^n = \frac{1}{\sqrt{2}}$$

$$\therefore \qquad \frac{1}{\sqrt{2}} = \left\{ \frac{1}{\sqrt{1+\left(\frac{f_1}{f_{1n}}\right)^2}} \right\}$$

When n stages are connected in cascade, the lower cut off frequency for nth stage is,

$$\left(\frac{A_L}{A_M}\right)^n = \frac{1}{\sqrt{2}}.$$

$$\left\{ \frac{1}{\sqrt{1+\left(\frac{f_1}{f_{1n}}\right)^2}} \right\}^n = \frac{1}{\sqrt{2}}$$

or $$\left[1+\left(\frac{f_1}{f_{1_n}}\right)^2\right]^{n/2} = \sqrt{2}$$

or $$1+\left(\frac{f_1}{f_{1_n}}\right)^2 = 2^{1/n}$$

$$\frac{f_1}{f_{1_n}} = \sqrt{2^{1/n}-1}$$

or $$f_{1_n} = f_1 \Big/ \sqrt{2^{1/n}-1}$$

2.3.5 MIDBAND FREQUENCY (F_0)

It is the geometric mean of the lower cut off frequency f_1 and upper cut off frequency f_2. It lies in the middle of the mid frequency range and has maximum gain. $f_0 = \sqrt{f_1 f_2}$.

f_β and f_T are the frequency constants of a Transistor in Common Emitter configuration.

For a Transistor in Common Emitter configuration, the plot of A_i V_S R_L is as shown Fig. 2.13. Short circuit current gain of Common Emitter Transistor varies with frequency. Short circuit current gain of transistor varies with frequency.

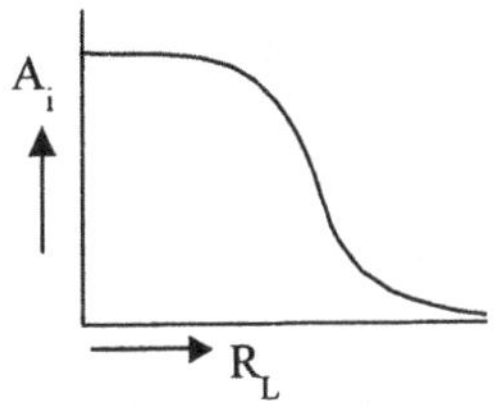

Fig. 2.13 Variation of A_i with R_L

The frequency f_β; It is the frequency at which short circuit current gain of the Transistor falls to $\frac{1}{\sqrt{2}}$ of its maximum value. This is called as the Bandwidth of the transistor.

$$h_{fe} \cdot f_\beta = f_T$$

2.3.6 CASCADING TRANSISTOR AMPLIFIERS

When the amplification of a single transistor is not sufficient for a particular purpose (say to deliver output to the speaker or to drive a transducer etc) or when the input or output impedance is not of the correct magnitude for the desired application, two or more stages may be connected in cascade. Cascade means in series i.e. The output of first stage is connected to the input of the next stage. (Fig. 2.14).

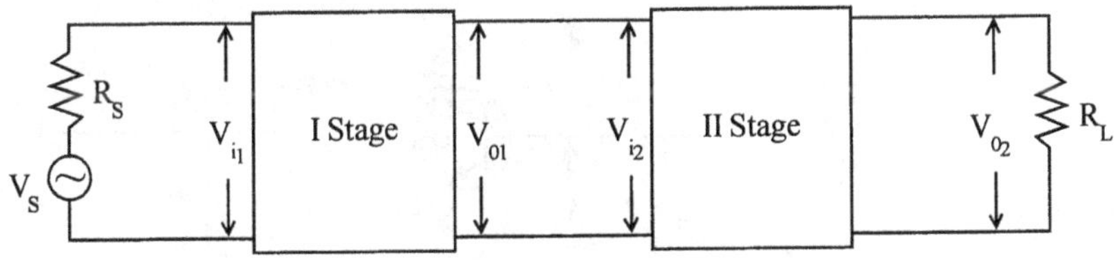

Fig. 2.14 Cascaded amplifier stages

Let us consider two stage cascaded amplifier. Let the first stage is in common emitter configuration. Current gain is high and let the II stage is in common collector configuration to provide high input impedance and low output impedance. So what are the expressions for the total current gain A_I of the entire circuit (i.e. the two stages), Zi, A_V and Y_0 ? To get these expressions, we must take the h-parameters of these transistors in that particular configuration. Generally manufactures specify the h-parameters for a given transistor in common emitter configuration. It is widely used circuit and also A_I is high. To get the transistor h-parameters in other configurations, conversion formulae are used.

2.3.7 The Two Stage Cascaded Amplifier Circuit

The Transistor Q_1 is in Common Emitter configuration (Fig. 2.15). The second Transistor Q_2 is in Common Collector (CC) configuration. Output is taken across 5K, the emitter resistance. Collector is at ground potential in the A.C. equivalent circuit.

Biasing resistors are not shown since their purpose in only to provide the proper operating point and they do not affect the response of the amplifier. In the low frequency equivalent circuit, since the capacitors have large value, and so is X_C low, and can be neglected. So the capacitive reactance is not considered, and capacitive reactance $X_c = \frac{1}{2\pi f C}$ is low when C is large and taken as short circuit.

The small signal Common Emitter configuration circuit reduces as shown in Fig 2.15.

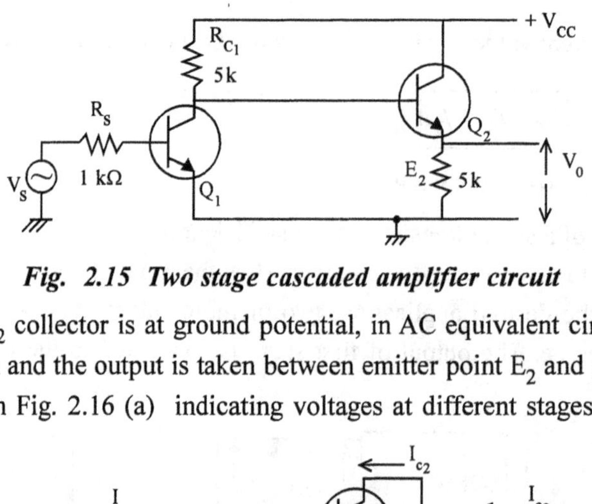

Fig. 2.15 Two stage cascaded amplifier circuit

In this circuit Q_2 collector is at ground potential, in AC equivalent circuit. It is in Common Collector configuration and the output is taken between emitter point E_2 and ground. So the circuit is redrawn as shown in Fig. 2.16 (a) indicating voltages at different stages and input and output resistances.

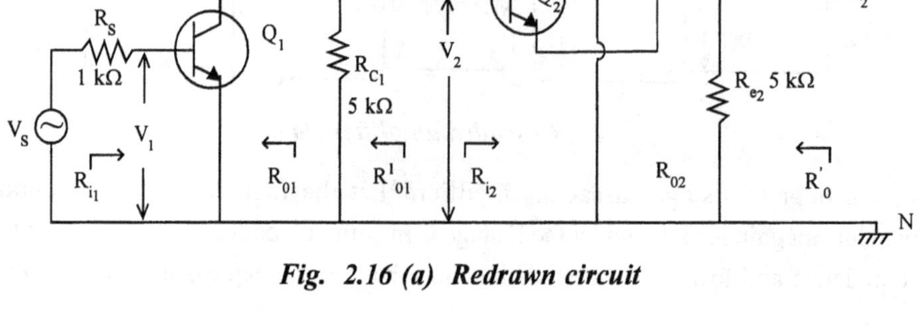

Fig. 2.16 (a) Redrawn circuit

2.3.8 Phase Response

Since the transistor is in Common Emitter configuration, there will be a phase shift of 180° between input and output. (Fig. 2.16(b)).

But in the L.F. range, $\theta = +\tan^{-1}\left(\frac{f_1}{f}\right)$

When $f = f_1$, $\theta = +\tan^{-1}(1) = 45°$. As f, increases, θ decreases.

$\therefore$ Total phase difference in the L.F. range is, +180°. Phase shift due to Common Emitter configuration $\theta = +\tan^{-1}\left(\frac{f_1}{f}\right)$

This phase lead decreases as frequency increases.

In the high frequency range $\theta = \tan^{-1}\left(\frac{f}{f_2}\right)$. When $f = f_2$, $\theta = \tan^{-1} 1 = 45°$

As f increases, θ decreases. Therefore Phase lag decreases

In the low frequency range, the signal is leading. Since θ is positive and +180°,

$$\text{Total Phase angle} = +180° + \theta;\ \phi = 180° + \tan^{-1}\left(\frac{f_1}{f}\right)$$

In the high frequency range the signal is lagging. Therefore $+180° - \theta$;

$$\phi = 180° - \tan^{-1}\left(\frac{f}{f_2}\right)$$

Phase Response : 180°, because, Common Emitter configuration introduces 180° phase shift. [Fig. 2.16(b)].

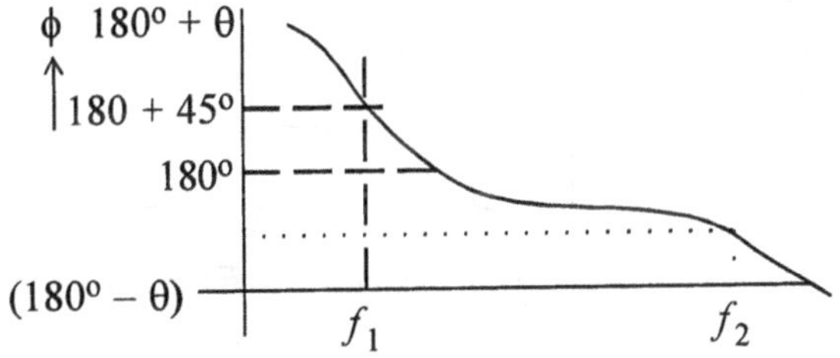

Fig. 2.16 (b) Variation of phase angle with frequency

In the low frequency range, Phase shift $\phi = (180° + \theta)$
In the high frequency range, Phase shift $\phi = (180° - \theta)$

2.3.9 Gain-Bandwidth Product

$$A_{MF} = \frac{-h_{fe}}{h_{ie}}\left(\frac{R_C R_L}{R_C + R_L}\right)$$

$$\text{Bandwidth} = f_2 - f_1 \simeq f_2 \quad \because\ f_1 \ll f_2$$

$$f_2 = \frac{1}{2\pi C_S.\left(\frac{R_C.R_L}{R_C + R_L}\right)}$$

$\therefore$ The product of these two, (A_{MF} and BW) is,

$$A \times f_2 = f_T = \frac{-h_{fe}}{h_{ie}} \times \frac{1}{2\pi C_S} = f_T$$

For a given value of C_S, this product is constant.

We can increase the voltage gain by increasing R_C, R_L parallel combination.

But $f_2 \propto \frac{1}{R_C \parallel R_L}$. Therefore f_2 will reduce if A_{MF} is increased. *Therefore if voltage gain increases, Bandwidth decreases and vice versa.*

f_2 can be increased without affecting voltage gain by reducing the value of C_S - But there is a limit to which C_S can be reduced. If higher gain is required then Bandwidth has to be sacrificed. It (B.W) will reduce.

The product of midband gain and Bandwidth is also known as the figure of Merit of the circuit.

Example : 2.5

If β = 150, what are the cutoff frequencies of the input and output lead networks of the given circuit ? (Fig. 2.17).

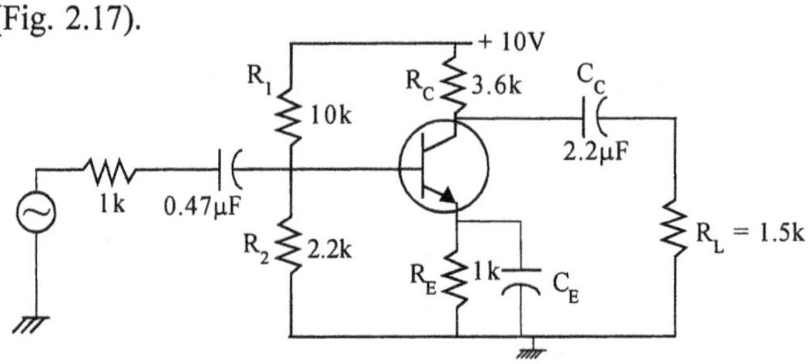

Fig. 2.17 Circuit for Ex : 2.5

Solution

When β value is given, and not h_{ie} of the transistor, the input impedance of the transistor can be determined.

$$Z_{in} = \frac{V_{in}}{i_b}$$

$$V_{in} = i_e r_e' \qquad i_e \simeq i_C \simeq \beta i_b$$

$$\therefore \quad V_{in} \simeq \beta i_b . r_e'$$

$$\therefore \quad Z_{in} = \frac{\beta i_b . r_e'}{i_b} = \beta r_e'$$

$$R_{in} = R_1 \parallel R_2 \parallel \beta_{re}'$$

$$\beta_{re}' = 150 \times 22.7 = 3.14\ k\Omega$$

$$R_{in} = 10k \parallel 2.2k \parallel 3.14\ k = 1.18k\Omega$$

$$f_{in} = \frac{1}{2\pi(R_S + R_{in})C_{in}}$$

Cutoff frequencies of input network (HPF)

$$f_0 = \frac{1}{2\pi(R_0 + R_L)C_0}$$

$$f_{in} = \frac{1}{2\pi(1\ k\Omega + 1.18\ k\Omega)(0.47\,\mu F)}$$

$$= 155 Hzs.$$

$$f_0 = \frac{1}{2\pi(3.6\ k + 1.5\ k\Omega)(2.2\,\mu F)}$$

$$= 14.2\ Hzs$$

LPF Cutoff frequency of output network.

2.3.10 Emitter Bypass Capacitor

Consider circuit shown in Fig. 2.18.

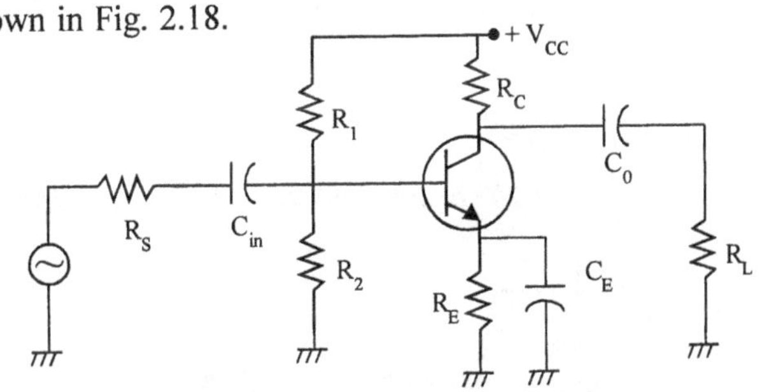

Fig. 2.18 CE amplifier circuit

C_E is the emitter bypass capacitor. This causes the frequency response of an amplifier to break at a cutoff frequency, designated f_E. To understand the effects of emitter bypass capacitor, suppose, C_{in} and C_0 (coupling capacitor) are shorted, then the frequency response will be, as shown in Fig. 2.19(a).

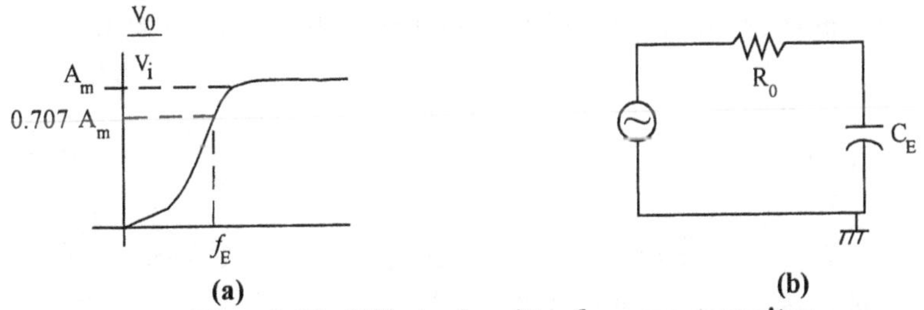

Fig. 2.19 Effect of emitter by pass capacitor

This means that frequency response breaks at f_E. Thevenin's equivalent resistance driving Common Emitter, R_{out} is the Thevenin's resistance facing the capacitor.

$$\therefore \quad R_{out} \simeq r_e' + \frac{R_S \| R_1 \| R_2}{\beta}$$

$$[i_e r_e' + i_e R_E - V_{in} + i_b (R_S \| R_1 \| R_2)] = 0$$

$$\because i_b = i_c/\beta, \ \cong i_e/\beta,$$

Solving for i_e,
$$i_e \simeq \frac{V_{in}}{R_E + r_e' + (R_s \| R_1 \| R_2)/\beta}$$

The emitter resistor R_E is driven by an AC source with an AC output resistance of

$$Z_O \text{ (emitter)} = r_e' + \frac{R_s \| R_1 \| R_2}{\beta}$$

2.4 EQUIVALENT CIRCUITS

The circuits are shown in Fig. 2.20 (a) and (b).

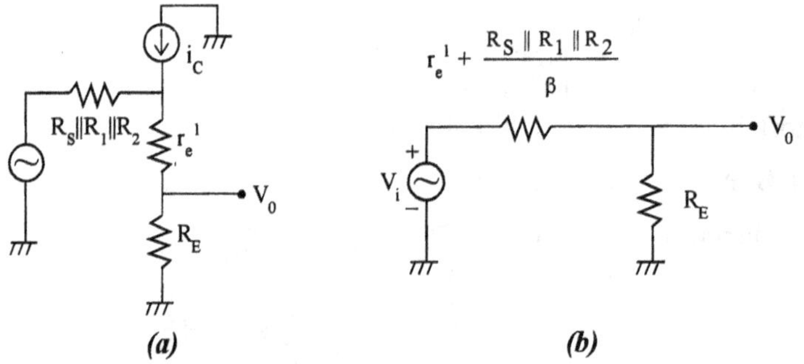

Fig. 2.20 Equivalent circuits

$$\therefore \quad f_E = \frac{1}{2\pi R_{out} C_E}$$

f_E = Cutoff frequency of emitter network.
R_{out} = Output resistance facing bypass capacitor.
C_E = Emitter bypass capacitance.

2.4.1 DECIBEL

In many cases it is convenient to compare two powers on a logarithmic scale rather than on a linear scale. The unit of this logarithmic scale is called the decibel abbreviated as db.

Suppose P_2 is the output power and P_1 is the input power, then the power gain in decibels is

$$N = 10 \log_{10} \frac{P_2}{P_1}$$

Where N is in dbs

If N is negative, it means P_2 is less than P_1.

Noise power is also expressed in decibels (dbs). It should be negative for a given device or amplifier i.e the output noise is less than what is present in the input.

If for a given amplifier circuit, the input and output resistances are same (as R), then

$$P_1 = \frac{V_1^2}{R}, \quad P_2 = \frac{V_2^2}{R}$$ where V_1 and V_2 are input and output voltages.

$$\therefore \quad N = 20 \log \frac{V_2}{V_1} = 20 \log A_V$$

But even though, the input and output impedances are not equal, this convention is followed for convenience i.e, $N = 20 \log A_V$. If $A_V = 10$, $N = 20 \log 10$ is 20 the decibel voltage gain of the amplifier. 20 is not the power gain because, the input resistances are not equal. Therefore 20 is the decibel voltage gain. If the output resistances are equal decibel voltage gain is equal, to power gain. Overall db V of a multistage amplifier is equal to sum of db V of individual stages.

2.5 MILLER'S THEOREM

Fig. 2.21(a) shows an amplifier with a capacitor between input and output terminals. It is called as feedback capacitor. When the gain K is large, the feedback will change the input Z and output Z of the circuit.

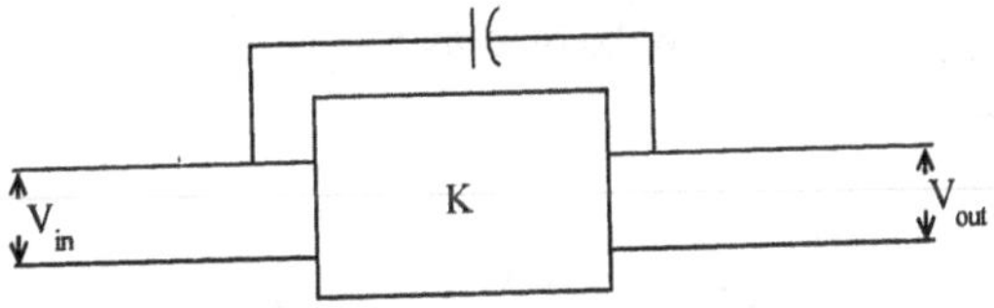

Fig. 2.21 (a) Feedback capacitor

A circuit as shown above is difficult to analyze, because of capacitor. So according to the Miller's theorem, the feedback capacitor can be split into two values, one as connected in the input side and the other on the output side, as shown in Fig. 2.21 (b).

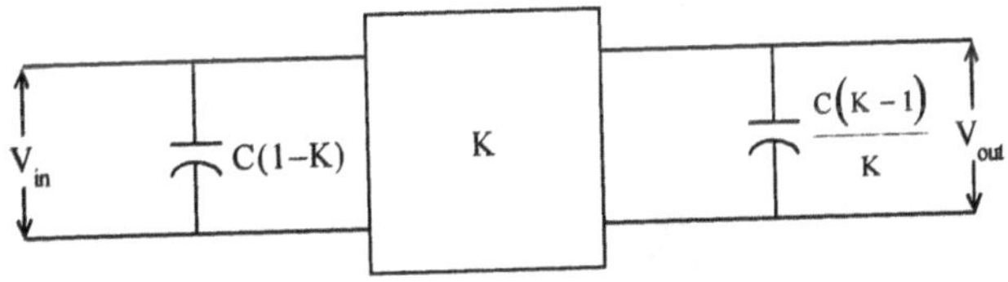

Fig. 2.21 (b) Splitting of feedback capacitor using miller's theorem.

2.5.1 MATHEMATICAL PROOF OF MILLER'S THEOREM

The AC current passing through capacitor (C) in Fig. 2.21 (a) is

$$I_C = \frac{V_{in} - V_{out}}{\left(\frac{1}{j\omega c}\right)} = \frac{(V_{in} - V_{out})}{-J\, X_C}$$

$$V_{out} = K\, V_{in}$$

$$\therefore \qquad I_C = \frac{(V_{in} - KV_{in})}{-jX_C} = \frac{V_{in}(1-K)}{-jX_C}$$

$$\frac{V_{in}}{I_C} = Z_{in} = \frac{V_{in}}{\dfrac{V_{in}(1-K)}{-jX_C}} = \frac{-jX_C}{(1-K)}$$

$$= \frac{-j}{2\pi f C(1-K)}$$

Since $\qquad X_C = \dfrac{1}{2\pi f C}$

$\dfrac{V_{in}}{I_C}$ is the input Z as seen from the input terminals.

$$\therefore \qquad Z_{in} = \frac{-j}{2\pi f[C(1-K)]}$$

$$\therefore \qquad C_{in} = C\,(1-K)$$

Similarly output capacitance can be derived as follows :

Current in the capacitor,

$$I_C = \frac{V_{out} - V_{in}}{-jX_C} = \frac{V_{out}\left(1 - \dfrac{V_{in}}{V_{out}}\right)}{-j\,X_C}$$

$$I_C = \frac{V_{out}\left(1 - \dfrac{1}{K}\right)}{-jX_C}$$

$$= \frac{V_{out}}{I_C} = Z_{out}$$

$$= \frac{V_{out}}{\dfrac{V_{out}\left(1 - \dfrac{1}{K}\right)}{-jX_C}}$$

$$Z_{out} = \frac{-jX_C}{\left(\dfrac{A-1}{A}\right)} = \frac{-j}{2\pi f C\left(\dfrac{K-1}{K}\right)}$$

$$\therefore \qquad C_{out}\,(\text{Miller}) = C\left(\frac{K-1}{K}\right)$$

2.6 FREQUENCY EFFECTS

2.6.1 Lead Network

The circuit is shown in Fig. 2.22.

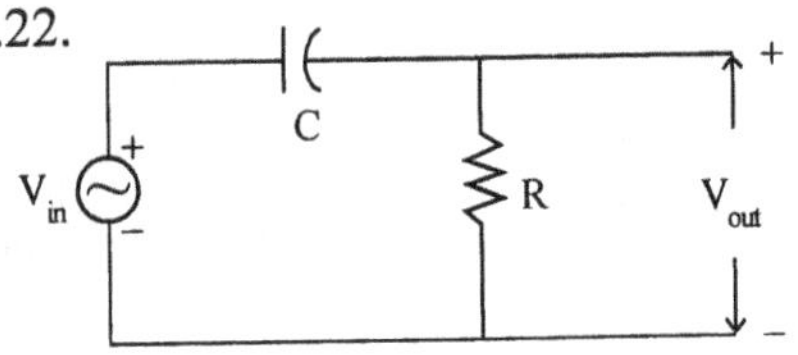

Fig. 2.22 Lead network

$$I = \frac{V_{in}}{R + \frac{1}{j\omega C}}$$

$$V_0 = I.\,R = \frac{V_{in}.R}{\left(R + \frac{1}{j\omega C}\right)}$$

V_0 leads with respect to V_{in}. So it is called as *lead network* for the above circuit, at low frequencies, $X_C = \infty$ (Fig. 2.22).

∴ V_0 is low since (I is low). As f increases X_c decreases. Hence I flows, and V_o increases.

∴ Gain increases. Hence the frequency response is as shown in Fig. 2.23.

2.6.2 Cut-off Frequency

Consider the frequency response shown in Fig. 2.23.

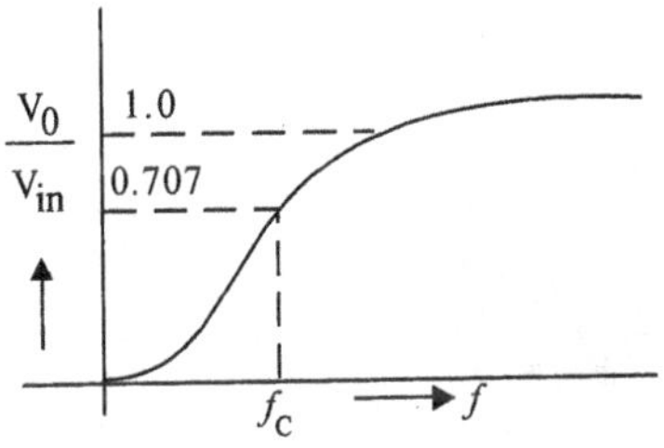

Fig. 2.23 Frequency response

$$V_{out} = \frac{R}{\sqrt{R^2 + X_C^{\;2}}}.V_{in}$$

$$\frac{V_{out}}{V_{in}} = \frac{R}{\sqrt{R^2 + X_C^2}}$$

Cut-off frequency is the frequency at which $\frac{V_{out}}{V_{in}} = \frac{1}{\sqrt{2}}$

This happens when, $X_c = R$

$$\frac{1}{2\pi f_c C} = R \quad \text{or} \quad \boxed{f_C = \frac{1}{2\pi RC}}$$

Lower cut-off frequency.

2.6.3 Half Power Point

At $f = f_C, \quad P = \frac{V_{out}^2}{2R}$

Power is half of maximum power when V_{out} is maximum. Therefore it is also called as half power point

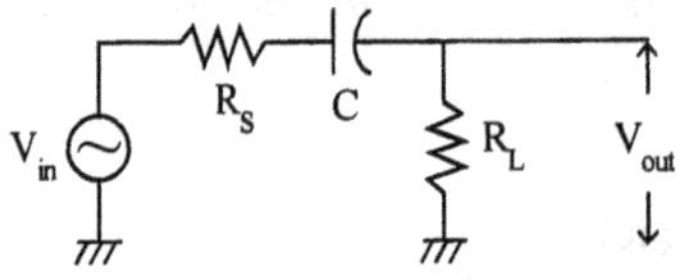

Fig. 2.24 Equivalent circuit

Considering Source Resistance : (Fig. 2.24).

$$\frac{V_{out}}{V_{in}} = \frac{R}{\sqrt{(R_S + R_L) + X_C^2}}$$

$$\text{At } f_C, \quad R_S + R_L = X_C$$

$$\therefore \quad f_C = \frac{1}{2\pi(R_S + R_L).C}$$

In the midband frequency region $X_C \cong 0$.

$$\therefore \quad \frac{V_{out}}{V_{in}} = \frac{R_L}{R_S + R_L}$$

$$A_{mid} \text{ (midband voltage gain)} = \frac{R_L}{R_S + R_L}$$

2.6.4 Stiff Coupling

If $X_c = 0.1\ (R_S + R_L)$ it is called *Stiff Coupling.*

The coupling capacitor must have (or bypass capacitor).

$$X_c = \frac{1}{10} R_E$$

This is known as *Stiff Coupling.*

2.7 AMPLIFIER ANALYSIS

Consider the circuit shown in Fig. 2.25.

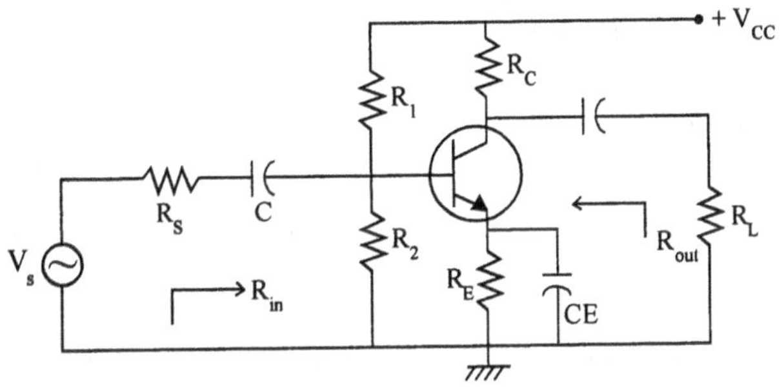

Fig. 2.25 Amplifier circuit

The equivalent circuit from input side, not considering R_S.

The equivalent circuit from output side, not considering R_L.

$$R_{in} = R_1 \parallel R_2 \parallel \beta r_e' \text{ (Not considering } R_S\text{) (See Fig. 2.26).}$$

$$R_{out} \cong R_C \text{ (Not considering } R_L\text{)}$$

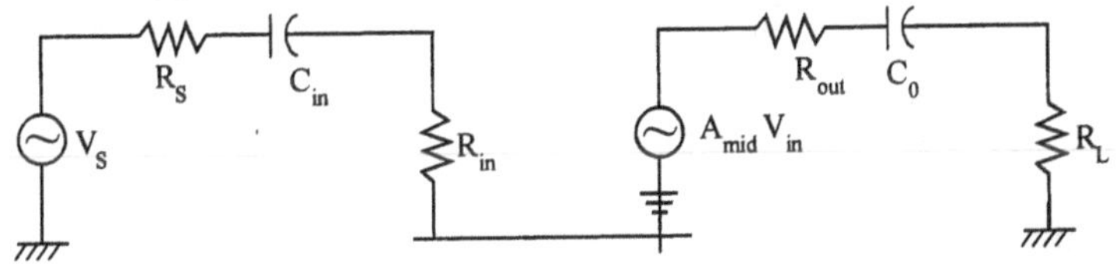

Fig. 2.26 Equivalent circuit

Considering the input side as a load network,

$$f_{in} = \frac{1}{2\pi(R_S + R_{in})C_{in}}$$

Similarly output lead network has cut-off frequency

$$f_{out} = \frac{1}{2\pi(R_{out} + R_L)C_{out}}$$

2.7.1 LAG NETWORKS

See Fig. 2.27 and 2.28.

$$V_0 = I.\, X_c$$

$$= \frac{V_i \cdot X_C}{\left(R + \dfrac{1}{j\omega C}\right)}$$

$$V_0 = \frac{\dfrac{1}{j\omega C} \cdot V_{in}}{\sqrt{R^2 + X_C{}^2}}$$

$$V_0 = \frac{-j(V_i/\omega C)}{\sqrt{R^2 + X_C^2}}$$ therefore it is lag network

V_0 lags with respect to V_i.

$$\frac{V_{out}}{V_{in}} = \frac{X_c}{\sqrt{R^2 + X_C^2}}$$

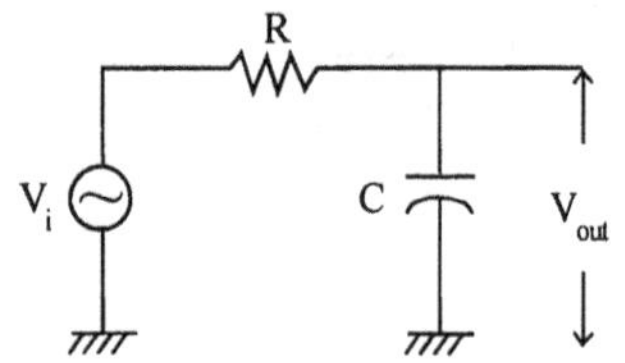

Fig. 2.27 Lag Network

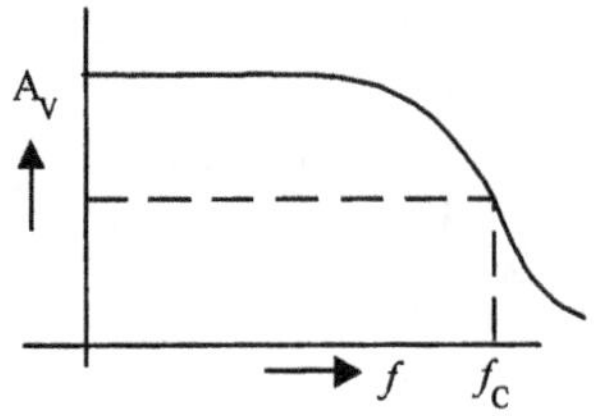

Fig. 2.28

at f_C, $R = X_C$

$\therefore$ $$\boxed{f_C = \frac{1}{2\pi RC}}$$

Midband gain $A_{mid} = \dfrac{R_L}{R_S + R_L}$

2.7.2 Decibel

Power gain = G = $\dfrac{P_2}{P_1}$

P_2 = Output power.

P_1 = Input power.

Decibel power gain = $G' = 10 \log_{10} G$.

If $G = 100$, $G' = 10 \log 100 =$ **20 db**.

If $G = 2$, $G' = 10 \log 2 = 3.01$db.

Usually, it is rounded off to 3.

2.7.3 Negative Decibel

If a < 1, a' will be negative.

If $G = \dfrac{1}{2}$,

$G' = 10 \log \dfrac{1}{2} =$ **– 3.01 db**

If

G	G′
1	0 db
10	10 db
100	20 db
1000	30 db
10, 000	40 db

Ordinary Gains Multiply :

If two stages are there, as shown in Fig. 2.29.

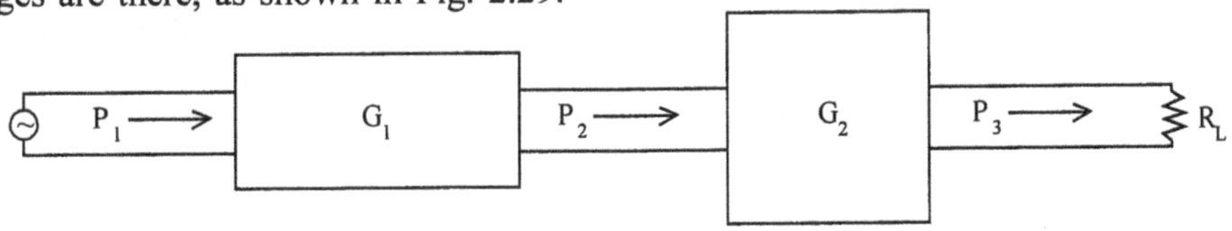

Fig. 2.29 Cascaded amplifier stages

$$G_1 = \frac{P_2}{P_1}; \; G_2 = \frac{P_3}{P_2}$$

Overall gain $$G = \frac{P_3}{P_1}$$

$$= \frac{P_2}{P_1} \cdot \frac{P_3}{P_2}$$

$$\mathbf{G = G_1 \, G_2}$$

Decibel gains add up.

2.7.4 Decibel Voltage Gain

If $A =$ Normal Voltage Gain $\frac{V_2}{V}$, decibel voltage gain $A' = 20 \log A$. $A = \frac{V_2}{V_1}$

If R_1 is output resistance,

Input power $$P_1 = \frac{V_1^2}{R_1}$$

If R_2 is output resistance, $P_2 = \frac{V_2^{\,2}}{R_2}$

$\therefore$ Power gain $$G = \frac{P_2}{P_1} = \frac{V_2^{\,2}}{V_1^{\,2}} \cdot \frac{R_1}{R_2}$$

If the impedances are matched, i.e., input resistance $R_1 =$ output resistance R_2,

Power gain $$G = \frac{V_2^{\,2}}{V_1^{\,2}} = A^2$$

In decibels, $10 \log G = G' = 10 \log A^2$

In decibels, $10 \log G = G' = 20 \log A$

2.7.5 Cascaded Stages

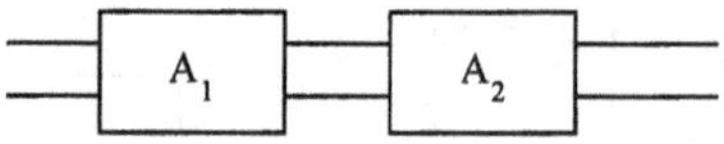

Fig. 2.30 Cascaded stages

$$A = A_1 \times A_2$$

$$A_1 = A_1' + A_2' \text{ (in decibels)}$$

2.7.6 Stiff Coupling

When the capacitor is chosen such that, $X_C = \frac{-R_E}{10}$, it is called as stiff coupling. (Fig. 2.31).

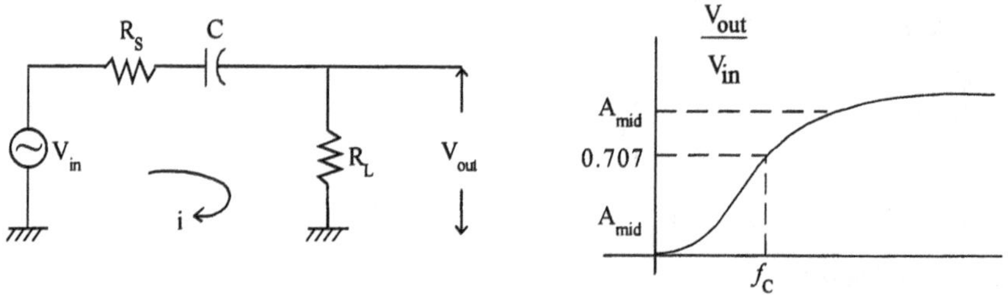

Fig. 2.31 Stiff coupling network

For the circuit, shown above, (Fig. 2.31) in the mid frequency range, X_C is negligible.

$$\therefore \qquad i = \frac{V_{in}}{(R_S + R_L)}$$

$$V_{out} = i.\ R_L = \left(\frac{V_{in}}{R_S + R_L}\right) R_L$$

$\therefore$ $A_{(mid)}$ = Voltage gain in the mid frequency range is,

$$A_{mid} = \frac{V_{out}}{V_{in}} = \frac{R_L}{R_S + R_L}$$

For the complete circuit, $\frac{V_{out}}{V_{in}} = \frac{R_L}{\sqrt{(R_S + R_L)^2 + X_C^2}}$

When $(R_S + R_L) = X_C, = \frac{1}{2\pi f_C C}$

$$\frac{V_{out}}{V_{in}} = \frac{R_L}{\sqrt{2}(R_S + R_L)}$$

$$= 0.707\ A_{mid}$$

$$\therefore \qquad f_C = \frac{1}{2\pi(R_S + R_L).C}$$

When $\qquad X_C = 0.1\ (R_S + R_L),$

$$\frac{V_{out}}{V_{in}} = \frac{R_L}{\sqrt{(R_S+R_L)^2 + [0.1\,(R_S+R_L)]^2}}$$

$$\frac{V_{out}}{V_{in}} = 0.995\ A_{mid}$$

X_C is made = 0.1 (R_S + R_L), at the lowest frequency. (f_C lower)

$\therefore$ At the frequency, A = 0.995 A_{mid}. So it is called as *Stiff Coupling.*

$$A = \frac{V_{out}}{V_{in}} = \frac{X_C}{\sqrt{R^2 + X_C^{\ 2}}} = \frac{1}{\sqrt{1+\left(\frac{R}{X_C}\right)^2}}$$

$$\because \qquad \frac{R}{X_C} = \frac{R}{\frac{1}{2\pi\, f_C}} = 2\pi f\ RC = \frac{f}{f_C}$$

$$\because \qquad f_C = \frac{1}{2\pi RC}$$

$$\therefore \qquad A = \frac{1}{\sqrt{1+\left(\frac{f}{f_C}\right)^2}}$$

$$\text{Decibel voltage gain } A' = 20 \log \frac{1}{\sqrt{1+\left(\frac{f}{f_C}\right)^2}}$$

f_C = cutoff frequency

When $\qquad \frac{f}{f_C} = 0.1,$

$$A' = 20 \log \frac{1}{\sqrt{1+(0.1)^2}} \cong 0\text{db}$$

When $\dfrac{f}{f_C} = 1,$

$$A' = 20 \log \frac{1}{\sqrt{1+1^2}} = -3.01 \text{ db} \cong 3 \text{ db}$$

When $\dfrac{f}{f_C} = 10,$

$$A' = 20 \log \frac{1}{\sqrt{1+10^2}} = -20 \text{ db}$$

When $\dfrac{f}{f_C} = 100,$

$$A' = 20 \log \frac{1}{\sqrt{1+100^2}} = -40 \text{ db}$$

When $\dfrac{f}{f_C} = 1000,$

$$A' = 20 \log \frac{1}{\sqrt{1+1000^2}} = 60 \text{ db}$$

$\therefore$ when $f = \dfrac{f_C}{10}$, $A' = 0$

$f = f_C$, $A' = -3$ db

$f = 10 f_C$, $A' = -20$ db

and so on.

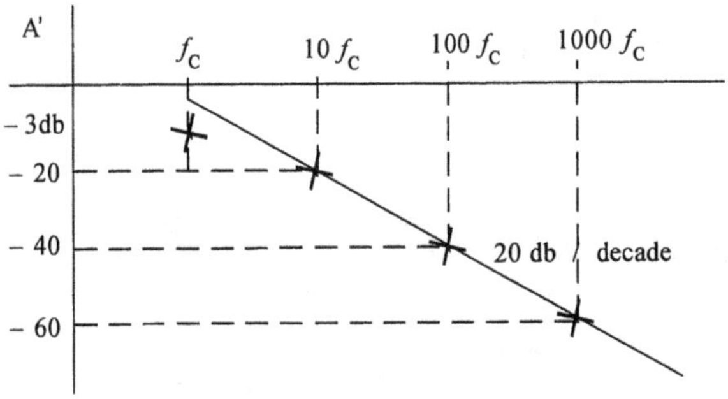

Fig. 2.32 Frequency roll-off

$\therefore$ A change of **20db** per decade *change* in frequency. (Fig. 2.32)

An Octave is a factor of 2 in frequency change

When *f changes from 100 to 200 Hzs, it has changed by one octave.*

When, *f* changes from 100 to 400 Hzs, it is two octaves for lead network, Bode Plot is as shown in Fig. 2.33.

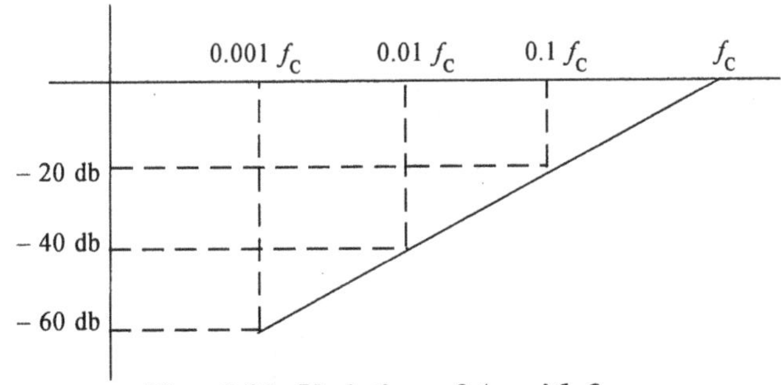

Fig. 2.33 Variation of A_V with f

2.8 HIGH INPUT RESISTANCE TRANSISTOR CIRCUITS

This topic is mentioned in previous chapter. For the convenience of reader and, continuity, it is given here again.

In some applications the amplifier circuit will have to have very high input impedance. Common Collector Amplifier circuit has high input impedance and low output impedance. But its $A_V < 1$. If the input impedance of the amplifier circuit is to be only 500 kΩ or less the Common Collector Configuration can be used. But if still higher input impedance is required a circuit shown in Fig. 2.34 is used. This circuit is known as the ***Darlington Connection*** (named after Darlington) or ***Darlington Pair Circuit.***

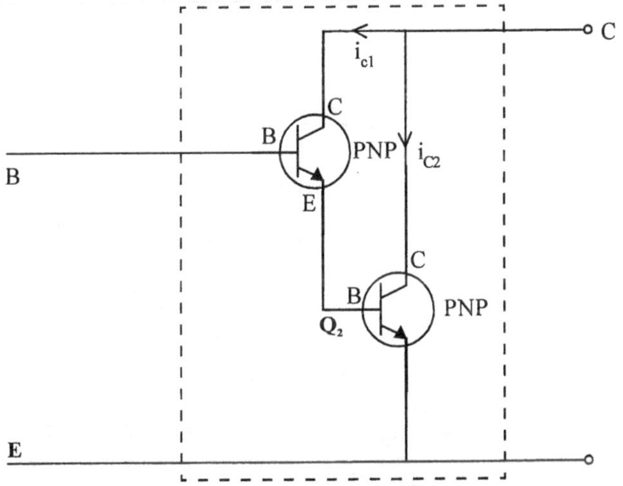

Fig. 2.34 Darlington pair circuit

In this circuit, the two transistors are in Common Collector Configuration. The output of the first transistor Q_1 (taken from the emitter of the Q_1) is the input to the second transistor Q_2 at the base. The input resistance of the second transistor constitutes the emitter load of the first transistor. So, Darlington Circuit is nothing but two transistors in Common Collector Configuration connected in series. The same circuit can be redrawn as AC equivalent circuit. So, DC is taken as ground shown in Fig.2.35. Hence, 'C' at ground potential. Collectors of transistors Q_1 and Q_2 are at ground potential.

The AC equivalent Circuit is shown in Fig. 2.35.

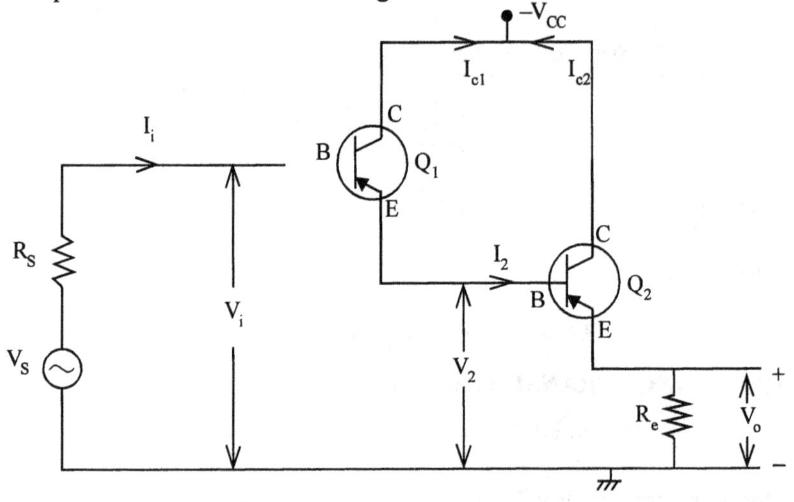

Fig. 2.35 AC equivalent circuit of darlington pair

There is no resistor connected between the emitter of Q_1 and ground i.e., Collector Point. So, we can assume that infinite resistance is connected between emitter and collector. For the analysis of the circuit, consider the equivalent circuit shown in Fig. 2.36 and we use Common Emitter ***h-parameters***, h_{ie}, h_{re}, h_{oe} and h_{fe}.

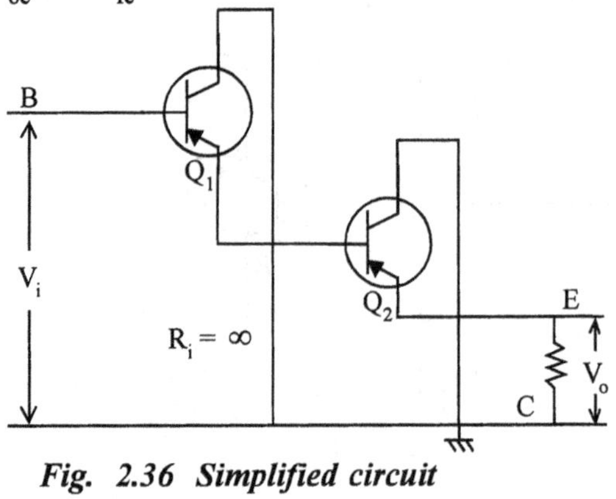

Fig. 2.36 Simplified circuit

For PNP transistor, I_c leaves the transistor, I_e enters the transistor and I_b leaves the transistor.

2.8.1 Current Amplification for Darlington Pair

$$I_c = I_{c1} + I_{c2}.$$

$$I_{c1} = I_{b1} h_{fe};$$

$$I_{c_2} = I_{b_2} h_{fe}.\text{ (Assuming identical transistor and } h_{fe} \text{ is same)}$$

But $\quad I_{b_2} = I_{e_1}$ (The emitter of Q_1 is connected to the base of Q_2)

$$\therefore \quad I_c = I_{b1} h_{fe} + I_{e1} h_{fe} \qquad \text{.....(2.13)}$$

$$I_{e_1} = I_{b_1} + I_{c_1} = I_{b_1}(1 + h_{fe}) \qquad \text{.....(2.14)}$$

$$\therefore \quad I_{c_1} = h_{fe} I_{b_1}$$

Substituting equation (2.14) in (2.13),

$$I_c = I_{b_1} h_{fe} + I_{b_1}(1 + h_{fe})h_{fe} = I_{b_1}(2h_{fe} + h_{fe}^2)$$

But $h^2_{fe} >> 2h_{fe}$

Since, h_{fe} is of the order of 100.

$$\therefore \quad I_c = I_{b_1} h_{fe}^2$$

It means that we get very large current amplification $\left(A_I = \frac{I_c}{I_{b_1}}\right)$ in the case of Darlington Pair Circuit, it is of the order h_{fe}^2 i.e. $100^2 = 10{,}000$.

$$\therefore \quad \boxed{A_I = \frac{I_c}{I_{b_1}} \cong (h_{fe})^2}$$

2.8.2 Input Resistance (R_i)

Input resistance R_{i_2} of the transistor Q_2 (which is in Common Collector Configuration) in terms of ***h- parameters*** in Common Emitter Configuration is,

$$R_{i_2} = h_{ie} + (1+h_{fe})\, R_L$$

But $\quad h_{ie} << h_{fe}R_L$, and $h_{fe}\, R_E >> h_{ie}$, $A_{I_2} = \frac{I_0}{I_2} = (1 + h_{fe})$

Here R_L is R_e, since, output is taken across emitter resistance.

$$\therefore \quad R_{i_2} \cong (1+ h_{fe})\, R_e$$

The input resistance Ri_1 of the transistor Q_1 is, since it is in Common Collector Configuration,

$$R_i = h_{ic} + h_{re}\, A_I.R_L.$$

Expressing this in term of Common Emitter ***h-parameters,***

$$h_{ic} \cong h_{ie};\ h_{rc} \cong 1.$$

(For Common Collector Reverse Voltage Gain is equal to 1) and R_L for transistor Q_1 is the input resistance of transistor Q_2.

$$\therefore \quad R_{i_1} = h_{ie} + A_{I_1}\, R_{i_2}.\ R_{i_2}, \text{ is large,}$$

Therefore, $\quad h_{oe}.R_{i_2} \leq 0.1.$ and $A_I \cong 1+ h_{fe}$

$$R_{i_1} \cong A_{I_1}\, R_{i_2}$$

$$R_{i_2} = (1 + h_{fe})\, R_e$$

But the expression for Common Collector Configuration in terms of Common Emitter ***h-parameters*** is

$$A_I = \frac{1+h_{fe}}{1+h_{oe}.R_L}$$

Here, $\quad R_L = R_{i_2}$ and $R_{i_2} = (1 + h_{fe})\, R_e.$

$$\therefore \quad A_{I1} = \frac{1+h_{fe}}{1+h_{oe}(1+h_{fe})R_e}$$

$h_{oe}\,R_e$ will be is less than 0.1 and can be neglected.

$\therefore$ h_{o_e} value is of the order of μ mhos (micro mhos)

$$\therefore \quad A_{I1} = \frac{1+h_{fe}}{1+h_{oe}h_{fe}R_e}$$

$$\therefore \quad R_{i1} \simeq A_{I1}\cdot R_{i2}$$

$$\boxed{R_i \simeq \frac{(1+h_{fe})^2 R_e}{1+h_{oe}h_{fe}R_e}}$$

This is a very high value. If we take typical values, of R_e = 4 kW, using ***h-parameters***,

R_{i_2} = 205 kΩ
R_i = 1.73 MΩ
A_I = 427

Therefore, Darlington Circuit has ***very high input impedance*** and very ***large current gain*** compared to Common Collector Configuration Circuit .

2.8.3 VOLTAGE GAIN

General expression for A_V for Common Collector in term of ***h-parameters*** is

$$A_v = 1 - \frac{h_{ie}}{R_i}; \; h_{ie} \cong h_{ic} \quad \text{or } A_{V1} = \frac{V_2}{V_i} = \left[1 - \frac{h_{ie}}{R_{i_1}}\right]$$

But $\quad R_i \cong A_{I1}\cdot R_2$

$$\therefore \quad A_{V1} = \left(-\frac{h_{ie}}{A_{I_1}\,R_{i_2}}\right)$$

$$\therefore \quad A_{V2} = \frac{V_o}{V_2} = \left(1 - \frac{h_{ie}}{R_{i2}}\right)$$

Therefore, over all Voltage Gain $A_v = A_{v1} \times A_{v2}$

$$= \left(1 - \frac{h_{ie}}{A_{I_1}.R_{i_2}}\right)\left(1 - \frac{h_{ie}}{R_{i_2}}\right)$$

$$A_v \cong \left(1 - \frac{h_{ie}}{Ri_2}\right) \qquad \because R_{i_2} >> h_{ie} \text{ and } A_{i_1} \text{ is } >> 1$$

Therefore, A_V is always less than 1.

2.8.4 Output Resistance

The general expression for R_0 of a transistor in Common Collector Configuration in terms of Common Emitter ***h-parameters*** is,

$$R_o = \frac{R_S + h_{ie}}{1 + h_{fe}}$$

$$\therefore \quad R_{o_1} = \frac{R_S + h_{ie}}{1 + h_{fe}}$$

Now for the transistor Q_2, R_s is R_{01}.

$$\therefore \quad R_{o_2} = \frac{\frac{R_S + h_{ie}}{1 + h_{fe}} + h_{ie}}{1 + h_{fe}}$$

Therefore, R_{o2} is the output resistance of the Darlington Circuit.

$$\therefore \quad R_{o_2} = \frac{R_s + h_{ie}}{(1 + h_{fe})^2} + \frac{h_{ie}}{1 + h_{fe}}$$

This is a small value, since, $1 + h_{fe}$ is >> 1.

Therefore, the characteristic of Darlington Circuit are

1. *Very High Input Resistance (of the order of $M\Omega$).*
2. *Very Large Current Gain (of the order of 10,000).*
3. *Very Low Output Resistance (of the order of few Ω).*
4. *Voltage Gain, $A_V < 1$.*

Darlington Pairs are available in a single package with just three leads, like one transistor in integrated form.

2.8.5 Disadvantages

We have assumed that the ***h-parameters*** of both the transistors are identical. But in practice it is difficult to make ***h-parameters*** depend upon the operating point of Q_1 and Q_2. Since the emitter current of transistor Q_1 is the base current for transistor Q_2, the value of $I_{c2} >> I_{c1}$

1. *The quiescent or operating conditions of both the transistors will be different. h_{fe} value will be small for the transistor Q_1. $\because h_{fe} = (I_c/I_b).I_{b_2}$ is less*

   CDIL make CIL997 is a transistor of Darlington Pair Configuration with $h_{fe} = 1000$.
2. *The second drawback is leakage current of the first transistor Q_1 which is amplified by the second transistor Q_2 ($\because I_{e_1} = I_{b_2}$).*

Hence overall leakage current is more. Leakage Current is the current that flows in the circuit with no external bias voltages applied

(a) *The **h-parameters** for both the transistors will not be the same.*

(b) *Leakage Current is more.*

Darlington transistor pairs are in single package available with h_{fe} as high as 30,000

2.8.6 Boot Strapped Darlington Circuit

The maximum input resistance of a practical Darlington Circuit is only 2 MΩ. Higher input resistance cannot be achieved because of the biasing resistors R_1, R_2 etc. They come in parallel with R_i of the transistors and thus reduce the value of R_i. The maximum value of R_i is only $\frac{1}{h_{ob}}$ since, h_{ob} is the resistance between base and collector. The input resistance can be increased greatly by boot strapping, the Darlington Circuit through the addition of C_0 between the first collector C_1 and emitter B_2.

What is Boot Strapping ?

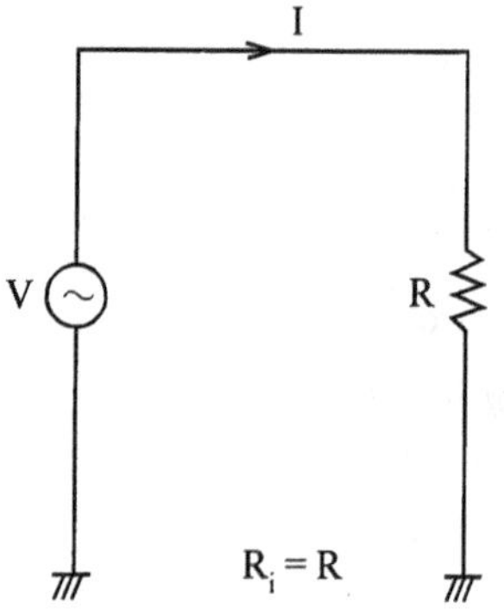

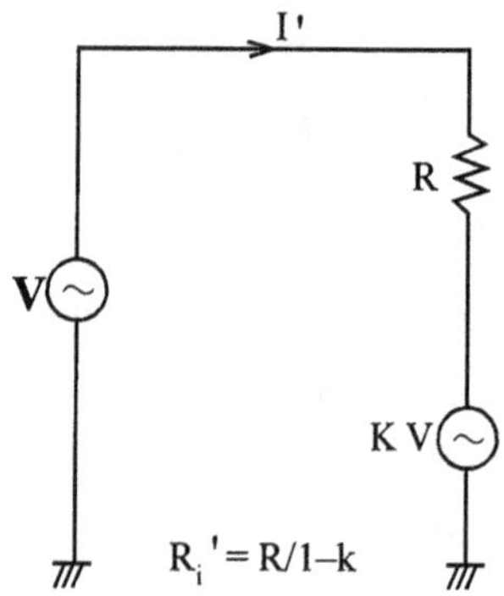

Boot strapping

Fig. 2.37 *Fig 2.38*

In Fig. 2.37, V is an AC signal generator, supplying current I to R. Therefore, the input resistance of the circuit as seen by the generator is $R_i = \frac{V}{I} = R$ itself. Now suppose, the bottom end of R is not at ground potential but at higher potential i.e. another voltage source of KV (K< 1) is connected between the bottom end of R and ground (Fig. 2.38). Now the input resistance of the circuit.

$$R_i' = \frac{V}{I'} \qquad I' = \frac{(V - KV)}{R}$$

or

$$R_i' = \frac{VR}{V(1-K)} = \frac{R}{1-K}$$

I' can be increased by increasing V. When V increases KV also increases. K is constant. Therefore the potential at the two ends of R will increase by the same amount, K is less than 1, therefore $R_i > R$. Now if K = 1, there is no current flowing through R (So V = KV there is no

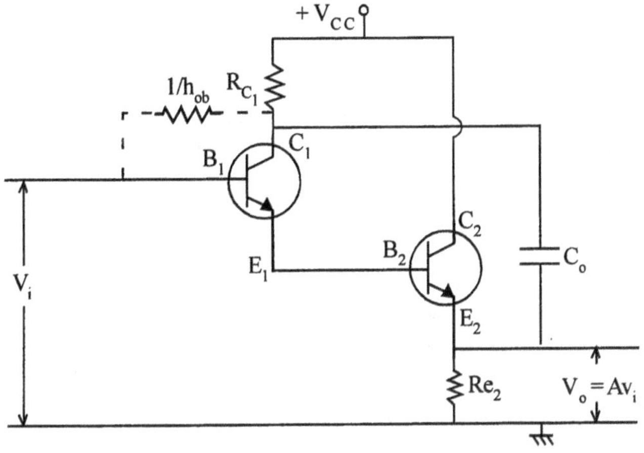

Fig. 2.39 Boot strap circuit

potential difference). So the input resistance $R_i = \infty$. Both the top and bottom of the resistor terminals are at the same potential. This is called as the Boots Strapping method which increases the input resistance of a circuit. If the potential at one end of the resistance changes, the other end of R also moves through the same potential difference. It is as if R is pulling itself up by its bootstraps. For CC amplifiers $A_v < 1 \cong 0.095$. So R_i can be made very large by this technique. $K = A_V \cong 1$. If we pull the boot with both the edges of the strap (wire) the boot lifts up. Here also, if the potential at one end of R is changed, the voltage at the other end also changes or the potential level of R_3 rises, as if it is being pulled up from both the ends. (Fig. 2.39).

For Common Collector Amplifier,

$$R_i = \frac{h_{ie}}{1-A_V}; \quad A_V \cong 1.$$

Therefore, R_i can be made large, since it is of the same form as

$$R_i = \frac{R}{1-K}$$

In the circuit shown in Fig. 2.39, capacitor C_0 is connected between C_1 and E_2. If the input signal changes by V_i, then E_2 changes by $A_v.V_i$ (assuming the resistance of C_0 is negligible).

Therefore, $\frac{1}{h_{ob}}$ is now effectively increased to $\frac{1}{h_{ob}(1-A_v)} \cong 400\ M\Omega$

2.8.7 AC Equivalent Circuit

Consider the circuit shown in Fig. 2.40.

The input resistance

$$R_i = \frac{V_i}{I_{b1}} \cong h_{fe_1} h_{fe_2} R_e$$

If we take h_{fe} as 50, $R_e = 4$ kΩ, we get R_i as 10MΩ. If a transistor with $h_{fe} = 100$ is taken, R_i will be much larger. The value of X_{C0} is chosen such that at the lower frequencies, under consideration X_{C_0} is a virtual short circuit. If the collector C_1 changes by certain potential, E_2 also changes by the same amount. So C_1 and E_2 are boot strapped. There is $\frac{1}{h_{ob}}$ between B_1 and C_1. (Fig. 2.40).

$$\therefore \quad R_{eff} = \frac{1}{h_{ob}(1 - A_v)}$$

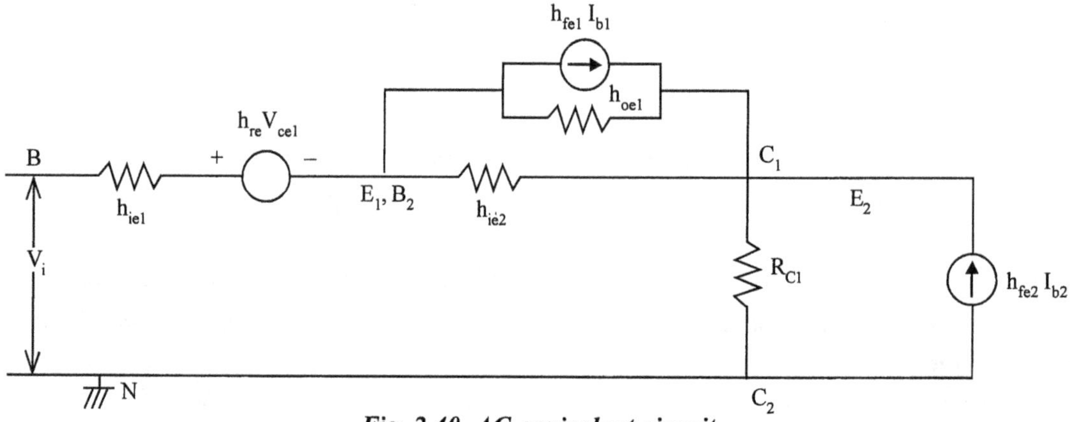

Fig. 2.40 AC equivalent circuit

Direct short circuit is not done between C_1 and E_2. Since, DC condition will change, X_{c_0} is a short only for AC signals and not for DC.

2.9 THE CASCODE TRANSISTOR CONFIGURATION

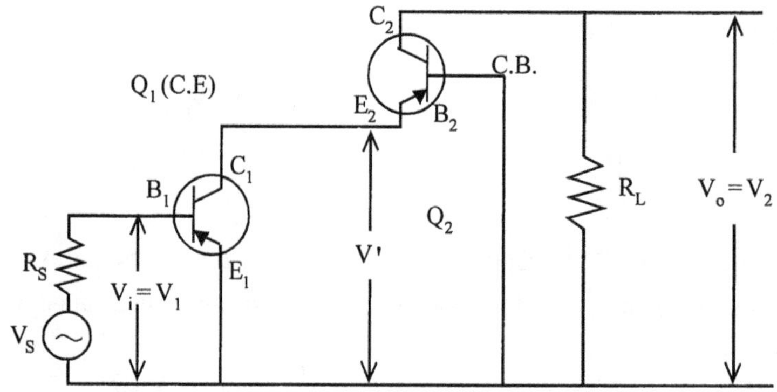

Fig. 2.41 CASCODE amplifier (C.E, C.B configuration)

The circuit is shown in Fig. 2.41. This transistor configuration consists of a ***Common Emitter Stage*** in cascade with a ***Common Base Stage***. The collector current of transistor Q_1 equals the emitter current of Q_2.

The transistor Q_1 is in Common Emitter Configuration and transistor Q_2 is in Common Base Configuration. Let us consider the input impedance (h_{11}) etc., output admittance (h_{22}) i.e., the ***h - parameters*** of the entire circuit in terms of the ***h - parameters*** of the two transistors.

2.9.1 Input |Z| (h_{11})

$$h_{11} = \text{Input } Z = \left.\frac{V_1}{I_1}\right|_{V_2 = 0}$$

If V_2 is made equal to 0, the net impedance for the transistor Q_1 is only h_{ib_2}. But h_{ib}, for a transistor in common emitter configuration is very small $\cong$ 20. We can conclude that the collector of Q_1 is effectively short circuited.

$$\therefore \qquad h_{11} \cong h_{ie}$$

When $V_2 = 0$, C_2 is shorted. Therefore, $h_{i2} = h_{ib_2}$. But h_{ib_2} is very small. Therefore C_1 is virtually shorted to the ground.

$$\therefore \qquad \boxed{h_i = h_{ie}}$$

2.9.2 Short Circuit Current Gain (h_{21})

$$h_{21} = \left.\frac{I_2}{I_1}\right|_{V_2 = 0}$$

$$h_{21} = \frac{I_2}{I_1} = \left.\frac{I}{I_1} \times \frac{I_2}{I_0}\right|_{V_2=0}$$

$$\frac{I'}{I_1} = h_{fe} \qquad \text{since, } I = I_{C_1}.\ I_1 = I_{B_1}$$

$$\frac{I_2}{I'} = -h_{fb} \qquad \text{since, } I = I_{E_2}.\ I_2 = I_{C_2}$$

$$\therefore \qquad h_{21} = -h_{fe}.\ h_{fb}$$

$$h_{fe} >> 1\ .\ -h_{fb} \simeq 1, \qquad \text{since } h_{fb} = \frac{I_C}{I_E}$$

$$\therefore \qquad \boxed{h_{21} \cong h_{fe}}$$

2.9.3 OUTPUT CONDUCTANCE (h_{22})

Output Conductance with input open circuited, for the entire circuit is,

$$h_{22} = \left.\frac{I_2}{V_2}\right|_{I_1 = 0}$$

when $I_1 = 0$, the output resistance of the transistor Q_1 is $\frac{1}{h_{oe}} \cong 40\ k\Omega$. (Since Q_1 is in Common Emitter configuration and h_{oe} is defined with $I_1 = 0$).

$$\therefore \qquad \frac{1}{h_{oe}} \cong 40\ k\Omega$$

is the source resistance for Q_2. Q_2 is in Common Base Configuration. What is the value of R_o of the transistor Q_2 with $R_s \cong 40$ k

It is $\cong 1/h_{ob}$ itself.

Since, h_{oe1} is very large, we can say that $I_1' = 0$ or between E_2 and ground there is infinite impedance. Therefore, output conductance of the entire circuit is $h_{22} \cong h_{ob}$.

2.9.4 REVERSE VOLTAGE GAIN (h_{12})

$$h_{12} = \left.\frac{V_1}{V_2}\right|_{I_1 = 0}$$

$$= \left.\frac{V_1}{V'} \times \frac{V^1}{V_2}\right|_{I_1 = 0}$$

$$\left.\frac{V_1}{V_0'}\right|_{I_i = 0} = h_{re}\cdot \left.\frac{V'}{V_2}\right| = h_{rb}$$

(Since, Q_2 is in Common Base configuration)

$\therefore \qquad h_{12} \cong h_{re}\, h_{rb}$.

$h_{re} \cong 10^{-4} \qquad h_{rb} = 10^{-4}$. $\because$ h_{12} is very small

$\because \qquad h_i = h_{11} \cong h_{ie}$. Typical value = 1.1 kΩ

$h_f = h_{21} \cong h_{fe}$. Typical value = 50

$h_o = h_{22} \cong h_{ob}$. Typical value = 0.49 μ A/V

$h_r = h_{12} \cong h_{re}\, h_{rb}$. Typical value = 7×10^{-8}.

Therefore, for a CASCODE Transistor Configuration, its input Z is equal to that of a single Common Emitter Transistor (h_{ie}). Its Current Gain is equal to that of a single Common Base Transistor (h_{fe}). Its output resistance is equal to that of a single Common Base Transistor (h_{ob}). The reverse voltage gain is very very small, i.e., there is no link between V_i (input voltage) and V_2 (output voltage). In otherwords, there is negligible internal feedback in the case of, a CASCODE Transistor Circuit, acts like a single stage C.E. Transistor (Since h_{ie} and h_{fe} are same) with negligible internal feedback ($\therefore$ h_{re} is very small) and very small output conductance, ($\cong h_{ob}$) or large output resistance ($\cong 2$ MΩ equal to that of a Common Base Stage). The above values are correct, if we make the assumption that $h_{ob} R_L < 0.1$ or R_L is < 200K. When the value of R_L is < 200 K. This will not affect the values of h_i, h_r, h_o, h_f of the CASCODE Transistor, since, the value of h_r is very very small.

CASCADE Amplifier will have

1. *Very Large Voltage Gain.*
2. *Large Current Gain (h_{fe}).*
3. *Very High Output Resistance.*

Example: 2.6

Find the voltage gains A_{Vs}, A_{V1} and A_{V2} of the amplifier shown in Fig.2.42.

Assume

$$h_{ie} = 1\ k\Omega, \quad h_{re} = 10^{-4}, \quad h_{fe} = 50$$

and $h_{oe} = 10^{-8}$ A/V.

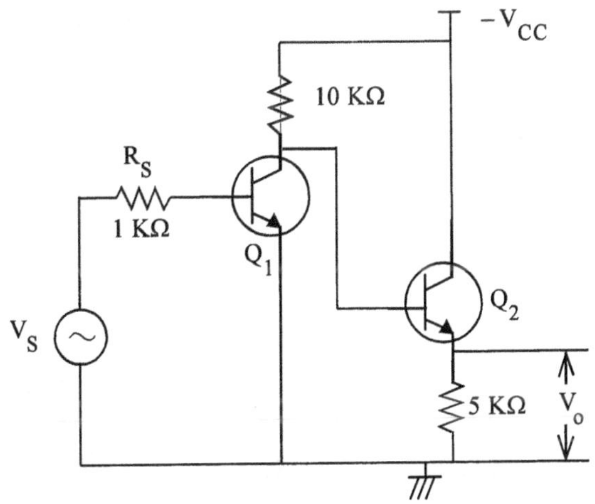

Fig. 2.42 Amplifier circuit Ex : 2.6

Solution :

The second transistor Q_2 is in Common Collector Configuration Q_1 is in Common Emitter Configuration. It is convenient if we start with the II stage.

II Stage :

$$h_{oe} \cdot R_{L_2} = 10^{-8} \times 5 \times 10^3 = 5 \times 10^{-5} < 0.1.$$

Therefore, $h_{oe}\, R_{L_2}$ is < 0.1, approximate analysis can be made, rigorous expression of C.C. and C.B Configuration need not be used.

$$\therefore \quad R_{i_2} = h_{ie} + (1+h_{fe})R_{L_2}$$

$$= 1\ k\Omega + (1 + 50)\ 5\ k\Omega = 256\ k\Omega.$$

Av_2 is the expression for Voltage Gain of the transistor in Common Collector Configuration in terms of Common Emitter ***h-parameters*** is

$$A_{V_2} = 1 - \frac{h_{ie}}{R_{i_2}}$$

$$= 1 - \frac{1\ k\Omega}{256\ k\Omega} = 1 - 0.0039 = 0.996$$

I Stage :

$$R_{L_1} = 10\ k \parallel R_{i_2} = 10\ k \parallel 256\ k\Omega = 9.36\ k\Omega.$$

$h_{oe} \cdot R_{L_1} < 0.1. \quad \because$ approximate equation can be used

$$A_{I_1} = -50$$

$$R_{i_1} = h_{ce} = 1\ k\Omega.$$

$$A_{v1} = -h_{fe}\frac{R_{L1}}{h_{ie}} = \frac{-50 \times 9.63k}{1\ k\Omega} = -48.$$

$$= \left(A_I \cdot \frac{R_{L1}}{R_i}\right)$$

$$\text{Overall Voltage Gain} = A_{v1} \cdot A_{v2} = -482 \times 0.996 = -480.$$

$$\therefore \quad A_{vs} = A_v \times \frac{R_S}{R_S + h_{ie}} = A_V \times \frac{1\ k\Omega}{2\ k} = -240.$$

2.10 CE – CC AMPLIFIERS

This is another type of two-stage BJT amplifier. The first stage in Common Emitter (CE) configuration provides voltage and current gains. The second stage in Common-Collector (CC) configuration provides impedance matching. This circuit is used in audio frequency amplifiers. The circuit is shown in Fig. 2.43.

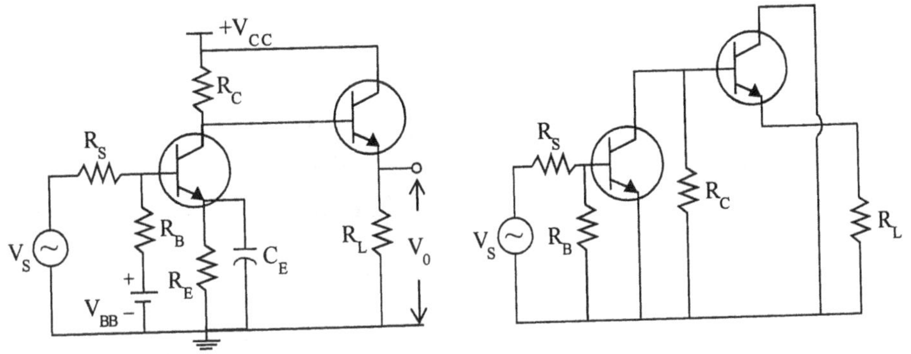

Fig. 2.43 CE, CC amplifiers

Analysis

Here biasing resistors are neglected for simplification of analysis.

II Stage in CC amplifier :

$$R_{L_2} \simeq R_L$$

$$h_{oc}\ R_{L_1} \leq 0.1$$

$$A_{I2} = A_{I'2} = (1 + h_{fe})$$

$$R_{i_2} = (1 + h_{fe})\ R_{L_2}$$

$$A_{V2} = A_{V2'} = \frac{A_{I_2} . R_{L_2}}{R_{i_2}}$$

$$= \frac{(1 + h_{fe}\ R_{L_2})}{h_{ie} + (1 + h_{fe})R_{L_2}}$$

$$= 1 - \frac{h_{ie}}{R_{i_2}};$$

$$\therefore \quad A_{V_2} < 1$$

for CE stage

$$A_{I_1} = A_{I_1}' = -h_{fe}$$

$$R_{i_1} = R_{i_1}' = h_{ie}$$

$$A_{V_1} = A_{V_1}' = (A_{I_1} . R_{L_1} / R_i)$$

Overall Characteristics

$$A_V = A_{V_1} . A'_{V_2}$$
$$R_i = R_{i_1}$$
$$R_0 = R_{0_1}$$
$$A_I = \frac{A_V . R_{i_1}}{R_{L_2}}$$

2.11 TWO STAGE RC COUPLED JFET AMPLIFIER (IN COMMON SOURCE CS CONFIGURATION)

The circuit for two stages of RC coupled amplifier in CS configuration is as shown in Fig. 2.44.

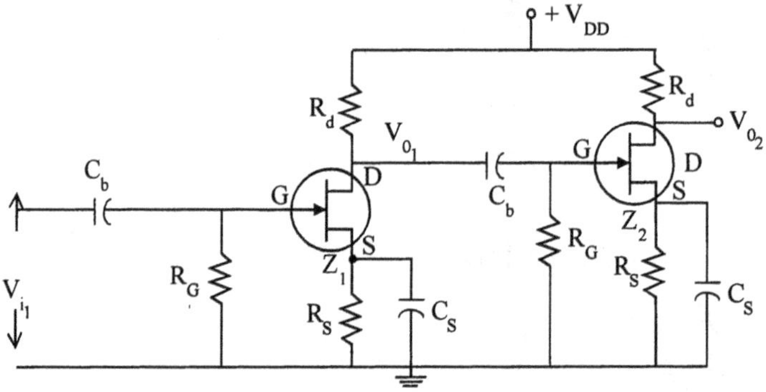

Fig. 2.44 Two stage RC coupled JFET amplifier

The output V_{0_1} of I Stage is coupled to the input V_{i_2} of II Stage through a blocking capacitor C_b. It blocks the DC components present in the output of I Stage from reaching the input of the II stage which will alter the biasing already fixed for the active device. Resistor R_g is connected between gate and ground resistor R_D is connected between drain and V_{DD} supply. C_S is the bypass capacitor used to prevent loss of gain due to negative feedback. The active device is assumed to operate in the linear region. So the small signal model of the device is valid.

Frequency Roll-off is the term used for the decrease in gain with frequency in the upper cut-off region. It is expressed as db/octave on db/decade. In the logarithmic scale of frequency,

octave is $\frac{f_2'}{f_2} = 2$ decade is $\frac{f_2'}{f_2} = 10$.

The purpose of multistage amplifiers is to get large gain. So with BJTs, Common Emitter Configuration is used. If JFETs are employed, common source configuration is used.

2.12 DIFFERENCE AMPLIFIER

This is also known as *differential amplifier*. The function of this is to amplify the difference between the signals. The advantage with this amplifier is, we can eliminate the noise in the input signals which is common to both the inputs. Thus S/N ratio can be improved. The difference amplifier can be represented as a blackbox with two inputs V_1 and V_2 and output V_0 where $V_0 = A_d (V_1 - V_2)$.

where A_d is the gain of the differential amplifier. But the above equation will not correctly describe the characteristic of a differential amplifier. The output V_0 depends not only on the difference of the two signals $(V_1 - V_2) = V_d$ but also on the average level called *common mode signal* $V_C = \frac{V_1 + V_2}{2}$.

If one signal is (V_1) 100 μV and the other signal (V_2) is –100 μV.

∴ V_0 should be A_d (200) μV.

Now in the second case, if $V_1 = 800$ μV, and $V_2 = 600$ μV, $V_d = 800 - 600 = 200$ μV and V_0 should be A_d (200) μV. So in both cases, for the same circuit. V_0 should be the same. But in practice it will not be so because the average of these two signals V_1 & V_2 is not the same in both the cases.

$$V_d = V_1 - V_2$$

$$V_c = \frac{1}{2}(V_1 + V_2)$$

from the equations above, we can write that,

$$V_1 = V_c + \frac{1}{2} V_d$$

[∵ If we substitute the values of V_c and V_d we get the same.

$$V_2 = V_c - \frac{1}{2} V_d \qquad V_1 = V_c + \frac{1}{2} V_d = \frac{V_1}{2} + \frac{V_1}{2} = V_1]$$

V_0 can be represented in the most general case as

$$V_0 = A\,V \;+ A_2\,V_2$$

Substituting the values of V_1 and V_2

$$V_0 = A_1\,[V_c + \frac{1}{2}\,v_d] + A_2\,[V_c - \frac{1}{2}\,V_d]$$

$$= A_1\,V_c + \frac{A_1}{2}\,V_d + A_2\,V_c - \frac{A_2}{2}\,V_d$$

$$V_0 = V_c\,(A_1 + A_2) + V_d\left[\frac{A_1 - A_2}{2}\right]$$

∴ $$V_0 = V_c A_c + V_d A_d$$

where $A_d = \frac{A_1 - A_2}{2}$ and $A_c = A_1 + A_2$.

for operational amplifiers, always input is given to the inverting node to get $\frac{A_1 - (-A_2)}{2}$ so that A_d is very large and A_c is very small.

A_1 and A_2 are the voltage gains of the two amplifier circuits separately.

The voltage gain from the difference signal is A_d.

The voltage gain from the common mode signal is A_c.

$$V_0 = A_d V_d + A_c V_c.$$

To measure A_d, directly set $V_1 = -V_2 = 0.5V$ so that

$$V_d = 0.5 - (-0.5) = 1V.$$

$$V_c = \frac{(0.5 - 0.5)}{2} = 0$$

$\therefore$ $$V_0 = A_d \cdot 1 = A_d \text{ itself}$$

$\therefore$ If $V_1 = -V_2$ and output voltage is measured,

Output voltage directly gives the value of A_d.

Similarly if we set $V_1 = V_2 = 1V$. then

$$V_a = 0, \quad V_c = \frac{V_1 + V_2}{2} = \frac{2}{2} = 1V.$$

$\therefore$ $$V_0 = 0 + A_c \cdot 1 = A_c.$$

$\therefore$ The measured output voltage directly gives Ac. We want A_d to be large and A_c to be very small because only the difference of the two signals should be amplified and the average of the signals should not be amplified. $\therefore$ The ratio of the these two gains $\rho = \left|\frac{A_d}{A_c}\right|$ is called the *common mode rejection ratio* (CMRR). This should be large for a good difference amplifier.

$$V_0 = A_d V_d + A_c V_c$$

$$\rho = \frac{A_d}{A_c} \qquad \therefore A_c = \frac{A_d}{\rho}$$

$\therefore$ $$V_0 = A_d V_d + \frac{A_d}{\rho} \cdot V_c$$

$$V_0 = A_d V_d \left(1 + \frac{1}{\rho} \cdot \frac{V_c}{V_d}\right)$$

2.13 CIRCUIT FOR DIFFERENTIAL AMPLIFIER

In the previous D.C amplifier viz., C.B, C.C and C.E, the output is measured with respect to ground. But in difference amplifier, the output is α to the difference of the inputs. So V_0 is not measured w.r.t ground but w.r.t to the output of one transistor Q_1 or output of the other transistor Q_2. (Fig. 2.45).

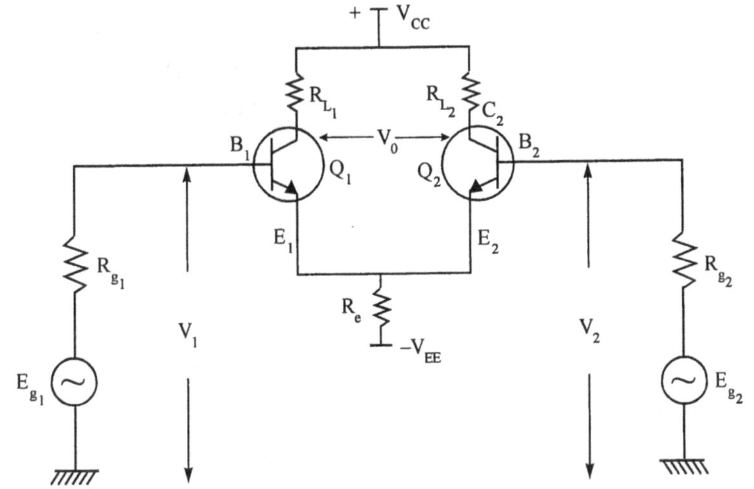

Fig. 2.45 Differential amplifier

Equivalent Circuit (Fig. 2.46)

The advantage with this type of amplifiers is the drift problem is eliminated. Drift means, even when there is no input, V_i there can be some output V_0 which is due to the internal thermal noise of the circuit getting amplified and coming at the output. Drift is reduced in this type of circuit, because, the two points should be exactly identical. Hence, I_{c_s}, h_{FE}, V_{BE} will be the same for the two transistors. Now if I_{c_1} rises across R_L (I_{c_1} R_{L_1}) increases with increase in I_{c_1}. So the voltage at collector of Q_1 decreases. If Q_2 is also identical to Q_1 its collector voltage also drops by the same amount. Hence V_0 which is the difference of these voltages remains the same thus the drift of these transistors gets cancelled.

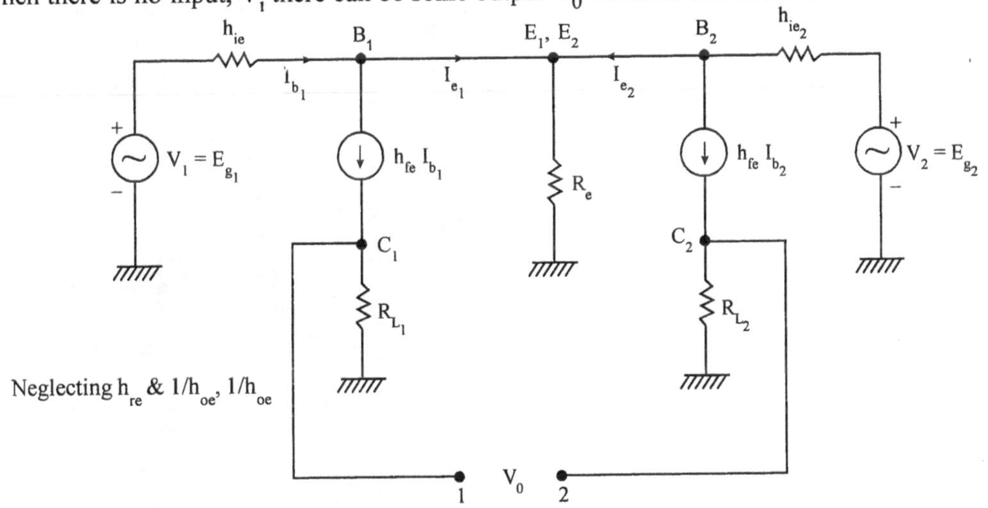

Fig. 2.46 Equivalent circuit

The input to a differential amplifier are of two types. 1. Differential mode 2. Common mode.

If V_1 and V_2 are the inputs, the differential mode input = $V_2 - V_1$.

Here two different a.c. signal are being applied V_1 & V_2. So these will be interference of these signals and so both the signals will be present simultaneously at both input points i.e., if V_1 is applied at point 1, it also prices up the signal V_2 and so the net input is $(V_1 + V_2)$. This is due to interference.

$$\text{Common node input} = \frac{V_1 + V_2}{2}$$

An ideal differential amplifier must provide large gain to the differential mode inputs and zero gain to command input.

$$\therefore \quad V_0 = A_2 V_2 - A_1 V_1 \qquad \text{.....(2.15)}$$

A_2 = voltage gain of the transistor Q_2

A_1 = voltage gain of the transistor Q_1

we can also express the output in term of the common mode gain A_c and differential gain A_d.

$$\therefore \quad V_0 = A_d (V_2 - V_1) + A_c \left(\frac{V_1 + V_2}{2}\right) \qquad \text{.....(2.16)}$$

$$= A_d V_2 - A_d . V_1 + A_c . \frac{V_1}{2} + A_c . \frac{V_2}{2} \qquad \text{.....(2.17)}$$

$$V_0 = V_2 \left(A_d + \frac{A_c}{2}\right) - V_1 \left(A_d - \frac{A_c}{2}\right) \qquad \text{.....(2.18)}$$

Comparing eqns. (2.18) and (2.15),

$$A_2 = A_d + \frac{A_c}{2}$$

$$A_1 = A_d + \frac{A_c}{2}$$

Solving these two eqns. $A_d = \dfrac{A_1 + A_2}{2}$

$$A_c = A_2 - A_1/2$$

μA 730 is an I.C differential amplifier. (Fig. 2.47).

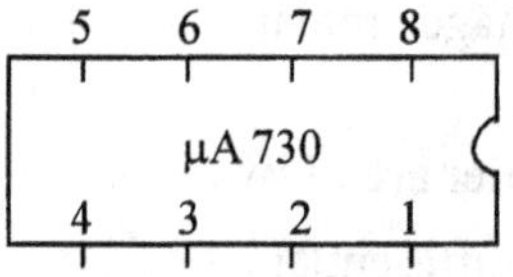

Fig. 2.47 IC mA 730 pin configuration

8 pins. Input is given to pins 2 and 3. V^+_{cc} to pin 7, 4 is ground. Output is taken at pin no 6. In the difference amplifier, the difference of the input voltages V_1 and V_2 is amplified. The collectors of the transistor Q_1 & Q_2 are floating. They are not at ground potential. So the output voltage is not at ground potential. Hence the output voltage is the difference of the collector voltages (a.c) of transistors Q_1 & Q_2. Difference amplifiers are used in measuring instants and instrumentations systems. The difference of V_{i_1} & V_{i_2} may be 1 μV which is difficult to measure. So if this is amplified to 1mV or 1V the measurement will be accurate. So difference amplifiers are used to measure very small increased voltages.

While computing A_1 & A_2 of individual transistors, the other input should be made zero. One while computing A_1, $V_2 = 0$. Because there should be no common mode signal, while computing A_1, A_1 is the actual gain. Not differential gain. ∴ the other input is made zero.

In the case of operational amplifiers for single ended operation, always the positive end is grounded (non inverting input) & input is applied to the inverting input (–). It is because, at this part the feedback current & input current get added algebraically. So this is known as the swimming junction. When sufficient negative feedback is used, the closed loop performance becomes virtually independent of the characters of the operational amplifiers & depends on the external passive elements, which is desired.

SUMMARY

- Based on type of coupling, amplifiers are classified as (a) R-C coupled (b) Transformer coupled (c) Direct coupled (d) Tuned circuit (R-L coupled) amplifiers.
- R-C coupled amplifier, equivalent circuit and analysis are given.
- When n-stages of amplifiers are cascaded, overall gain $A_n = A_1, A_2, A_3$ .... A_n.
- Millers theorem and mathematical proof for the same are given.
- Darlington and Boot strapped Darlington pair circuits are analysed.
- Two stage R-C coupled JFET amplifier circuit analysis is given.
- Circuit for difference amplifier and analysis are given.

  Common Mode Rejection Ratio (CMRR) = $\dfrac{A_d}{A_c}$ is explained

- CASCODE amplifier which is a combination of CE and CB amplifiers is given and the circuit is analysed. CASCODE configuration gives large R_0, high A_v and A_I.

OBJECTIVE TYPE QUESTIONS

1. Based on the type of Coupling, the Amplifiers are classified as __________________.
2. Based on Bandwidth, the amplifiers are classified as __________________.
3. Different types of Distortion in amplifiers are __________________.
4. When n stages with gains A_1, A_2 .... A_n are cascaded, overall voltage gain A_{Vn} = __________________.
5. Expression for voltage gain A_v in the Mid Frequency range, interms of h_{fe}, R_L, R_S and h_{ie} is, A_V(M.F) = __________________.
6. Expression for A_v (H.F) in terms of f, f_2 and A_v (M.F) is, A_v (H.F) = __________________.
7. Phase shift ϕ interms of f_1 and f is, __________________.
8. Expression for A_v (L.F) in terms of f, f_1 and A_v (M.F) is __________________.
9. When n stages are cascaded, the relation between f_{2n}, and f_2 is __________________.
10. When h stages are cascaded, the relation between f_{1n} and f_1 is, __________________.
11. The relation between h_{fe}, f_β and f_T is __________________.
12. Figure of Merit of an amplifier circuit is __________________.
13. CMRR (ρ) = __________________.
14. Expression for V_0 interms of A_d, V_d, V_c and ρ is, __________________.

15. Expression for A_d and A_c interms of A_1 and A_2 are ____________________.
16. Stiff coupling is ____________________.
17. When frequency change is octave $\dfrac{f_2}{f_1}$ = ____________________.
18. In cascade form, ordinary gains ____________________ decibal gains ____________________.
19. Phase response is a plot between ____________________.
20. According to Miller's theorem the feedback capacitance when referred to input side, with gain A is ____________________.
21. Interms of h_{fe}, current gain in Darlington Pair circuit is approximately ____________________.
22. The disadvantage of Darlington pair circuit is ____________________.
23. Compared to Common Emitter Configuration R_i of Darlington pair circuit is ____________________.
24. In CASCODE amplifier, the transistors are in ____________________ configuration.
25. The salient featuers of CASCODE Amplifier are ____________________.
26. What is distortion ?
27. What are the types of distortion ? Define them.
28. How does the amplifier behave for low frequencies and high frequencies ?
29. Which configuration is the best in cascade for an output stage and for an intermediate stage ?
30. What is the darlington pair ? What is its significance ?
31. How is the bandwidth of a cascade affected compared to the bandwidth of a single stage ?
32. Why is the emitter bypass capacitor used in an RC coupled amplifier ?
33. What is the affect of emitter bypass capacitor on low frequency response ?
34. What are types of cascade ?
35. How would you differentiate an interacting stage from a non-interacting stage ?
36. What is the expression for the upper 3dB frequency for a n-stage non interacting cascade ?
37. What is the expression for lower 3dB cutoff frequency in n-stage interacting cascade ?
38. What is the expression for upper 3dB cutoff frequency in a n-stage interacting cascade ?
39. What is the slope of the amplitude response for an n-stage amplifier ?
40. Why do we go for multistage amplifier ?

ESSAY TYPE QUESTIONS

1. Explain about the classification of Amplifiers based on type of coupling and bandwidth.
2. What are the different types of distortions possible in amplifiers outputs ?
3. Obtain the expression for the voltage gain A_v in the L.F, M.F and H.F ranges, in the case of single stage BJT amplifier.
4. When h-identical stages of amplifiers are cascaded, derive the expressions for overall gain A_{vn}, lower cutoff frequency f_{1n} and upper cutoff frequency f_{2n}.
5. With the help of necessary waveforms, explain about the step response of amplifiers.
6. What is the significance of square wave testing in amplifiers ?
7. Draw the circuit for differential amplifier and derive the expression for CMRR.
8. Explain about the characteristics of operational amplifiers.
9. Draw the circuit for Darlington pair and device the expressions for A_I, A_V, R_i and R_O.
10. Draw the circuit for CASCODE Amplifier. Explain its working, obtaining overall values of the circuit for h_i, h_f, h_O and h_r.

ANSWERS OF OBJECTIVE TYPE QUESTIONS

1. (a) R-C coupled (b) Transformer coupled
   (c) Direct coupled (d) Tuned circuit
2. (a) Narrow band (b) Untuned amplifiers
3. (a) Frequency distortion (b) Phase distortion (c) Amplitude distortion
4. $A_{vn} = A_1, A_2, A_3 \ldots\ldots A_n$
5. $A_v\,(M.F) = \dfrac{-h_{fe}.R_L}{R_S + h_{ie}}$
6. $A_v\,(H.F) = A_V\,(M.F) \Big/ 1 + j\left(\dfrac{f}{f_2}\right)$
7. $\phi = 180^0 - \tan^{-1}\left(\dfrac{f_1}{f}\right)$
8. $A_v\,(L.F) = \dfrac{A_V(M.F)}{1 - j\left(\dfrac{f_1}{f}\right)}$
9. $\dfrac{f_{2n}}{f_2} = \sqrt{\left(2^{1/n} - 1\right)}$
10. $f_{1_n} = f_1 \Big/ \sqrt{2^{1/n} - 1}$
11. $h_{fe}, f_\beta = f_T$
12. Figure of merit = Gain × Bandwidth
13. CMRR = (A_d / A_c)
14. $V_0 = A_d V_d \left(1 + \dfrac{1}{\rho}.\dfrac{V_C}{V_d}\right)$
15. $A_d = \dfrac{(A_1 + A_2)}{2}$; $A_c = (A_2 - A_1)/2$
16. When capacitor is chosen such that $X_C = \dfrac{R_E}{10}$
17. 2
18. Multiply, Add up

19. Phase angle and frequency
20. $C_{in} = C (1 - A)$
21. $(h_{fe})^2$
22. Leakage current is more
23. Very high
24. CE – CB
25. Large voltage and current gains
26. If the output waveform is not the replica of the input waveform, it is called distortion.
27. (i) Non linear distortion : If harmonic frequencies are generated at the o/p, it is called non linear or amplitude distortion.

    (ii) Freqeuncy distortion : If different frequency components are amplified differently, it is called frequency distortion.

    (iii) Phase distortion : If the output is shifted by different phased each time, it is called phase distortion.
28. Low frequencies - high pass filter.

    High frequencies - low pass filter.
29. Output stage - CC

    Intermediate stage - CE
30. CC-CC cascade is called Darlington pair.

    Significance : (i) It has very high input impedance. (ii) It behaves life a constant current source.
31. It decreases.
32. To decrease the loss in gain due to negative feedback
33. The tilt is more
34. Interacting and non interacting.
35. If the input impedance of next stage loads the previous stage, it is called interacting stages. If the input impedance of next stage does not load the previous stage, it is called non interacting stage.
36. $f_L = \dfrac{f}{\sqrt{2^{1/n} - 1}}$
37. $f_H = f.\ \sqrt{2^{1/n} - 1}$
38. $f_H = 0.94\, f_D$.
39. 6n dB/octave or 20n dB/decade.
40. For high gain.

3 BJT - Amplifiers, Frequency Response

In this Unit,

- Cascading of single stage amplifiers is discussed.
- Expressions for overall voltage gain are derived.
- Lower cut-off frequency, upper cut-off frequency, when n-stages are cascaded are given.
- Phase response of an amplifier, decibel voltage gain, stiff coupling terms are explained.
- Other types of transistor circuits, Darlington pair circuits, Boot strapped sweep circuit, Cascode amplifiers are also discussed.
- Numerical examples are also given.

3.1 LOGARITHMIC DECIBELS

In many cases it is convenient to compare two powers on a logarithmic scale rather than on a linear scale. The unit of this logarithmic scale is called the decibel abbreviated as db.

Suppose P_2 is the output power and P_1 is the input power, then the power gain in decibels is

$$N = 10 \log_{10} \frac{P_2}{P_1}$$

Where N is in dbs

If N is negative, it means P_2 is less than P_1.

Noise power is also expressed in decibels (dbs). It should be negative for a given device or amplifier i.e the output noise is less than what is present in the input.

If for a given amplifier circuit, the input and output resistances are same (as R), then

$$P_1 = \frac{V_1^2}{R}, \quad P_2 = \frac{V_2^2}{R}$$ where V_1 and V_2 are input and output voltages.

$$\therefore \quad N = 20 \log \frac{V_2}{V_1} = 20 \log A_V$$

But even though, the input and output impedances are not equal, this convention is followed for convenience i.e, $N = 20 \log A_V$. If $A_V = 10$, $N = 20 \log 10$ is 20 the decibel voltage gain of the amplifier. 20 is not the power gain because, the input resistances are not equal. Therefore 20 is the decibel voltage gain. If the output resistances are equal decibel voltage gain is equal, to power gain. Overall db V of a multistage amplifier is equal to sum of db V of individual stages.

3.2 FREQUENCY RESPONSE

The graph of voltage gain versus frequency is as shown in Fig. 3.1.

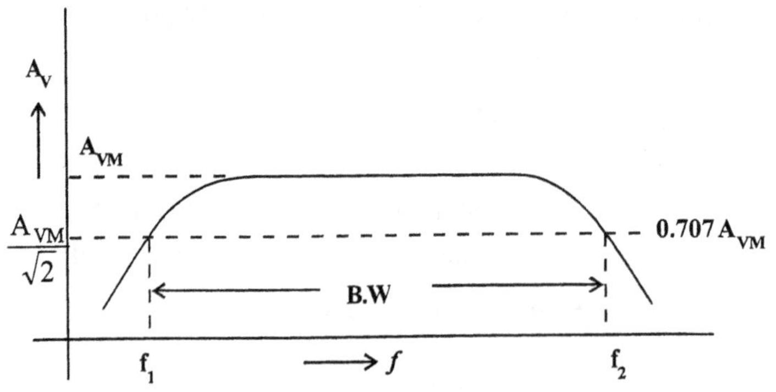

Fig 3.1 Frequency response.

Initially as frequency increases, capacitance reactance due to C_b, $\alpha_{Cb} = \frac{1}{2\pi f C_b}$ decreases. So the drop across it also decreases and hence net input at the base of transistor increases and so V_0 increases. Hence $A_V = \frac{V_0}{V_i}$ also increases.

In the mid frequency range, X_{Cb} is almost a ***short circuit,*** since its reactance decreases with increasing frequency. In the low frequency range, $X_{C_C} = \frac{1}{2\pi f C_C}$ is open circuit (in the A.C equivalent circuit).

In the mid frequency range, the capacitance reactance of both C_b and C_c are negligible. So the net change with frequency, remains constant as shown in the figure.

As the frequency is further increased, capacitance reactance due to C_C decreases. Hence V_0 decreases because, V_0 is taken across C_c. So as frequency increases, V_0 decreases and hence A_V decreases. So the shape of the curve is as shown in the Fig. 3.1.

$$\text{Expression } A_V \text{ (M.F)} = \frac{-hf_e R_L}{R_S + h_{ie}}$$

$$f_1 \simeq \frac{1}{2\pi C_b R_c}$$

f_1 is also known as lower cut-off frequency or lower 3 db frequency.

$$f_2 \simeq \frac{1}{\pi C_C R_C}$$

f_2 is also known as upper cut-off frequency or upper 3 db point.

PHASE RESPONSE

Phase shift between V_0 and V_i varies as shown in Fig. 3.2.

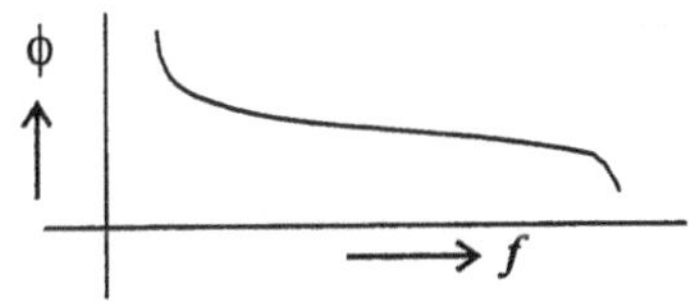

Fig 3.2 Phase response.

3.3 ANALYSIS AT LOW AND HIGH FREQUENCIES

In order to obtain some idea of a transistor's high frequency capability and what transistor to choose for a given application, we examine how transistor's CE short-circuit forward-current gain varies with frequency. When $R_L = 0$, the approximate high-frequency equivalent circuit is drawn. (Fig. 3.3).

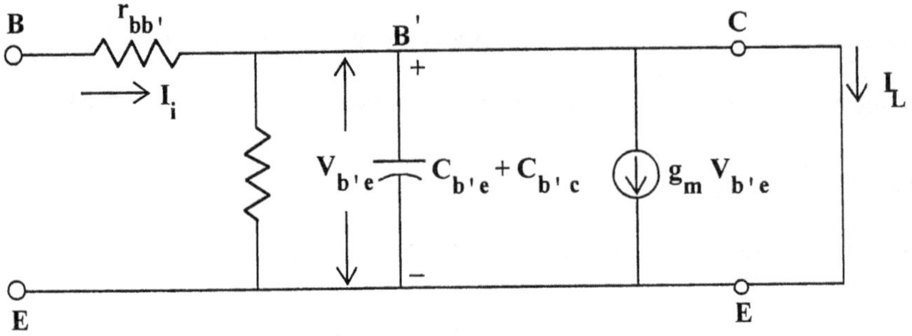

Fig 3.3 Approximate high-frequency circuit.

The current gain, $A_I = \frac{I_L}{I_i} = \frac{-g_m V_{b'e}}{I_i}$

$$I_i = V_{b'e}\, g_{b'e} + V_{b'e}\, [j\omega\, (C_{be} + C_{b'c})]$$

$$A_I = \frac{-g_m V_{b'e}}{V_{b'e}\left[g_{b'e} + j\omega(C_{b'e} + C_{b'c})\right]} = \frac{-g_m}{g_{b'e} + j\omega(C_{b'e} + C_{b'c})}$$

$$\because g_{b'e} = \frac{g_m}{h_{fe}}$$

$$A_I = \frac{-g_m h_{fe}}{g_m + j2\pi f(C_{b'e} + C_{b'c})h_{fe}}$$

$$= \frac{-h_{fe}}{\left[(j + j2\pi f\ (C_{b'e} + C_{b'c})\frac{h_{fe}}{g_m}\right]} = \frac{-h_{fe}}{1 + j2\pi f\ (C_{b'e} + C_{b'c})}$$

$$\left[\because r_{b'e} \frac{h_{fe}}{g_m}\right]$$

$$A_I = \frac{-h_{fe}}{1 + j\left(\frac{f}{f_\beta}\right)}$$

Where $f_\beta = \frac{1}{2\pi r_{b'e}(C_{b'e} + C_{b'c})}$

h_{fe} = CE small signal short-circuit current gain.

β Cut-off Frequency 'f_β' :

The 'β' cut-off frequency f_β, also referred as f_{hfe}, is the CE short-circuit small-signal forward-current transfer ratio cutoff frequency. It is the frequency at which a transistors CE short-circuit current gain drops 0.707 from its value at low frequency. Hence, f_β represents the maximum attainable bandwidth for current gain of a CE amplifier with a given transistor.

α Cut-off Frequency 'f_α' :

A transistor used in the CB connection has a much higher 3-dB frequency. The expression for the current gain by considering approximate high frequency circuit of the CB with output shorted is given by

$$A_I = \frac{I_L}{I_i} = \frac{-h_{fb}}{1 + j\left(\frac{f}{f_\alpha}\right)}$$

Where $$f_\alpha = \frac{1}{2\pi r_{b'e}(1 + h_{fb})C_{b'e}} \approx \frac{h_{fe}}{2\pi\, r_{b'e}C_{b'e}}$$

$$\boxed{f_\alpha = \frac{h_{fe} f_\beta (C_{b'e} + C_{b'c})}{C_{b'e}}}$$

f_α is the alpha cutoff frequency at which the CB short-circuit small-signal forward current transfer ratio drops 0.707 from its 1KHZ value.

Gain Bandwidth Product :

Although f_a and f_b are useful indications of the high-frequency capability of a transistor, are even more important characteristics is f_T.

The f_T is defined as the frequency at which the short-circuit CE current gain has a magnitude of unity. The 'f_T' is lies in between f_β and f_α.

$$|A_I| = \frac{h_{fe}}{\sqrt{1 + (f / f_\beta)^2}}$$

at $$f = f_T,\ |A_I| = 1,\ \left(\frac{f_T}{f_\beta}\right)^2 >> 1,$$

we obtain $$f_T / f_\beta \approx h_{fe}$$

$$\boxed{f_T \approx h_{fe}\, f_\beta}$$

$\therefore$ f_T is the product of the low frequency gain, h_{fe}, and CE bandwidth f_β. f_T is also the product of the low frequency gain, h_{fb}, and the CB bandwidth, f_α . i.e $f_T = h_{f\beta} \cdot f_\alpha$. The value of f_T ranges from 1MHz for audio transistors upto 1GHZ for high frequency transistors.

The Fig. 3.4 indicates how short-circuit current gains for CE and CB connections vary with frequency.

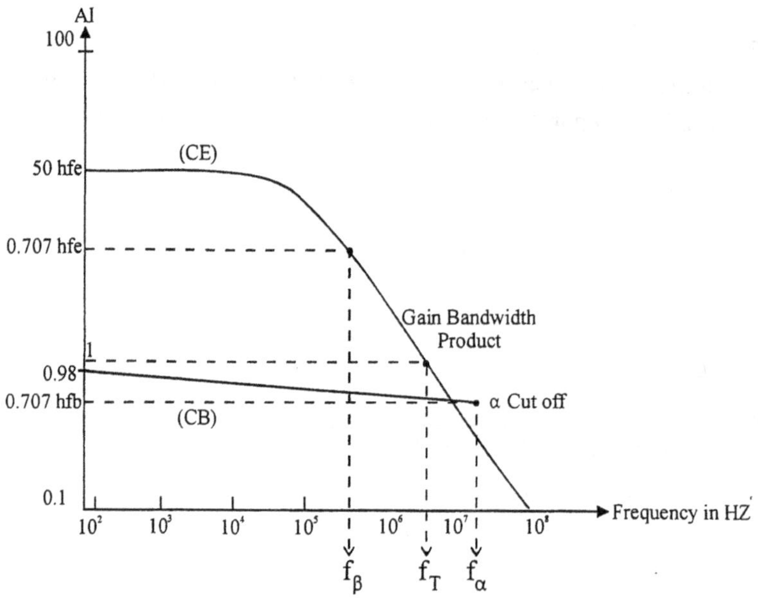

Fig. 3.4

3.4 EFFECT OF COUPLING AND BYPASS CAPACITOR

The purpose of coupling capacitor is explained from the Fig. 3.5 (a). In this arrangement, the signal developed across collector resistor R_C of the first stage in coupled to the base of second stage through the coupling-capacitor C_C.

The coupling capacitor C_C transmits ac signal but blocks the dc voltage of the first stage from reaching the base of the second stage. Thus the dc biasing of the next stage is not interfered with. For this reason, the coupling capacitor C_C is also called the blocking capacitor.

3.4.1 Effect of Coupling Capacitor on Different Frequency Ranges

(a) Mid frequencies:

At Mid frequencies, (50 HZ - 20 kHz) the impedance offered by the coupling capacitor C_C is so small that it can be effectively considered as a short circuit and its effect can be neglected.

- Equivalent circuit of Fig 3.5(a) is shown in Fig 3.5(b) and thevenin's equivalent circuit is given Fig 3.5(c).

$$\text{Current, } I = \frac{h_{fe}\, I_b\, R_C}{R_C + h_{ie}}$$

$$\text{So current gain, } A_{in} = \frac{I}{I_b} = \frac{h_{fe}\, R_C}{R_C + h_{ie}}$$

Output voltage, $V_{out} = h_{ie}\, I = \dfrac{h_{ie}\, h_{fe}\, I_b\, R_C}{R_C + h_{ie}}$

Input voltage, $V_{in} = h_{ie}\, I_b$.

So voltage gain $A_{vm} = V_{out}/V_{in} = \dfrac{h_{ie} h_{fe} I_b R_C/(R_C + h_{ie})}{h_{ie} I_b}$

$$= \frac{h_{fe}\, R_C}{R_C + h_{ie}}$$

- Here the voltage gain of the amplifier is constant, with the increase in the frequency in this range. The reactance of C_C reduces there by gain increases but at the same time lower capacitive reactance causes higher loading resulting in lower voltage gain. Thus the two effects cancel each other and uniform gain is obtained in mid-frequency range.

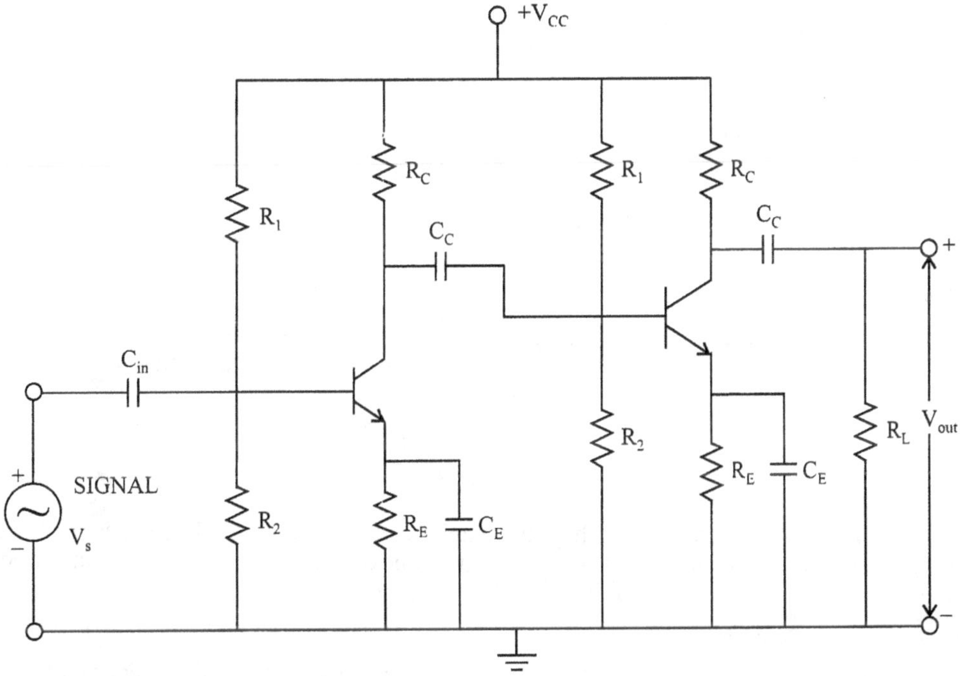

Fig. 3.5(a) Two stage RC coupled transistor amplifier.

(b) Low-frequency Range:

In low-frequency range (< 50 Hz), the impedance offered by coupling capacitor C_c is comparable to collector resistance R_C. C_C largely affects the current amplification. So C_C included in equivalent Circuit.

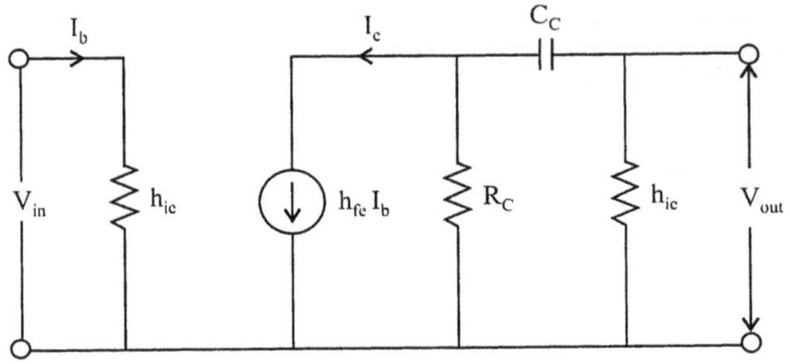

Fig. 3.5 (b) Approximate model of on R-C coupled transistor amplifier.

From the circuit,

$$\text{current,} \quad I = \frac{h_{fe} I_b R_C}{h_{ie} + R_C - j/\omega C_C}$$

$$\text{so current gain,} \quad A_{il} = \frac{I}{I_b} = \frac{h_{fe} R_C}{h_{ie} + R_C + j/\omega C_C}$$

$$\text{output voltage,} \quad V_{out} = h_{ie} I = \frac{h_{fe} h_{ie} I_b R_C}{h_{ie} + R_C - j/\omega C_C}$$

$$\text{Input voltage,} \quad V_{in} = h_{ie} I_b$$

$$\text{so voltage gain,} \quad A_{vl} = V_{out}/V_{in} = \frac{\dfrac{h_{ie} h_{fe} I_b R_C}{h_{ie} + R_C - j/\omega C_C}}{h_{ie} I_b} = \frac{h_{fe} R_C}{h_{ie} + R_C - j/\omega C_C}$$

$$= \frac{h_{fe} R_C}{h_{ie} + R_C - j/\omega C_C}$$

- At low frequencies (below 50 Hz) higher capacitive reactance of coupling capacitor C_C allows very small part of signal to pass one stage to the next and higher C_E and R_E not effectively shunted. This the voltage gain falls off at low frequencies.

(c) High Frequency Range:

In high frequency range, exceeding 20 kHz, the reactance offered by C_c is very small and it is considered as short circuit. Equivalent circuit as shown in Fig 3.5(c) and Fig 3.5(g).

From Thevenin's equivalent circuit,

$$\text{current } I = \frac{h_{fe} I_b \dfrac{R_C h_{ie}}{R_C + h_{ie}}}{\dfrac{R_C h_{ie}}{R_C + h_{ie}} + \dfrac{1}{j\omega C_d}} = \frac{h_{fe} I_b R_C h_{ie}}{R_C h_{ie} + \dfrac{1}{j\omega C_d}(R_C + h_{ie})}$$

$$\text{current gain} = A_{in} = \frac{I}{I_b} = \frac{h_{fe}\, h_{ie}\, R_C}{h_{ie}\, R_C + \dfrac{1}{j\omega C_d}(h_{ie} + R_C)}$$

$$\text{output voltage, } V_{out} = I \times \frac{1}{j\omega C_d} = \frac{h_{fe}\, h_{ie}\, I_b\, R_C/(h_{ie} + R_C)}{\dfrac{h_{ie}\, R_C}{h_{ie} + R_C} + \dfrac{1}{j\omega C_d}} \times \frac{1}{j\omega C_d}$$

$$= \frac{h_{fe}\, h_{ie}\, I_b\, R_C}{h_{ie}\, R_C\, j\omega C_d + h_{ie} + R_C}$$

Input voltage, $V_{in} = h_{ie}\, I_b$

$$\therefore \quad \text{Vtg gain}, A_{v_h} = \frac{V_{out}}{V_{in}} = \frac{h_{fe}\, R_C}{h_{ie}\, R_C\, j\omega C_d + h_{ie} + R_C}$$

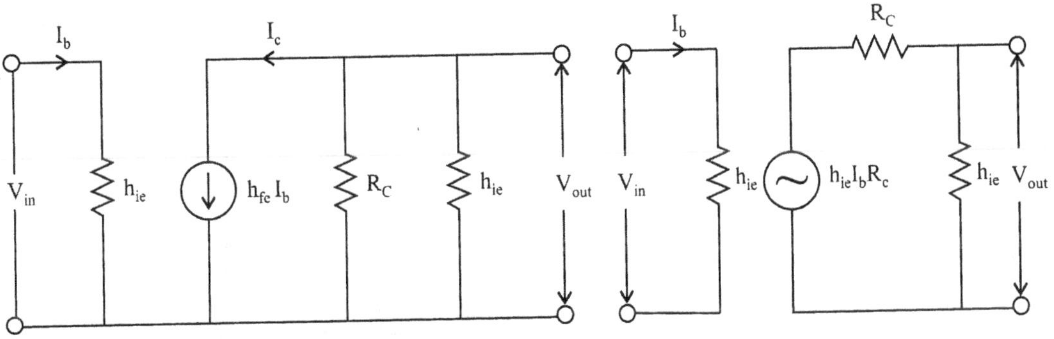

Fig. 3.5 (c) Equivalent circuit for mid-frequency range *Thevenin's equivalent circuit*

At high frequencies, the gain of the amplifier decreases with the increase in frequency. Several factors are responsible for the reduction in gain. Here reactance of C_C becomes very small and behaves as a short circuit. This increases the loading of next stage and reduces voltage gain.

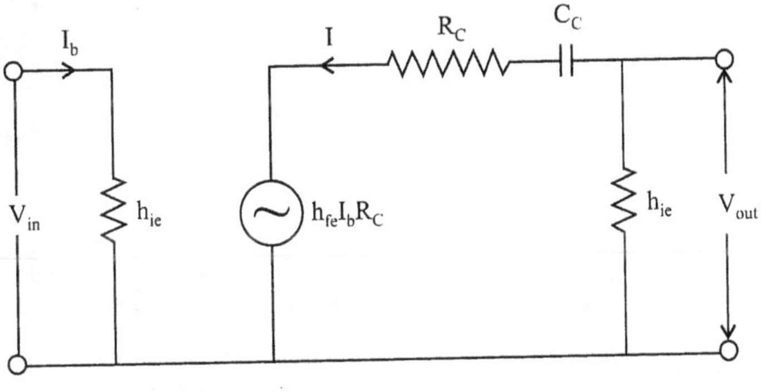

Fig.3.5 (d) Thevenin's equivalent circuit for low-frequency range.

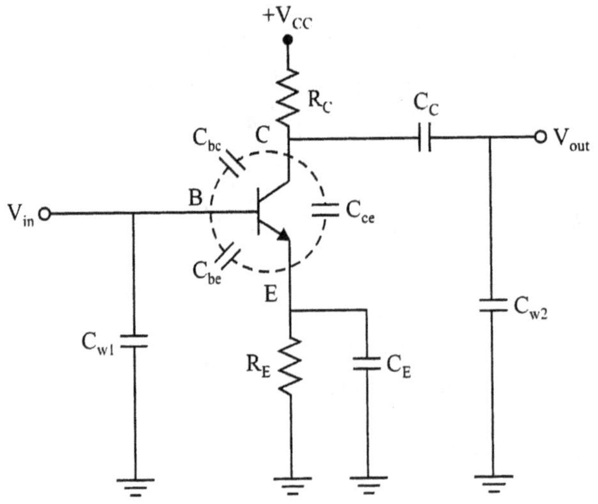

Fig. 3.5 (e) R-C coupled amplifier at high frequencies

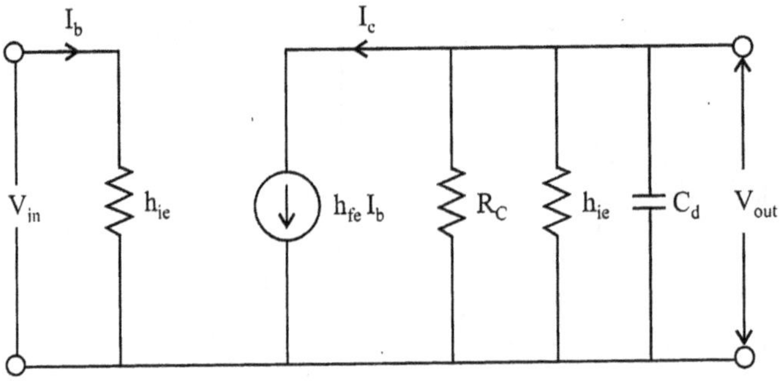

Fig. 3.5 (f) Equivalent circuit for high frequency range

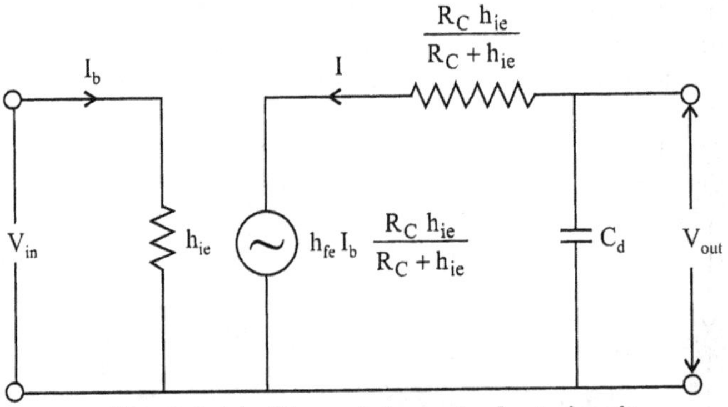

Fig. 3.5 (g) Thevenin's equivalent circuit

The emitter-bypass capacitor C_E, offers low reactance path to the signal. If it is not present, then the voltage drop across R_E will reduce the effective voltage available across the base-emitter terminals and thus reduces gain.

Effect of Emitter Capacitor on Different Frequency Ranges

For analysis for different frequencies that is for three categories

(a) Mid frequencies

(b) Low frequencies

(c) High frequencies the reactance of C_E is so small so that parallel combination of R_E and C_E can be considered as a short-circuit.

General Frequency Considerations

- At low frequencies reactance of capacitance increases so that the coupling capacitors and bypass capacitors can no longer be replaced by the short-circuit approximation.
- The frequency dependent parameters of the small signal equivalent circuits and the stray capacitive elements associated with the active device and the network will limit the high frequency response of the system.
- An increase in the number of stages of a cascaded system will also limit both the high and low frequency response.
- The horizontal scale of frequency curve represents logarithmic scale starts from low to high frequency.

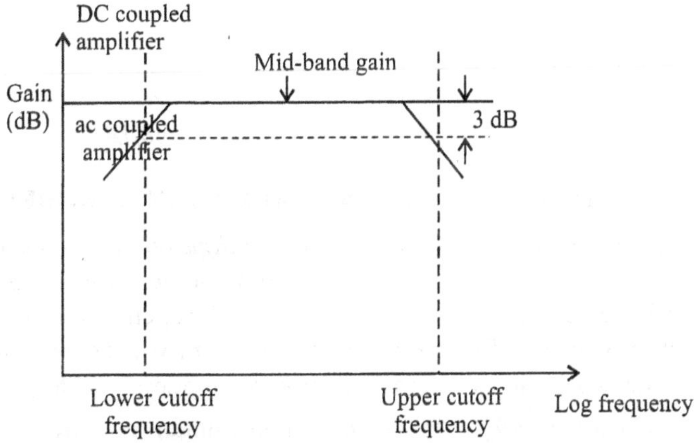

Fig. 3.6 (a) Frequency response.

- When we consider RC coupled amplifier, the drop of low frequencies is due to the increasing reactance C_C and C_E where the upper frequency limit is determined by either parasitic capacitive elements of the network.
- In the frequency curve magnitude of gain is equal to midband value.
- The corresponding frequencies f_1 and f_2 are also called as corner, cut off, band, break, or half-power frequencies.

  Here output power is half of the mid band power output,

$$P_{O_{mid}} = |V_o|^2 / R_o = |A_{V_{mid}} v_i|^2 / R_o$$

At cut off frequencies, (half-power frequencies)

$$P_{OHPF} = |0.707\ A_{Vmid} V_i|^2 / R_o$$

$$= 0.5 |A_{Vmid} V_i|^2 / R_o$$

$$P_{OHPF} = 0.5\ P_{o_{mid}}$$

- The bandwidth is the difference of frequencies that is $f_2 - f_1$.
- The decibel plot can be obtain from the equation,

$$\left(A_V / A_{V_{mid}}\right) dB = 20\ \log_{10}\left(A_V / A_{V_{mid}}\right) \quad .....(A)$$

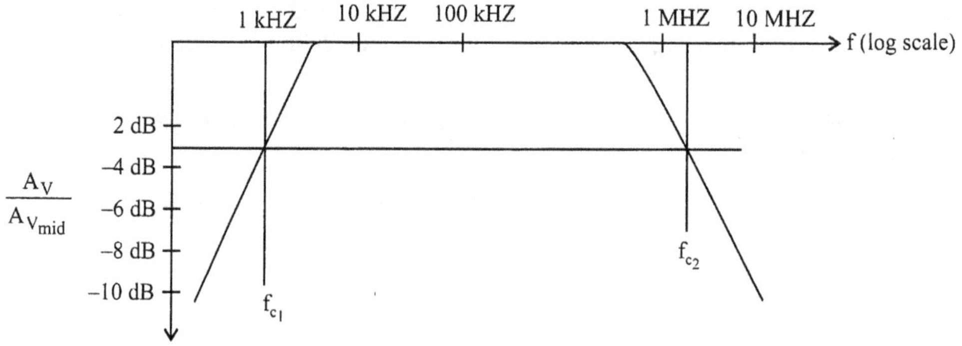

Fig. 3.6 (b) Decibel plot for the equation (A)

3.5 HYBRID - π COMMON EMITTER TRANSCONDUCTANCE MODEL

For *Transconductance* amplifier *circuits Common Emitter configuration is preferred. Why* ? Because for Common Collector ($h_{rc} < 1$). For Common Collector Configuration, voltage gain $A_V < 1$. So even by cascading you can't increase voltage gain. For Common Base, current gain $h_{fb} < 1$. So overall voltage gain is < 1. But for Common Emitter, $h_{fe} >> 1$. Therefore Voltage gain can be increased by cascading Common Emitter stage. So Common Emitter configuration is widely used.

The *Hybrid-π* or *Giacoletto Model* for the Common Emitter amplifier circuit (single stage) is as shown in Fig. 3.7.

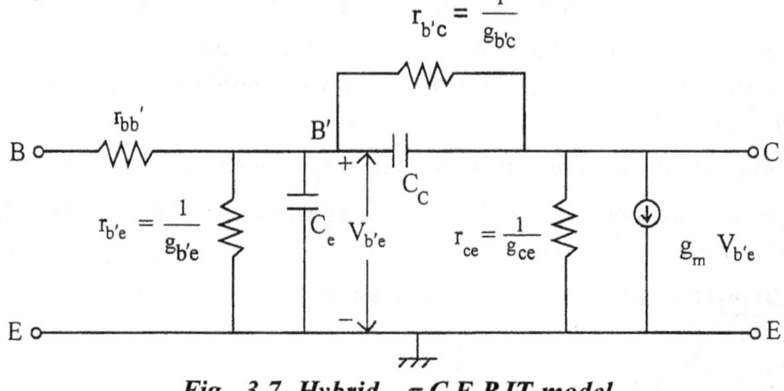

Fig. 3.7 Hybrid - π C.E BJT model

Analysis of this circuit gives satisfactory results at *all* frequencies not only at *high frequencies* but also at *low frequencies*. All the parameters are assumed to be independent of frequency.

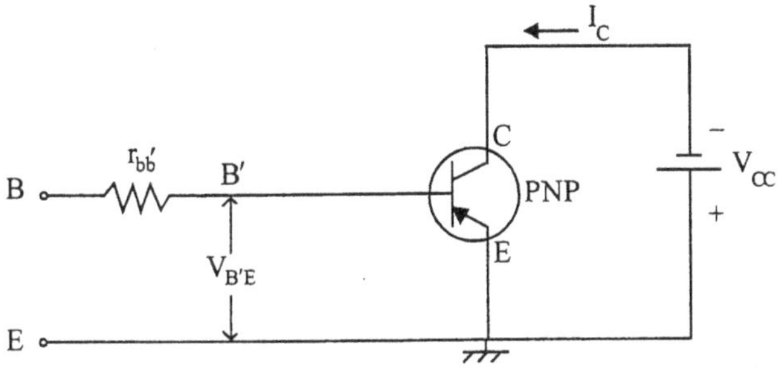

Fig. 3.8 PNP transistor amplifier

3.5.1 Circuit Components

Consider the circuit shown in Fig. 3.8.

B′ is the internal node of the base of the Transconductance amplifier. It is not physically accessible. The base spreading resistance $r_{b'b}$ is represented as a *lumped* parameter between base B and internal node B′. (g_m $V_{b'e}$) is a current generator. $V_{b'e}$ is the input voltage across the emitter junction. If $V_{b'e}$ increases, more carriers are injected into the base of the transistor. So the increase in the number of carriers is α $V_{b'e}$. This results in small signal current (since we are taking into account changes in $V_{b'e}$). This effect is represented by the current generator g_m $V_{b'e}$. This represents the current that results because of changes in $V_{b'e}$, when C is shorted to E.

When the number of carriers injected into the base increase, base recombination also increases. So this effect is taken care of by $g_{b'e}$. As recombination increases, base current increases. *Minority carrier storage* in the *base is represented* by C_e, *the diffusion capacitance.*

According to *Early Effect*, the change in voltage between Collector and Emitter changes the base width. So base width will be modulated according to the voltage between Collector and Emitter. When base width changes, the minority carrier concentration in base changes. Hence the current which is proportional to carrier concentration also changes. So I_E changes and hence I_C changes. This feedback effect [I_E on input side, I_C on output side] is taken into account by connecting $g_{b'c}$ between B', and C. The conductance between Collector and Base is g_{ce}.

C_C represents the collector junction barrier capacitance.

3.5.2 Hybrid - π Parameter Values

Typical values of the hybrid-π parameter at I_C = 1.3 mA are as follows :

g_m = 50 mA/V $r_{bb'}$ = 100 Ω $r_{b'e}$ = 1 kΩ

r_{ce} = 80 kΩ C_c = 3 pf C_e = 100 pf

$r_{b'c}$ = 4 MΩ

These values depend upon :

1. Temperature 2. Value of I_C

DETERMINATION OF HYBRID-π CONDUCTANCES

Transconductance or Mutual Conductance (g_m)

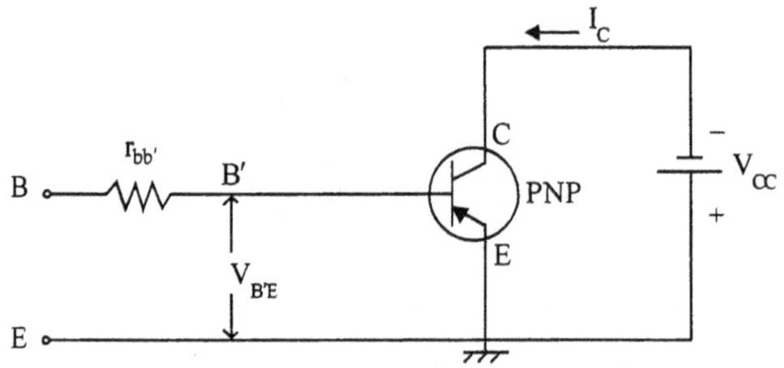

Fig. 3.9 PNP transistor amplifier

The circuit shown in Fig. 3.9. PNP transistor amplifier in Common Emitter configuration for AC purpose, Collector is shorted to Emitter.

$$I_C = I_{CO} - \alpha_0 \cdot I_E \qquad .....(3.1)$$

I_{CO} opposes I_E. I_E is negative. Hence $I_C = I_{CO} - \alpha_0\ I_E\ \alpha_0$ is the normal value of α at room temperature.

In the hybrid - π equivalent circuit, the short circuit current = $g_m\ V_{b'e}$

Here only transistor is considered, and other circuit elements like resistors, capacitors etc, are not considered.

$$g_m = \left.\frac{\partial I_C}{\partial V_{b'e}}\right|_{V_{CE} = K}$$

Differentiate (3.1) with respect to $V_{b'e}$ partially. I_{CO} is constant.

$$\therefore \qquad g_m = 0 - \alpha_0 \frac{\partial I_E}{\partial V_{b'e}} \text{ for a PNP transistor, } V_{b'e} = -V_E$$

Since, for PNP transistor, base is n-type. So negative voltage is given,

$$\therefore \qquad g_m = \alpha_0 \frac{\partial I_E}{\partial V_E}$$

If the emitter diode resistance is r_e, then $r_e = \dfrac{\partial V_E}{\partial I_E}$

$$\therefore \qquad g_m = \frac{\alpha_0}{r_e}$$

But for a diode, $r = \dfrac{\eta . V_T}{I}$ $\quad \because I = I_0\ (e^{V/\eta\ VT} - 1) \quad I \simeq I_0 \cdot e^{V/\eta\ VT}$

$$\left[g = \frac{dI}{dV} = \frac{J_0 . e^{V/\eta V_T}}{\eta . V_T} \cong \frac{I}{\eta V_T} \right]$$

$\therefore \quad r = \dfrac{\eta . V_T}{I} \quad \eta = 1, \qquad I = I_E \qquad r = \dfrac{V_T}{I_E}$

$\therefore \quad g_m = \dfrac{\alpha_0 . I_E}{V_T} \qquad \alpha_0 \simeq 1, \qquad I_E \simeq I_C$

$$I_E = I_{CO} - I_C$$

$\therefore \quad g_m = \dfrac{I_{CO} - I_C}{V_T}$

Neglecting I_{CO},

$g_m = \dfrac{|I_C|}{V_T} \qquad g_m$ = (Transconductance or Mutual Conductance)

g_m is directly proportional to I_C. g_m is also $\alpha \dfrac{1}{T}$. For PNP transistor, I_C is negative.

$\therefore \quad \boxed{g_m = \dfrac{-I_C}{V_T}}$

will become positive. Since I_C is negative. So g_m is always positive.

V_T is volt equivalent of temperature $V_T = T/11,600$

At room temperature, $T \simeq 300$ K,

$g_m = \dfrac{|I_C|}{26}$, I_C is in mA.

If $\quad I_C = 1.3$ mA, $g_m = 0.05$ A/V

If $\quad I_C = 10$ mA, $g_m = 400$ mA/V

INPUT CONDUCTANCE ($g_{b'e}$)

At low frequencies, *capacitive reactance* will be *very large* and can be considered as *Open circuit*. So in the hybrid-π equivalent circuit which is valid at low frequencies, all the capacitances can be neglected. The equivalent circuit is as shown in Fig. 3.10.

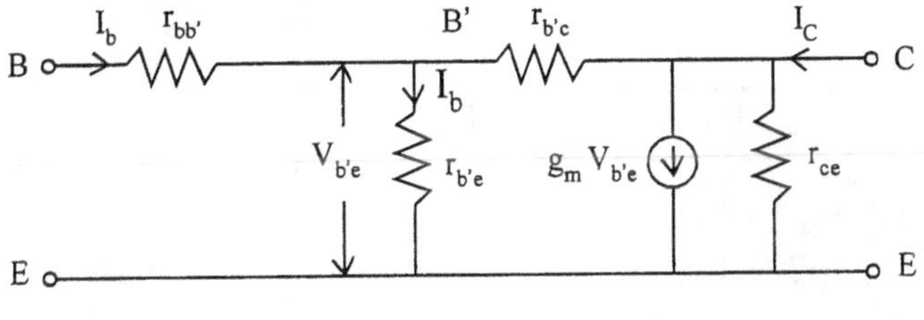

Fig. 3.10 Equivalent circuit at low frequencies

The value of $r_{b'c} >> r_{b'e}$ (Since Collector Base junction is Reverse Biased)

So I_b flows into $r_{b'e}$ only. [This is I_b. $(I_E - I_b)$ will go to collector junction]

$$\therefore \quad V_{b'e} \simeq I_b \cdot r_{b'e}$$

The short circuit collector current,

$$I_C = g_m \cdot V_{b'e}; \qquad V_{b'e} = I_b \cdot r_{b'e}$$

$$I_C = g_m \cdot I_b \cdot r_{b'e}$$

$$h_{fe} = \left.\frac{I_C}{I_B}\right|_{V_{CE}} = g_m \cdot r_{b'e}$$

or

$$\boxed{r_{b'e} = \frac{h_{fe}}{g_m}}$$

But

$$g_m = \frac{|I_C|}{V_T}$$

$$\therefore \quad r_{b'e} = \frac{h_{fe} \cdot V_T}{|I_C|}$$

$$\therefore \quad g_{b'e} = \boxed{\frac{|I_C|}{h_{fe} V_T}} \text{ or } \boxed{\frac{g_m}{h_{fe}}}$$

FEEDBACK CONDUCTANCE ($g_{b'c}$)

h_{re} = reverse voltage gain, with input open or $I_b = 0$

$$= \frac{V_{b'e}}{V_{ce}} = \frac{\text{Input voltage}}{\text{Output voltage}}$$

$$h_{re} = \frac{r_{b'e}}{r_{b'e} + r_{b'c}}$$

[With input open, i.e., I_b– 0, V_{ce} is output. So it will get divided between $r_{b'e}$ and $r_{b'c}$ only]

or

$$h_{re}(r_{b'e} + r_{b'c}) = r_{b'e}$$

$$r_{b'e}[1 - h_{re}] = h_{re} r_{b'c}$$

But

$$h_{re} << 1$$

$$\therefore \quad r_{b'e} = h_{re} \; r_{b'c}; \quad r_{b'c} = \frac{r_{b'e}}{h_{re}}$$

or

$$\boxed{g_{b'c} = h_{re} \; g_{b'e}} \quad \frac{1}{r_{b'c}} = g_{b'c} = \frac{h_{re}}{r_{b'e}}$$

$$h_{re} = 10^{-4}$$

$$\therefore \quad r_{b'c} >> r_{b'e}$$

BASE SPREADING RESISTANCE ($r_{bb'}$)

The input resistance with the output shorted is h_{ie}. If output is shorted, i.e., Collector and Emitter are joined, $r_{b'e}$ is in parallel with $r_{b'c}$.

But we have seen that $r_{b'e} = h_{re} \cdot r_{b'c}$

h_{oe} in very small and $r_{b'c} >> r_{b'e}$

$\therefore$ $r_{b'e}$ is parallel with $r_{b'c}$ is only $r_{b'e}$ (lower value)

$$h_{ie} = r_{bb'} + r_{b'e}$$

or

$$\boxed{r_{bb'} = h_{ie} - r_{b'e}}$$

$$h_{ie} = r_{bb'} + r_{b'e}$$

But we know that

$$r_{b'e} = \frac{h_{fe} \cdot V_T}{|I_C|}$$

$$h_{ie} = r_{bb'} + \frac{h_{fe} \cdot V_T}{|I_C|}$$

$r_{bb'}$ is small, few Ωs, to few hundred Ωs

$$\therefore \qquad h_{ie} \cong \frac{h_{fe} \cdot V_T}{|I_C|}$$

OUTPUT CONDUCTANCE (g_{ce})

This is the conductance with input open circuited. In h-parameters it is represented as h_{oe}. For $I_b = 0$, we have,

$$\therefore \qquad I_C = \frac{V_{ce}}{r_{ce}} + \frac{V_{ce}}{r_{b'c} + r_{b'e}} + g_m \, V_{b'e}$$

But $$h_{re} = \frac{V_{b'e}}{V_{ce}} \qquad \therefore \qquad V_{b'e} = h_{re} \cdot V_{ce}$$

$$\therefore \qquad I_C = \frac{V_{ce}}{r_{ce}} + \frac{V_{ce}}{r_{b'c} + r_{b'e}} + g_m \cdot h_{re} \cdot V_{ce}$$

But $$h_{oe} = \frac{I_C}{V_{ce}}$$

So dividing by V_{ce} and that $r_{b'c} >> r_{b'e}$, $r_{b'c} + r_{b'e} \simeq r_{b'c}$

$$\therefore \quad h_{oe} = \frac{1}{r_{ce}} + \frac{1}{r_{b'c}} + g_m \cdot h_{re}$$

$$= g_{ce} + g_{b'c} + g_m\, h_{re}$$

But $$g_{b'e} = \frac{g_m}{h_{fe}}$$

$$\therefore \quad g_m = g_{b'e} \cdot h_{fe}$$

$$h_{re} = \frac{r_{b'e}}{r_{b'e} + r_{b'c}} \approx \frac{r_{b'e}}{r_{b'c}} = \frac{g_{b'c}}{g_{b'e}}$$

$$\therefore \quad h_{oe} = g_{ce} + g_{b'c} + g_{b'e}\, h_{fe} \cdot \frac{g_{b'c}}{g_{b'e}}$$

or $$g_{ce} = h_{oe} - (1 + h_{fe}) \cdot g_{b'c}$$

If $$h_{fe} >> 1, \ 1 + h_{fe} \approx h_{fe}$$

$$\therefore \quad \mathbf{g_{ce} = h_{oe} - h_{fe} \cdot g_{b'c}}$$

But $$\boxed{g_{b'c} = h_{re} \cdot g_{b'e}}$$

$$\therefore \quad g_{ce} = h_{oe} - h_{fe} \cdot h_{re} \cdot g_{b'e}$$

But $$h_{fe} \cdot g_{b'e} = g_m$$

$$\therefore \quad g_{ce} = h_{oe} - g_m \cdot h_{re}$$

3.5.3 HYBRID - π CAPACITANCES

In the hybrid - π equivalent circuit, there are two capacitances, the capacitance between the Collector-Base junction is the C_C or $C_{b'c}$. This is measured with input open i.e., $I_E = 0$, and is specified by the manufacturers as C_{0b}. 0 indicates that input is open. Collector junction is reverse biased.

$$C_C \propto \frac{1}{(V_{CE})^n}$$

where $$n = \frac{1}{2} \text{ for abrupt junction}$$

$$= 1/3 \text{ for graded junction.}$$

C_e = Emitter diffusion capacitance C_{De} + Emitter junction capacitance C_{Te}

C_T = Transition capacitance.

C_D = Diffusion capacitance.

$C_{De} >> C_{Te}$

$\therefore \quad C_e \simeq C_{De}$

$C_{De} \alpha I_E$ and is independent of Temperature T.

3.5.4 THE DIFFUSION CAPACITANCE

For pnp transistor, base is n-type and emitter is p-type. So if E-B junction is forward biased, holes are injected into the base. The distribution of these injected holes between E and C is as shown in Fig. 3.11.

The collector is reverse biased. So the injected charge concentration p' at the collector junction is zero since they are attracted because of the negative potential at the collector. The base width W is assumed to be small compared to the diffusion length L_S of the minority carriers. If $W << L_B$, then p' varies almost linearly from the value $p'(0)$ at the emitter to zero at the collector. The stored charge in the base is the average concentration $\frac{p'(0)}{2}$ times the volume of the base W_A.

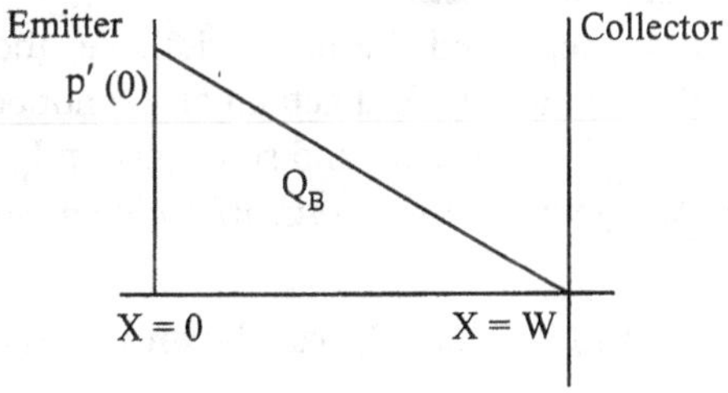

Fig. 3.11 Variation of charge in the base region

(where A is Cross Sectional Area), times charge Q.

W.A = Volume ;

W = Base width ;

A = cross sectional area of the junction.

Q = Charge

$\frac{p'(0)}{2}$ = Average concentration.

Diffusion current $I = -AQ \cdot D_B \cdot \frac{dp'}{dx} = -A\,Q\,D_B\,\frac{p'(0)}{W}$

D_B = diffusion constant for minority carriers in the base

$$\therefore \quad p'(0) = \frac{I.W}{AQD_B}$$

$$\therefore \quad Q_B = \frac{1}{2} \times \frac{1 \times W}{AQD_B} . AQW$$

$$Q_B = \frac{W^2}{2D_B}$$

Emitter diffusion capacitance $C_{de} = \frac{dQ_B}{dV}$

$$C_{de} = \frac{W^2}{2D_B} \cdot \frac{dI}{dV} = \frac{W^2}{2D_B} \cdot \frac{1}{r_e}$$

But $$\frac{dV}{dI} = \frac{V_T}{I_E}, \quad I_E = I_0 \left(e^{\frac{V}{\eta V_T}} \right)$$

$$\therefore \quad \boxed{C_{De} = \frac{W^2 . I_E}{2 D_B \ V_T} = g_m . \frac{W^2}{2 D_B}}$$

In the above derivation, we have neglected the base width i.e, there is no possibility for recombination in the base hence $I_b = 0$. Then $I_E = I_C$. But actually it cannot be so. Hence, the hybrid-π model is valid only when change in V_{BE} is very small and so change in I_E is equal to change in I_C. i.e., *Hybrid-π model is valid only under dynamic conditions, wherein change in I_B is negligible.* So change in I_E is equal to change in I_C.

Giacoletto who proposed the hybrid - π model, has shown that the hybrid parameters are independent of frequency when

$$2\pi f . \frac{W^2}{6 D_B} << 1 \qquad .....(3.2)$$

Where W = Base width

D_B = Diffusion constant for minority carriers in the base

f = Frequency of the input signal

But $$C_e = g_m \cdot \frac{W^2}{2 D_B} \qquad .....(3.3)$$

$$\frac{C_e}{g_m} = \frac{W^2}{2D_B}$$

$$\therefore \quad \frac{W^2}{6D_B} = \frac{C_e}{3g_m}$$

$$C_e = \frac{g_m}{2\pi f_T}, \quad \frac{C_e}{g_m} = \frac{1}{2\pi f_T}$$

$$\therefore \quad \frac{W^2}{6D_B} = \frac{1}{6\pi f_T}$$

$\therefore$ Equation becomes, $\dfrac{2\pi f}{6\pi f_T} << 1$

or $\quad f << 3 f_T$

$\therefore$ *Hybrid π model is valid for frequencies upto $\simeq f_T/3$.*

VARIATION OF HYBRID PARAMETERS WITH $|I_C|$, $|V_{CE}|$ AND T

TRANSCONDUCTANCE AMPLIFIER OR MUTUAL CONDUCTANCE (g_m)

$$g_m = \frac{|I_C|}{V_T}$$

$\therefore \quad g_m$ is $\propto I_C$

$V_T = T/11,600$

$$\therefore \quad g_m \propto \frac{1}{T}$$

g_m is independent of V_{CE}

Since in the active region of the transconductance, I_C is independent of V_{CE}.

3.5.5 BASE EMITTER RESISTANCE ($r_{b'e}$)

$$r_{b'e} = \frac{h_{fe} \cdot V_T}{I_C}$$

$$\therefore \quad r_{b'e} \propto \frac{1}{I_C}$$

$r_{b'e}$ increases as T increases since $r_{b'e} \propto V_T$

3.5.6 EMITTER CAPACITANCE (C_e)

$$C_e \simeq g_m \cdot \frac{W^2}{2D_B}$$

$g_m \propto I_C \qquad \therefore C_e \propto I_C$

As V_{ce} increases, W the effective base width decreases

$\therefore$ *C_e decreases with V_{CE} (increasing)*

COLLECTOR CAPACITANCE (C_C)

C_C depends on V_{CE}

$C_c \propto (V_{CE})^{-\eta}$

$\therefore$ *C_C is independent of I_C*

$\therefore$ *C_C is independent of T*

C_C decreases with increase in V_{CE}

3.5.7 BASE SPREAD RESISTANCE ($r_{bb'}$)

$r_{bb'}$ decreases with increase in I_C.

Since as I_C increases, conductivity increases. So $r_{bb'}$ decreases, because of conductivity modulation. But $r_{bb'}$ increases with increase in Temperature. Because as T increases, mobility of the carriers decreases. So conductivity decreases. So $r_{bb'}$ increases.

3.6 CURRENT GAIN WITH RESISTANCE LOAD

$$f_T = f_\beta \cdot h_{fe} = \frac{g_m}{2\pi(C_e + C_c)}$$

Considering the load resistance R_L,

$V_{b'e}$ is the input voltage and is equal to V_i

V_{ce} is the output voltage and is equal to V_2

$$K_2 = \frac{V_{ce}}{V_{b'e}}$$

This circuit is still complicated for analysis.

Because, there are two time constants associated with the input and the other associated with the output. The output time constant will be much smaller than the input time constant. So it can be neglected.

K = Voltage gain. It will be >> 1

$$\therefore \quad g_{b'c}\left(\frac{K-1}{K}\right) \simeq g_{b'c}$$

$$g_{b'c} < g_{ce} \quad \because \quad r_{b'c} \simeq 4\ M\Omega, \qquad r_{ce} = 80\ k \ \text{(typical values)}$$

So $g_{b'c}$ can be neglected in the equivalent circuit.

In a wide band amplifier R_L will not exceed 2kΩ, since $f_H \propto \frac{1}{R_L}$. If R_L is small f_H is large.

$$f_H = \frac{1}{2\pi C_S (R_C \| R_L)}$$

Therefore g_{ce} can be neglected compared with R_L

Therefore the output circuit consists of current generator gm $V_{b'e}$ feeding the load R_L so the Circuit simplifies as shown in Fig. 3.12.

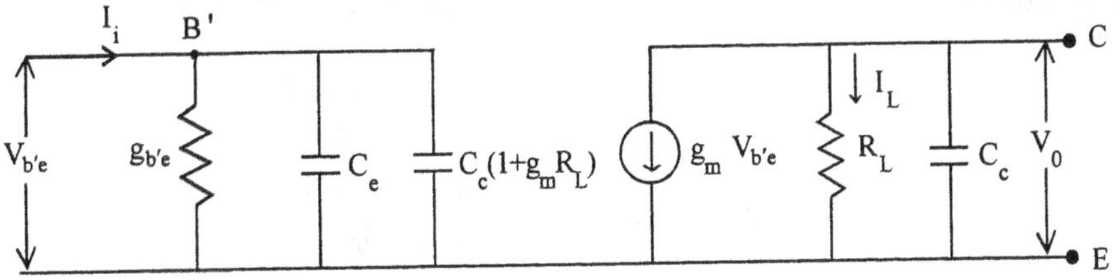

Fig. 3.12 Simplified equivalent circuit

$$K = \frac{V_{ce}}{V_{b'e}} = -g_m R_L ; \quad g_m = 50 \text{ mA\V}, \quad R_L = 2k\Omega \text{ (typical values)}$$

$$K = -100$$

So the maximum value is $g_{b'c}$ $(1 - k) \simeq 0.02595$. So this can be neglected compared to $g_{b'e} \simeq 1$ mA/V.

R_L should not exceed 2kΩ, therefore if $R_L > 2K\Omega$, C_c $(1 + g_m R_L)$ becomes very large and so band pass becomes very small.

$$C_c = \left[\frac{k-1}{k}\right] \simeq C_c$$

when $R_L = 2k\Omega$,

the output time constant is,

$$R_L \cdot C_c = 2 \times 10^3 \times 3 \times 10^{-12} = 6 \times 10^{-9} \text{ s} = 6 \text{ μ.sec. (typical values)}$$

Input time constant is,

$$r_{b'e} \ [C_e + C_c \ [1 + g_m R_L]] = 403 \text{ μ.sec. (typical values)}$$

So the band pass of the amplifier will be determined by the time constant of the input circuit.

The 3db frequency $f_M = \dfrac{1}{2\pi r_{b'e} C} = \dfrac{g_{b'e}}{2\pi C}$

where $C = C_e + C_c \ (1 + g_m R_L)$

3.7 CE SHORT CIRCUIT CURRENT GAIN

This is the circuit of transistor amplifier in common emitter configuration. (Fig. 3.13).

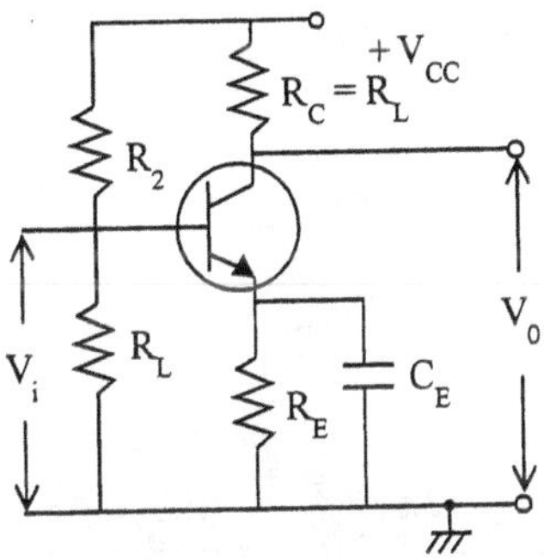

Fig. 3.13 CE amplifier circuit

The approximate equivalent circuit at high frequencies, with output shorted is shown in Fig. 3.14.

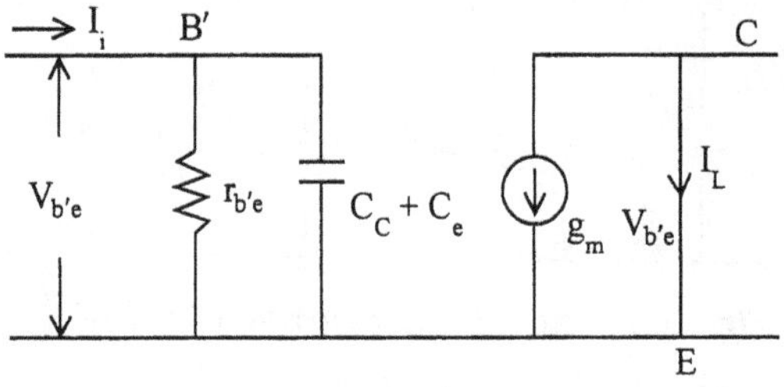

Fig. 3.14 Simplified equivalent circuit

$r_{b'e}$ is assumed to be very large. So it is open circuit.

r_{ce} disappears since it is in shunt with short circuited output.

$$I_L = -g_m V_{b'e}$$

Negative sign taking the direction of current into acount. I_L is contributed by the current source only. (Fig. 3.15).

$$V_{b'e} = I \times Z = I \times \frac{1}{Y}$$

$$V_{b'e} = \frac{I_i \times 1}{g_{b'e} + j\omega C_e}$$

$$\therefore \quad I_L = \frac{-g_m \cdot I_i}{g_{b'e} + j\omega C_e}$$

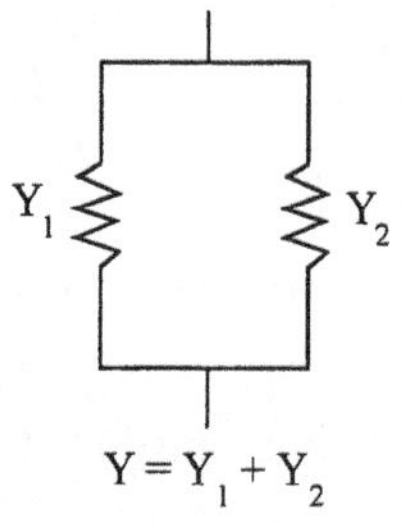

Fig. 3.15 Conductances in parallel

Conductances in parallel get added. (Fig. 3.15).

Therefore current gain under short circuit conditions is,

$$A_i = \frac{I_L}{I_i} = \frac{-g_m}{g_{b'e} + j\omega C_e}$$

But

$$g_{b'e} = \frac{g_m}{h_{fe}}$$

$$C_e = \frac{g_m}{2\pi f_T}$$

$$\therefore \quad A_i = \frac{I_L}{I_i} = \frac{-g_m}{\frac{g_m}{h_{fe}} + \frac{j\omega g_m}{2\pi f_T}}$$

$$\therefore \quad A_i = \frac{-1}{\frac{1}{h_{fe}} + \frac{j2\pi f}{2\pi f_T}} = \frac{-h_{fe}}{1 + \frac{jh_{fe} \cdot f}{f_T}}$$

$$\frac{f_T}{h_{fe}} = f_\beta$$

$$\therefore \quad \boxed{A_i = \frac{-h_{fe}}{1 + j\left(\frac{f}{f_\beta}\right)}}$$

when $f = f_\beta$, A_i falls by $\frac{1}{\sqrt{2}}$, or by 3db. The frequency range f_β is called Bandwidth of the amplifiers.

$\therefore$ f_β : Is the frequency at which the short circuit gain in common emitter configuration falls by 3 db.

f_T : This is defined as the frequency at which the common emitter shunt circuit current gain becomes 1.

$$A_i = \frac{-h_{fe}}{1 + j\left(\frac{f}{f_\beta}\right)}$$

Let $f = f_T$, $A_i = 1$

$$\therefore \quad 1 = \frac{h_{fe}}{\sqrt{1 + \left(\frac{f_T}{f_\beta}\right)^2}}$$

$$\therefore \quad 1 + \left(\frac{f_T}{f_\beta}\right)^2 = h_{fe}^2$$

$$\left(\frac{f_T}{f_\beta}\right)^2 = h_{fe}^2 - 1 \simeq h_{fe}^2 \qquad \because h_{fe} >> 1.$$

$$\therefore \quad f_T = f_\beta \cdot h_{fe}$$

f_β is the Bandwidth of the transistor

h_{fe} is the current gain

$\therefore$ f_T is the current gain, Bandwidth product.

In Common Emitter configurations, $A_i >> 1$. But as frequency increases A_i falls. Why should A_i decrease ? $A_i = \frac{I_L}{I_i}$. As frequency increases, X_c increases. So, X_e increases. i.e., more and more number of carriers will be stored in the base region itself. Due to this, less number of carriers will reach collector. Storage of carriers at the base and emitter increases. So I_C decreases. Therefore I_L decreases i.e., g_m decreases. So g_m $V_{b'e}$ decreases or I_L decreases. Hence A_i decreases.

f_T depends on the operating point of the transistor. The graph of f_T V_s I_c for a transistor is as shown in Fig. 3.16.

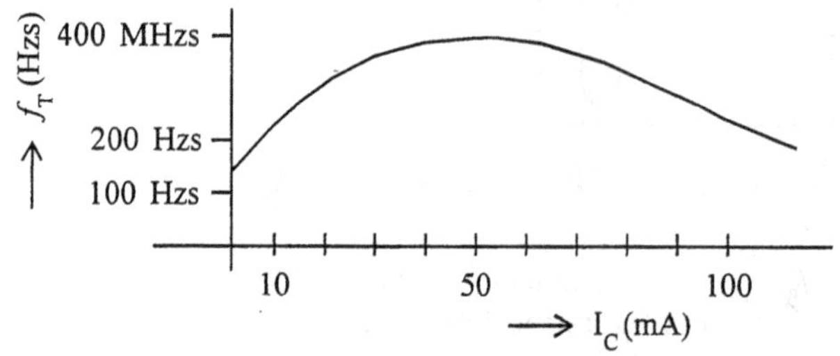

Fig. 3.16 Variation of I_C with frequency

For a typical transistor, f_T = 80 MHz

f_β = 1.6 MHz

3.8 EMITTER FOLLOWER AT HIGH FREQUENCIES

Broad band transistor (BJT) amplifiers use shunt feedback and series feedback methods so as to get large Gain – Band width (BW) product. If number of amplifier stages are cascaded, the output current is made up from several transistors and the individual device operating point canbe made to correspond to the optimem Gain - B.W. product of emitter current.

A single section of a multi stage amplifier is shown in Fig. 3.17. The input impedance of the amplifier circuit is low compared to the impedance of the parallel R-C network.

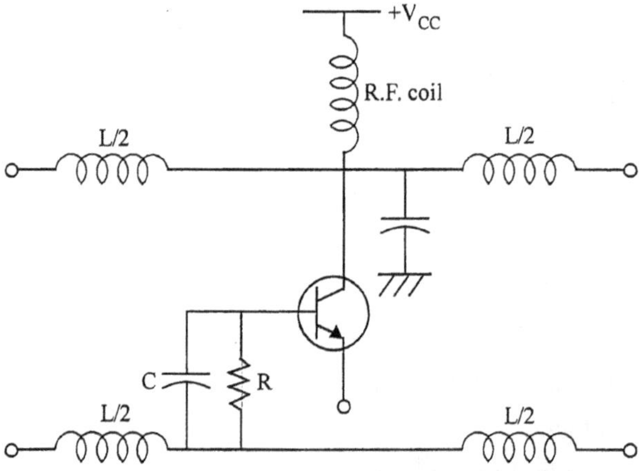

Fig. 3.17 One stage of multistage amplifier.

$$f_\beta = \frac{1}{2\pi RC}$$

where f_β is the β cut-off frequency of the transistor.

Now consider 3 such stages cascaded as shown in Fig. 3.18. R is the series resistor and h_{ie} is the transistor low frequency input impedance. h_{re} of the BJT is small and so it can be neglected.

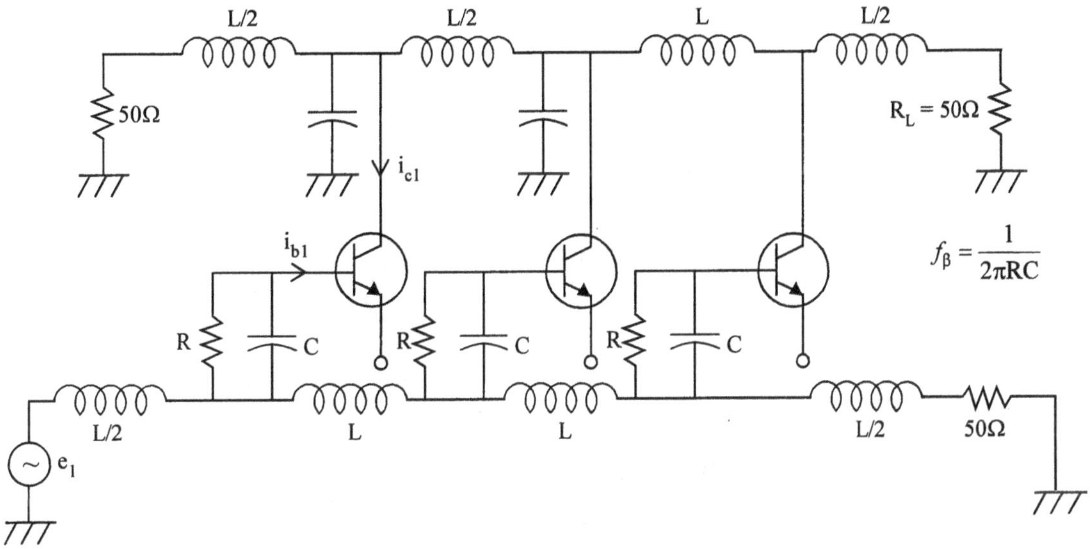

Fig. 3.18 Three stage cascaded amplifier.

e_1 is the input voltage

$$i_{b1} = \frac{e_1}{(R + h_{ie})} \simeq \frac{e_1}{R}$$

Since h_{ie} of the BJT is small compared to R.

R_L is a 50 Ω load for the circuit.

The voltage gain gets multiplied for each stage. So for n-stages,

$$e_0 = (n.\, i_c)\left(\frac{R_L}{2}\right) = n.\, \beta_o\, i_b\left(\frac{R_L}{2}\right)$$

$$= \frac{n.\, \beta_0 .\, e_1 .\, R_L}{2R}$$

where β_0 is the low frequency current gain (β) of the BJT.
n = number of cascaded stages.

The Voltage Gain for 'm' cascaded stages is, (V.G) m

$$(V.\, G)_m = \left[\frac{n.\, R_L .\, \beta_0}{2R}\right]^m$$

Typically, a single stage will have a 3-db frequency of 200 MHzs in high frequency amplifiers.

Effect of Emitter Degeneration :

The approximate high frequency equivalent circuit for a BJT operating in C.E. configuration is shown in Fig. 3.19.

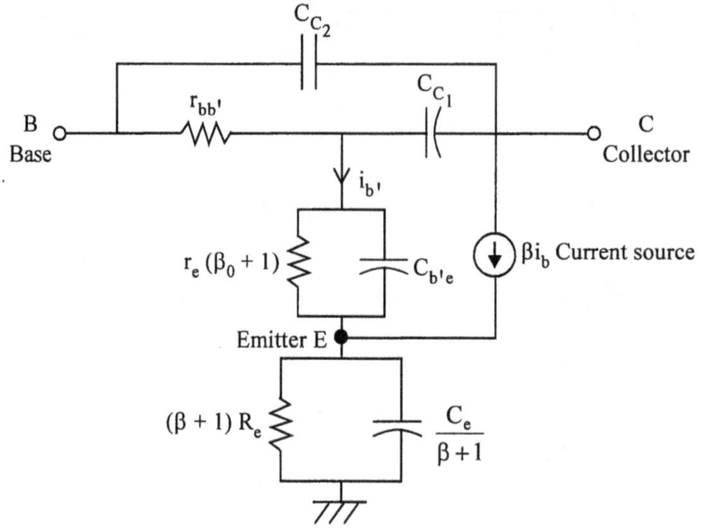

Fig. 3.19 High frequency equivalent circuit for a BJT.

$r_{bb'}$ = Base spread resistance

C_{C_1} = Collector emitter capacitance

C_{C2} = Collector-base capacitance

r_e = Emitter resistance = $\frac{VT}{J_e} = \left(\frac{25}{I_e}\right)$ Ω at 25 °C.

T = Temperature in OK; (273 + 25) = 298 °K for 25 °C.

k = Boltzmans' constant.

$C_{b'e}$ = Diffusion capacitance

Let β be the current gain at any frequency *f*.

Then $$\beta = \frac{\beta_0}{1 + j\left(\frac{f}{f_\beta}\right)}$$

where β_0 = Low frequency current gain

f_β = Beta cutoff frequency

Typical values of these parameters are, for $V_{CE} = 10V$, $I_E = 10mA$

$r_{bb'} = 70\ \Omega$

$C_{C1} = 1pf$

$C_{C2} = 2\ pf$

$\beta_0 = 100$

$f_\beta = 6$ MHZs

$r_e = 4\ \Omega$

$C_{b'e} = 130\ pf$

If $f >> f_\beta$, the input circuit can be simplified as, (Fig. 3.20)

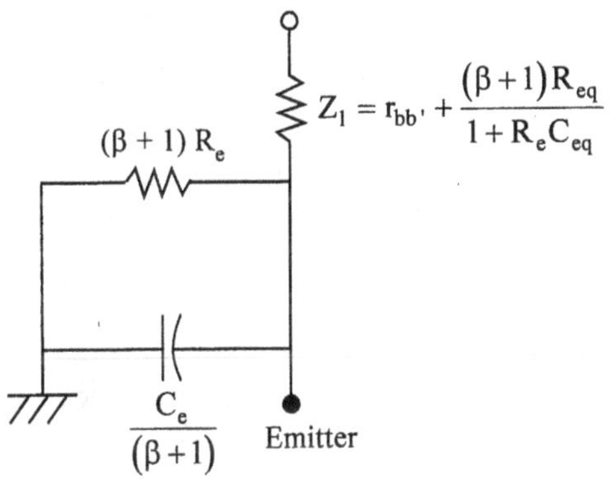

Fig. 3.20 Simplified input circuit.

The equivalent input impedance of the R-C combination on the input side along with transistor capacitances is shown in Fig. 3.20.

If the frequency range of consideration is such that X_{Ce_1} and X_{Ce_2} can be neglected and 'f' is greater than f_β,

$f > f_\beta$ but X_{Ce_1} and X_{Ce_2} are negligible,

$$Z_1 \simeq r_{bb'} + \frac{(\beta+1)R_e}{(1+j\omega R_eC_e)}$$

But,
$$\beta = \frac{\beta_0}{\left(1+j\frac{f}{f_\beta}\right)}$$

$$= \frac{\beta_0}{\left(1+j\frac{\omega}{\omega_\beta}\right)}$$

Upper cut-off frequency $f_2 = \dfrac{1}{2\pi R_e C_e}$

for frequencies below f_2,

$$Z_1 \simeq \frac{(r_{bb'} + R_e) + R_e \beta_0}{1 + \dfrac{j\omega}{\omega_\beta}}$$

with a parallel R-C network inserted in the base,

we have, $$Z_T = \left[\frac{R}{1 + j\omega CR}\right] + (r_{bb'} + R_e) + \frac{\beta_0 R_e}{1 + j\dfrac{\omega}{\omega_\beta}}$$

If the R-C combination has the same 3-dB points as f_β, then $f_\beta = \dfrac{1}{2\pi RC}$

so that $$Z_T = \frac{R + \beta_0 R_e}{1 + \left(\dfrac{j\omega}{\omega_\beta}\right)} \cdot [r_{bb'} + R_e]$$

If $f > f_\beta$ but less than f_2, $\left(f_2 = \dfrac{1}{2\pi R_e C_e}\right)$, the input impedance of the delay line comprises a resistance $(r_{bb'} + R_e)$ in series with a parallel combination of R_{eq} and C_{eq}.

Calculation of R_e and C_e for a required Gain - B.W value (for uncompensated case)

The voltage gain A_v of one section at low frequencies is given by,

$$A_V = \frac{\beta_0 R_L}{2[(R + \beta_0 R_e) + (r_{bb'} + R_e)]}$$

$$= \frac{\beta_0 R_L}{2(R + \beta_0 R_e)}$$

For uncompensated case, $C_e = 0$.

The Bandwidth B.W is given as,

$$B.W = \frac{1}{2\pi C_{eq}(R_e + r_{bb'})}$$

where $$C_{eq} = \frac{1}{2\pi f_\beta (R + \beta_0 R_e)}$$

$\therefore$ $$B.W \simeq f_2$$

$$\text{Gain Band width product} = \frac{f_{\beta}.\ \beta_0\ R_L}{2(r_{bb'} + R_e)}$$

For compensated case, $C_e \neq 0$

$$\therefore \qquad Z_T = \frac{R}{\left(1 + j\frac{\omega}{\omega_\beta}\right)} + r_{bb'} + \frac{R_e}{1 + j\omega\, R_e C_e} + \frac{\beta_0\, R_e}{\left(1 + j\frac{\omega}{\omega_\beta}\right)(1 + j\omega\, R_e C_e)}$$

In order to have good compasation, $R_e\ C_e$ must be chosen such that the break frequency is same as in the case of uncompensated Band width.

$\therefore$ we have, $\quad f_{2u} = f_2$ of uncompensated circuit,

$$f_{2u} = \frac{1}{2\pi\ C_e R_e}$$

The value of f_{2c}, the upper cut-off frequency for compensated circuit can be obtained by equating the real and imaginary parts of Z_T at f_{2c}.

Since $\quad f_{2c} > f_\beta$,

We have $\quad 1 + \left(\frac{j\omega}{\omega_\beta}\right) \simeq \frac{j\omega}{\omega_\beta}$

Also, $\quad \omega\ C_e\ R_e = 1$

$\therefore \quad Z_T$ at $f = f_{2c}$

$$= -j\left[R + \{(\beta_0 R_e (1 + j))\}\left(\frac{\omega_\beta}{\omega_{2c}}\right) + r_{bb'} + \left(\frac{R_e}{1 + j}\right)\right]$$

$$= \left[r_{bb'} + \left(\frac{R_e}{2}\right) - \left(\frac{j\beta\ \beta_0\ R_e}{2f_{2c}}\right)\right] - j\left[\frac{f_\beta\,(\beta_0\ R_e + 2R)}{2f_{2c}} + \left(\frac{R_e}{2}\right)\right]$$

Equating real and Imaginary parts,

$$f_{2c} = \frac{f_\beta\,(R_0 + \beta_0\ R_e)}{r_{bb'}}$$

The B.W improvement factor 'k' is,

$$k = \frac{f_{2c}}{f_{2u}} = 1 + \left(\frac{R_e}{r_{bb'}}\right)$$

L_0 and C_0 can be determined, using the transmission line equation,

$$Z_0 = \sqrt{(L_0 / C_0)\left[1 - \left(\frac{f}{f_{cut-off}}\right)^2\right]}$$

$$f_{cut\text{-}off} = \frac{1}{2\pi\sqrt{L_0C_0}}$$

C is usually chosen as 2 or 3 times C_0

R is computed to correspond to $R = \dfrac{1}{2\pi\, Cf_\beta}$.

R_e can be calculated from the equation for f_{2c}.

C_e is obtained from the equation for f_{2u}.

Typical values are :

$f_{cut\text{-}off} = 250$ MHz

$Z_0 = 48\ \Omega$

$L_0 = 0.05\ \mu H$

$C_0 = 19$ p*f*

$(R + \beta_0 R_e) = 2.5\ k\Omega$

$R_e = 30\ \Omega$

$C_e = 50$ p*f*

Problem 3.1

Find Z_i, Z_0 and A_v in the case of an emitter follower given that,

$C_{be'} = 1000$ pf $\quad C_{b'c} = 10$ pF

$r_{b'e} = 100\ \Omega \quad r_{bb'} = 30\ \Omega$

$h_{fe} = 100 \quad R'_e = 100\ \Omega$

$R_i^1 = 190\ \Omega$

Solution

Expression for midband input impedance is,

$$Z_i\ (\text{midband}) = r_{bb'} + r_{b'e} + (1 + h_{fe})\, R_{e'}$$
$$= 30 + 100 + (100 + 1)\ 100$$
$$= 130 + (101)\ (100)$$
$$Z_i\ (\text{midband}) \simeq 10k\Omega$$
$$(r_{b'}e + h_{fe}\, R_{e'}) >> R_{e'}$$

We have

$$C^1 = \frac{C_{b'e}}{1 + g_m R_{e'}}$$
$$= \frac{1000 \times 10^{-12}}{(1 + 100)}$$
$$\simeq 10\text{pF}$$

$$\omega_1 = \frac{1}{(r_{b'e} + h_{fe}R_e')(C_{b'c} + C')}$$

$$= \frac{1}{(100 + 10^4)\ (10 + 10)10^{-12}}$$

$$\omega_1 = 5 \times 10^6 \text{ rad/sec}$$

for emitter follower $A_v < 1$.

Problem 3.2

Design a single stage I.F. amplifier to have carrier frequency f_c = 455 kHZs, B.W = 10 kHZs, V_{cc} = – 9V, I_c = – 1mA. The small signal hybrid π parameters are :

$r_{bb'} = 70\ \Omega$	$C_{b'e} = 1550\ pf$
$g_{b'e} = 800\ \mu v$	$C_{b'e} = 9\ pf$
$g_{ce} = 8.6\ \mu\Omega$	$g_m = 38.6\ m\mho$
$g_{b'e} = 0.25\ \mu\Omega$	

Solution

The circuit diagram is shown in Fig. 3.21.

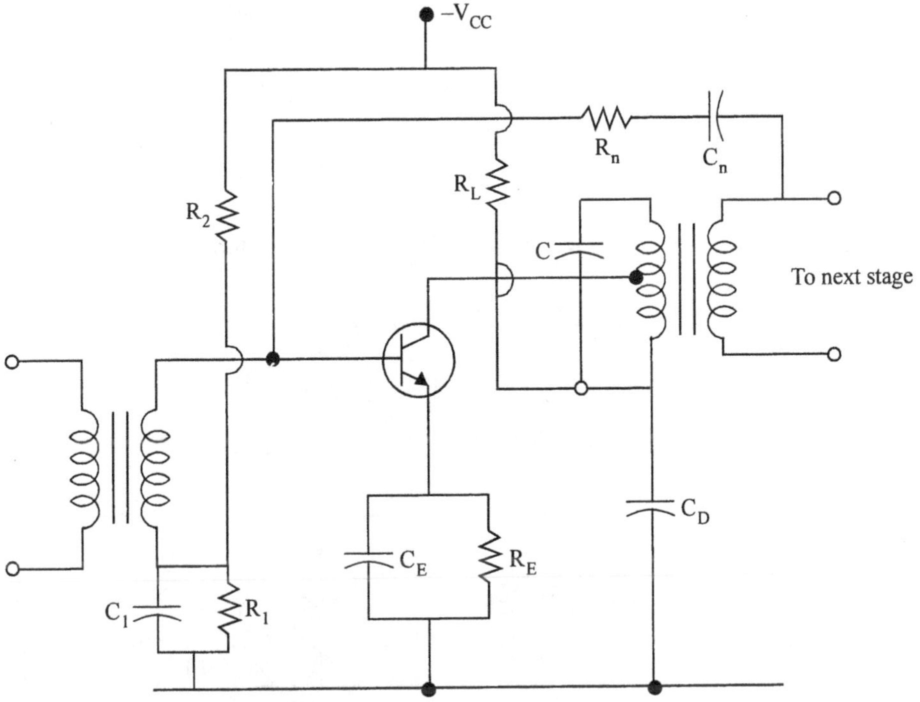

Fig. 3.21 I.F. amplifier single stage

$$r_{b'e} = \frac{1}{g\,b'e} = 1.25\text{ k}\Omega$$

$$r_{ce} = \frac{1}{g_{ce}} = 120\text{ k}\Omega$$

$$r_{b'c} = \frac{1}{g\,b'c} = 4\text{ M}\Omega$$

$$g_{bb'} = 0.133\text{ m℧}$$

$$\omega_r = 2\pi f_r = 2.8\text{ rad/sec}$$

Choose R_1 and R_2 such that I_c = 1mA.

Since $X_{C1} << R_1$, if R_1 = 5 kΩ, than C_1 = 0.05 μF

If R_1 = 1 kΩ, then C_1 = 0.1μF.

Let R_L = 500 Ω and C_D = 0.05 μF

Substituting in the expression for R_i and R_o,

$$R_i = 526\ \Omega \qquad R_o = 32.4\text{ k}\Omega.$$

$$C_i = 1268\text{ pF} \quad C_o = 33\text{ pF}$$

$$G_m = 34.6\text{ mA/V}.$$

R_y is calculated as 12.4 kΩ. C_y is calculated as 9 pF. R_o must be matched with the load (R_i of the next stage). So the transformer turns ratio must be,

$$n = \sqrt{\frac{R_o}{R_i}} = \sqrt{\frac{32.4\times10^3}{526}}$$

$$= 7.8 : 1$$

The feedback components, resistor R_n and capacitor C_n can be determined as,

$$R_n = \left(\frac{R_y}{n}\right) = \frac{12.7\text{ k}\Omega}{7.8} = 1640\ \Omega$$

$$C_n = n.\,C_y = 7.8 \times 9 \times 10^{-12} = 70.2\text{ pF}$$

The equivalent circuit is shown in Fig. 3.22.

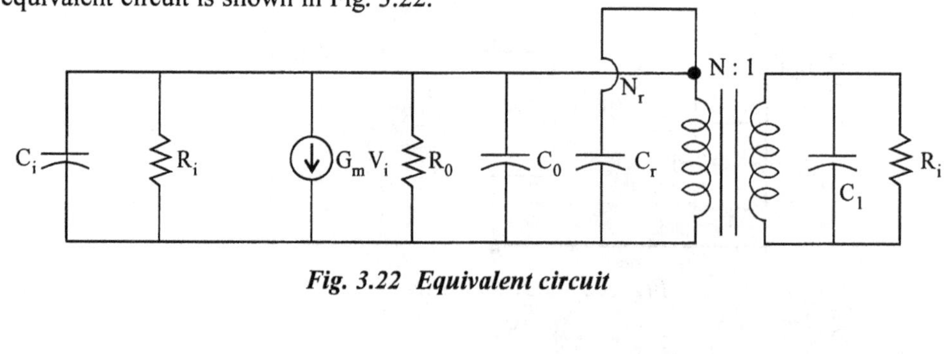

Fig. 3.22 Equivalent circuit

$$P_i = \frac{V_i^2}{R_i}; \quad R_L = R_o = n^2 R_i$$

$$I_L = I_o$$

$$\therefore \quad P_o = I_o^2 \cdot R_L = I_o^2 R_o = \left(\frac{G_m V_i}{2}\right)^2 \cdot R_o$$

$$\text{Power gain } A_p = \frac{P_o}{P_i} = \left(\frac{G_m^2 \cdot R_o R_i}{4}\right)$$

$$= \frac{\left(34.8\times10^{-3}\right)^2 \cdot 32.4\times10^3\times528}{4}$$

$$= 5200$$

Power gain in dB = 10 log (5200) = 37.16 dB

$$\text{Q factor} = \frac{f_r}{B} = \frac{455\times10^3}{10\times10^3} = 45.5$$

with a coil having $Q_c = 100$,

The inductance of the coil is,

$$L = \frac{R_o(Q_c - Q)}{2\omega r\, Q\, Q_c}$$

$$= \frac{32.4\times10^3(100-45.5)}{2\times2.86\times10^6\times45.5\times100}$$

$$= 67\ \mu H$$

Parallel tuning capacitance

$$C = \frac{1}{\omega r^2 L}$$

$$= \frac{1}{\left(8.18\times10^{12}\right)^2 \cdot \left(68\times10^{-6}\right)} \simeq 1800\text{pF}$$

The total transformer primary inductance L_T is,

$$L_T = \frac{1}{\omega r^2 c} = \frac{1}{\left(8.18\times10^{12}\right)^2 \left(206.2\times10^{-12}\right)}$$

$$L_T = 590\ \mu H$$

Problem 3.3

Design a JFET Single Tuned Narrow Band amplifier with a centre frequency of 5.0 MHZs, Q = 40, B.W = 100 KHZs midband gain A_{mid} = 150. Given, for the JFET, C_{DG} = 20 PF, C_{GS} = 50PF, V_{DD} = 15V, V_P = – 2V.

Solution

Given

f_0 = 5 MHZs L = ?
Q = 40 C = ?
A_{mid} = 150 R_D = ?
C_{DG} = 20 pF
C_{GS} = 50 pF
V_{DD} = 15 V
V_p = –2 V
g_n = 10 m℧
r_0 = 1 MW

Circuit Diagram is shown in Fig. 3.23.

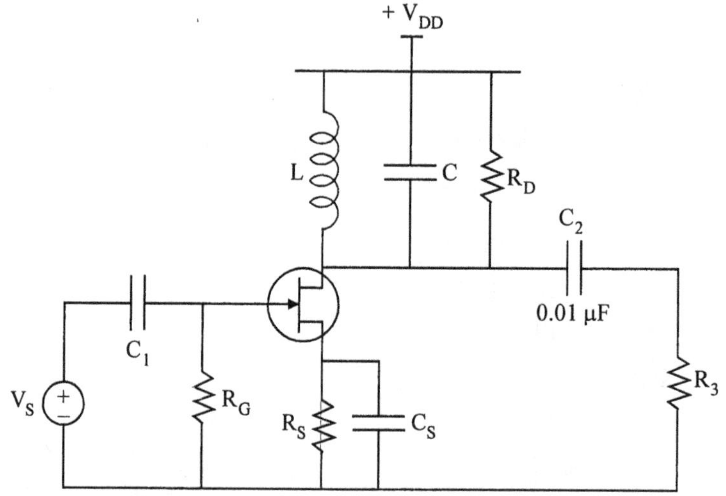

Fig. 3.23 Circuit diagram for problem 3.3

$$\text{Expression for } A_v(s) \simeq A_{mid} \frac{s\left(\frac{\omega_0}{Q}\right)}{s^2 + s\left(\frac{\omega_0}{Q}\right) + \omega_0^2}$$

$$\text{Centre frequency } \omega_0 = \frac{1}{\sqrt{L(C + C_{GD})}}$$

Quantity factor $Q = \omega_0 R_p (C + C_{GD}) = \frac{R_p}{\omega_0 L}$; $A_{mid} = g_m \cdot R_p$

where R_0 is the parallel resistance,

$$R_p = (r_0 \parallel R_D \parallel R_3)$$

The maximum value of $R_p = r_0$.

$$f_0 = 5 \text{ MHZs}$$

$$\omega_0 = 2\pi f_0 = 2 \times 3.14 \times 5 \times 10^6 = 31.4 \text{ MHZs}$$

$$\omega_0 = \frac{1}{\sqrt{L(C + C_{GD})}}$$

$$31.4 \times 10^6 = \frac{1}{\sqrt{L(C + 20 \times 10^{-12})}} \quad .....(3.4)$$

$$Q = \frac{R_p}{\omega_0 L}$$

$$40 = \frac{R_p}{31.4 \times 10^6 \times L} \quad .....(3.5)$$

$$A_{mid} = g_m \cdot R_p \quad .....(3.6)$$

$$150 = 10 \times 10^{-3} \times R_p$$

$$\therefore \quad R_p = \frac{150}{10 \times 10^{-3}} = 15 \text{ k}\Omega;$$

$$R_p = 15 \text{ k}\Omega$$

$$40 = \frac{15 \times 10^3}{31.4 \times 10^6 L}$$

$$\therefore \quad L = \frac{15 \times 10^3}{31.4 \times 10^6 \times 40} = 0.012 \times 10^{-3} \text{ H} = 0.012 \text{mH}$$

$$\omega_0 = \frac{1}{\sqrt{L(C + C_{GD})}}$$

$$\therefore \quad 31.4 \times 10^6 = \frac{1}{\sqrt{0.012 \times 10^{-3}(C + 20pf)}}$$

$$(31.4 \times 10^6)^2 = \frac{1}{0.012 \times 10^{-3}(C + 20pf)}$$

$$0.012 \times 10^{-3}\,(C + 20\text{ pF}) = \frac{1}{\left(31.4\times10^{6}\right)^{2}} = 0.001 \times 10^{-12} = 10^{-15}$$

$$(C + 20\text{ pF}) = \frac{10^{-15}}{0.012\times10^{-3}} = 83.3 \times 10^{-12} = 83.3\text{ pF}$$

$\therefore$ $C = 83.3 - 20 = 63.3$ pF

$\therefore$ $L = 0.012$ mH

$C = 83.3$ pF

Let $R_s = 100\text{ k}\Omega \quad R_p = r_0 \parallel R_D \parallel R_3$

$r_0 = 1\text{ M}\Omega$

$R_p = 15\text{ k}\Omega \;\therefore\; R_D = ?$

$$r_0 \parallel R_3 = \frac{1\text{M}\Omega\times100\text{ k}\Omega}{(1\text{M}\Omega+100\text{ k}\Omega)} = \frac{10\times10^{5}\times10^{5}}{\left(10\times10^{5}+10^{5}\right)}$$

$$r_0 \parallel R_3 = \frac{10^{11}}{11\times10^{5}} = \frac{10^{6}}{11}$$

$$R_p = \frac{10^{6}}{11} \parallel R_D$$

$$15 \times 10^{3} = \frac{\frac{10^{6}}{11}\times R_D}{\frac{10^{6}}{11}+R_D} = \frac{10^{6}R_D}{\left(10^{6}+11R_D\right)}$$

$15 \times 10^{9} + 165 \times 10^{3}\, R_D = 10^{6}\, R_D$

$15 \times 10^{9} = 10^{3}\, R_D\,(1000 - 165)$

$15 \times 10^{9} = 10^{3} \times R_D \times 835$

$$R_D = \frac{15\times10^{9}}{10^{3}\times835} = 0.012 \times 10^{6} = 12\text{ k}\Omega$$

$\therefore$ $R_D = 12\text{ k}\Omega$

Gain Bandwidth Product :

Although f_a and f_b are useful indications of the high-frequency capability of a transistor, are even more important characteristics is f_T.

The f_T is defined as the frequency at which the short-circuit CE current gain has a magnitude of unity. The 'f_T' is lies in between f_β and f_α.

$$|A_I| = \frac{h_{fe}}{\sqrt{1+(f/f_\beta)^2}}$$

at $\quad f = f_T,\ |A_I| = 1,\ \left(\frac{f_T}{f_\beta}\right)^2 >> 1,$

we obtain $\quad f_T / f_\beta \approx h_{fe}$

$$\boxed{f_T \approx h_{fe}\, f_\beta}$$

$\therefore$ f_T is the product of the low frequency gain, h_{fe}, and CE bandwidth f_β. f_T is also the product of the low frequency gain, h_{fb}, and the CB bandwidth, f_α . i.e $f_T = h_{f\beta} \cdot f_\alpha$. The value of f_T ranges from 1MHz for audio transistors upto 1GHZ for high frequency transistors.

The following Fig. 3.24 indicates how short-circuit current gains for CE and CB connections vary with frequency.

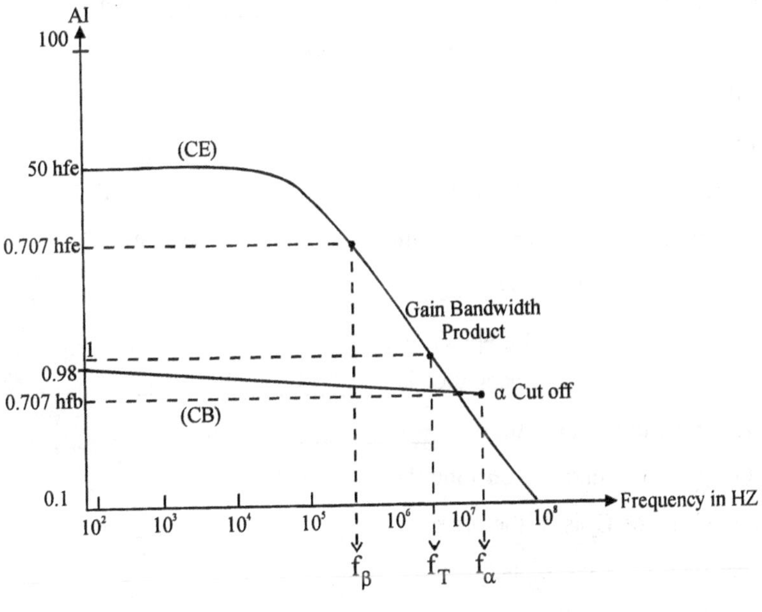

Fig 3.24

SUMMARY

- The frequency response of amplifier circuit is given. Expressions for lower cut-off frequency f_L and upper cut-off frequency f_H are given.
- B-cutoff frequency α-cut off frequency and f_T terms are explained. $f_T = h_{fe}\ f_B$.
- f_B is called C.E short circuit small signal forward current transfer ratio cut-off frequency.
- Hybrid-π C.E transistor model at high frequencies is given.
- Expressions for hybrid-π parameters are derived and typical values are given.
- Variation of hybrid-p parameter values with I_c, T and V_{CE} are given.

OBJECTIVE TYPE QUESTIONS

1. Hybrid - π model is also known as model ____________________ model.
2. Hybrid - π circuit is so named because ____________________.
3. b' to denote Base spread resistance $r_{bb'}$ is ____________________ terminal of the transistor.
4. Transconductance g_m in Hybrid - π model is defined as ____________________.
5. Typical value of $r_{bb'}$ is ____________________.
6. Expression for $g_{b'e}$ = ____________________.
7. Expression for h_{ie} in terms of h_{fe} and I_C is, ____________________.
8. f_T is the frequency at which Common Emitter short circuit current gain ____________________.
9. Relation between f_T, h_{fe} and f_β is ____________________.
10. Expression for C_e interms of g_m and f_T is, ____________________.
11. Typical value of $C_{ob'}$ output capacitance is ____________________.
12. r_π in Hybrid - π equivalent circuit is ____________________.
13. If h_{fe} = 100, g_m = 0.5 mhos, determine the value of $r_{b'e}$.
14. Hybrid π capacitance C_π is of the order of ____________________.
15. Relation between f_T, f_β and h_{fe} is, ____________________.
16. Classify amplifiers depending on the position of the quiescent point of each amplifier.
17. Draw the hybrid -π model for a transistor in CE configuration.
18. What is f_T?

19. What is the significance of the gain bandwidth product ?
20. What would you neglect while drawing a low frequency model ?
21. How does the trans conductance (g_m) depend on current ?
22. How does the trans conductance (g_m) depend on temperature ?
23. Write the expression for r_{be} in terms of g_m and h_{fe}.
24. How does the diffusion capacitance depend on current and temp ?
25. Write the expression for C_{De} in terms of g_m, W, D_B.
26. When is the hybrid - π model valid ? (at what frequencies)
27. What is an emitter follower ?
28. Which time constant is considered for the bandwidth ?
29. What is the expression for β cutoff frequency ?
30. How are f_β and f_T related ?

ESSAY TYPE QUESTIONS

1. Draw the high frequency equivalent circuit of a BJT and explain the same.
2. Give the typical values of various Hybrid - π parameters.
3. Derive the expressions for Hybrid - π parameters., C_e, $r_{bb'}$, $r_{b'e}$, C_C
4. Derive the expression for the Hybrid - π parameters g_m, r_{ce}, C_e and $r_{b'c}$, g_{ce}.
5. Explain about Hybrid - π capacitances. How do Hybrid - π parameters vary with temperature ?
6. Obtain the expressions for f_β and f_T of a transistor.
7. Draw the circuit and derive the expression for CE short circuit current gain A_i interms of at any frequency 'f' and f_β of the BJT.
8. Explain how f_β and f_T of a BJT can be determined ? Obtain the expression for the Gain - Bandwidth product of a transistor.

ANSWERS TO OBJECTIVE TYPE QUESTIONS

1. Giacoletto model
2. The parameters are of different units (Hybrid) Ω , , constants etc. The shape of the circuit is in π - shape.
3. Fictitious terminal
4. $g_m = \frac{\delta I_C}{\delta VB'E}\bigg|_{V_{CE}} = K.$
5. 100 Ω
6. $g_{b'e} = g_m / h_{fe}$
7. $h_{ie} = \frac{h_{fe} V_T}{|I_C|}$
8. becomes unity
9. $f_T = h_{feo} f_\beta$
10. $C_e = g_m / 2\pi f_T$.
11. 2.5 pf.
12. Incremental resistance.
13. $r_{b'e} = h_{fe} / g_m = 200\ \Omega$.
14. Picofarads
15. $f_T = h_{fe} f_\beta$
16. Class A, B, AB, C
17.

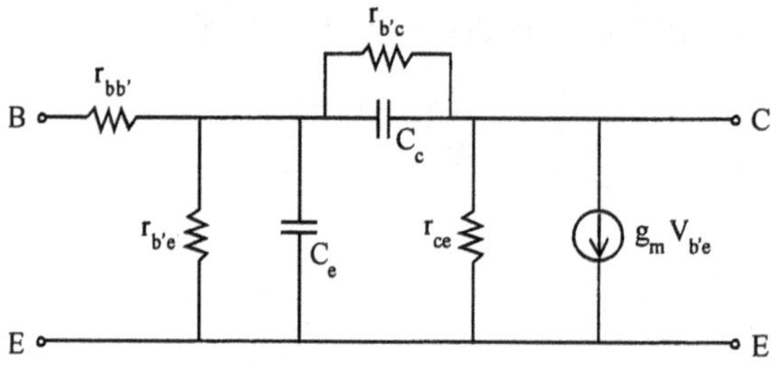

18. Frequency at which short circuit current gain is unity.

19. Tradeoff between gain and BW.

20. Capacitances.

21. g_m directly depends on current. or $g_m = \frac{|I_C|}{V_T}$ or $g_m = \frac{|I_C|}{26}$ mt.

22. g_m inversely proportional to temp.

23. $g_{b'e} = \frac{g_m}{h_{fe}} \Rightarrow r_{b'e} = \frac{h_{fe}}{g_m}$.

24. C_{de} α current α T^n

25. $C_{De} = g_m \frac{W^2}{2D_B}$

26. $2\pi f \frac{W^2}{6D_B} << 1$ or $f << 3f_T$ or $f = \frac{f_T}{3}$

27. CC

28. input time constant.

29. $f_\beta = \frac{1}{h_{fe}} \frac{g_m}{2\pi(C_e + C_c)}$

30. $f_T = h_{fe} f_\beta$

4 MOS Amplifiers

In this Chapter,

- The principle of working of Bipolar Junctions Transistors, (which are also simply referred as Transistors) and their characteristics are explained.
- The Operation of Transistor in the three configurations, namely Common Emitter, Common Base and Common Collector Configurations is explained.
- The variation of current with voltage, in the three configurations is given.
- The structure of Junction Field Effect Transistor (JFET) and its V-I characteristics are explained.
- The structure of Metal Oxide Semiconductor Field Effect Transistor (MOSFET) and its V-I characteristics are explained.
- JFET amplifier circuits in Common Source (CS) Common Drain (CD) and Common Gate (CG) configurations are given.

4.1 FIELD EFFECT TRANSISTOR (FET)

The Field Effect Transistor is a semiconductor device which depends for its operation on the control of current by an electric field. Hence, the name FET.

There are two types (FET) :

1. Junction Field Effect Transistor (JFET) or simply FET.
2. Insulated Gate Field Effect Transistor IGFET. It is also called as Metal Oxide Semiconductor (MOS) transistor or MOST or MOSFET.

The principle of operations of these devices, their characteristic are given in this chapter.

Advantages of FET over BJT (Transistor)

1. JFET is unipolar device. Its operation depends upon the flow of majority carriers only. Vacuum tube is another example of unipolar device. Because the current depends upon the flow of electrons emitted from cathode. Transistor is a bipolar device. So recombination noise is more.
2. JFET has high input resistance (MΩ). BJT (Transistor) has less input resistance. So loading effect will be there.
3. It has good thermal stability.
4. Less noisy than a tube or transistor. Because JFET is unipolar, recombination noise is less.
5. It is relatively immune to radiation.
6. JFET can be used as a symmetrical bilateral switch.
7. By means of small charge stored or internal capacitance, it acts as a memory device.
8. It is simpler to fabricate and occupies less space in IC form. So packing density is high.

Disadvantage

1. Small gain - bandwidth product.

4.1.1 JFET

There are two types;

1. n - channel JFET.

2. p - channel JFET.

If n - type semiconductor bar is used it is n - channel JFET.

If p - type semiconductor bar is used it is p - channel JFET.

Ohmic contacts are made to the two ends of a semiconductors bar of n-type or p-type semiconductor. Current is caused to flow along the length of the bar, because of the voltage supply connected between the ends. If it is n - channel JFET, the current is due to electrons only and if it is p-channel JFET, the current is due to holes only.

The three lead of the device are **1.** Gate **2.** Drain **3.** Source. They are similar to **1.** Base **2.** Collector **3.** Emitter of BJT respectively.

4.1.2 n-Channel JFET

The arrow mark at the gate indicates the direction of current if the Gate source junction is forward biased. (Fig. 4.1).

On both sides of the n - type bar heavily doped (p^+) region of acceptor impurities have been formed by alloying on diffusion. These regions are called as Gates. Between the gate and source a voltage V_{GS} is applied in the direction to reverse bias the p-n junction.

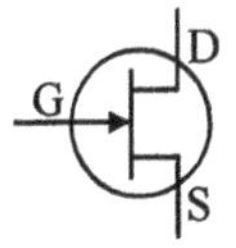

Fig 4.1 (n-channel JFET)

4.1.3 p-Channel JFET

The symbol for p-channel FET is given in Fig. 4.2. The semiconductor bar is p-type or source is p-type. Gate is heavily doped n-region. If G - S junction is forward biased, electrons from gate will travel towards source. Hence, the conventional current flows outside. So, the arrow mark is as shown in Fig. 4.2. indicating the direction of current, if G-S junction is forward biased.

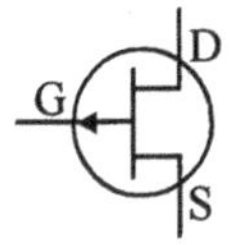

Fig 4.2 (p-channel JFET).

The FET has three terminals, Source, drain and Gate.

Source

The source S is the terminal through which the majority carriers enter the bar (Fig. 4.3). Conventional current entering the bar at S is denoted as I_S.

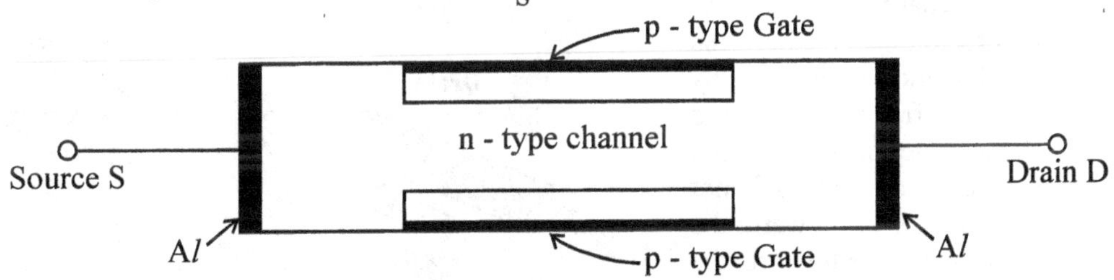

Fig 4.3 Structure of JFET.

Drain

The drain D is the terminal through which the majority carriers leave the bar. Current entering the bar at D is designated by I_D. V_{DS} is positive if D is more positive than S, source is forward biased.

Gate

In the case of n - channel FET, on both sides of the bar heavily doped (p+) region of acceptor impurities have been formed by diffusion or other techniques. The impurity regions form as Gate. Both these region (p+ regions) are joined and a lead is taken out, which is called as the gate lead of the FET. Between the gate and source, a voltage V_{GS} is applied so as to reverse bias the gate source p-n junction. ***Because of this reason JFET has high input impedance. In BJT Emitter Base junction is forward biased. So it has less Z.*** Current entering the bar at G is designated as I_G.

CHANNEL

The region between the two gate regions is the channel through which the majority carriers move from source to drain. By controlling the reverse bias voltage applied to the gate source junction the channel width and hence the current can be controlled.

4.1.4 FET STRUCTURE

FET will have gate junction on both sides of the silicon bar (as shown in Fig. 4.4).

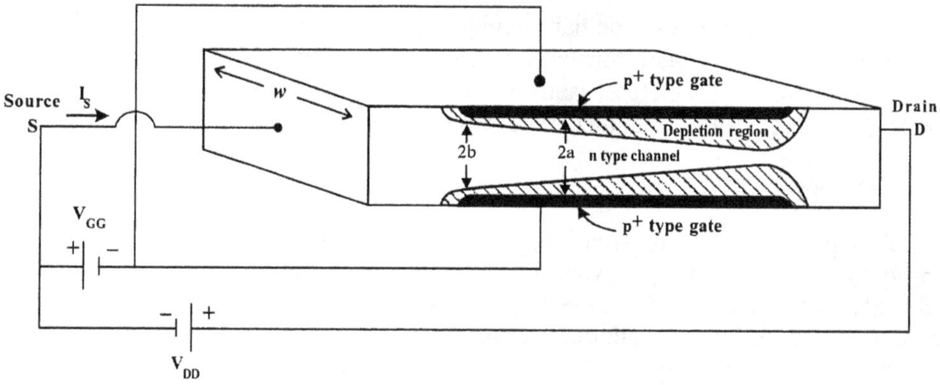

Fig 4.4 JFET structure.

But it is difficult to diffuse impurities from both sides of the wafer. So the normal structure that is adopted is, a p-type material is taken as substrate. Then n-type channel is epitaxially grown. A p-type gate is then diffused into the n-type channel. The region between the diffused p-type impurity and the source acts as the drain.

2b (x) = Actual channel width, at any point x, the spacing between the depletion regions at any point x from source end because the spacing between the depletion regions is not uniform, it is less towards the drain and more towards the source (see Fig.4.5). 2b depends upon the value of V_{GS}.

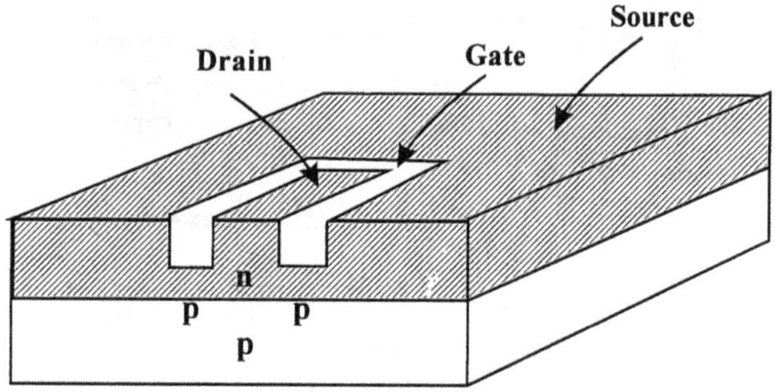

Fig 4.5 Structure

2a = Channel width, the spacing between the doped regions of the gate from both sides. This is the channel width when $V_{GS} = 0$ (and the maximum value of channel width).

4.1.5 THE ON RESISTANCE r_{DS} (ON)

Suppose a small voltage V_{DS} is applied between drain and source, the resulting small drain current I_D will not have much effect on the channel profile. So the effective channel cross section A can be assumed to be constant, throughout its length.

$\therefore$ $A = 2b\,w$

Where, 2b = Channel width corresponding to zero I_D (Distance between the space charge regions)

w = Channel dimension perpendicular to b.

∴ Expressions for,

$$I_D = A \cdot e \cdot N_D \, \mu_n \, \varepsilon_x$$

∵ $J = n\, e \cdot \mu \cdot \varepsilon \; (J = \sigma.\varepsilon.\ \sigma = ne.\mu)$

$I = A \times J,$

But $A = 2b\,(x) \cdot w$

Considering b along x-axis, since 'b' depends upon the voltage V_{GS}, b(x) is the width at any point 'x'.

∴ $I_D = 2b\,(x)\, w.\ q.\ N_D \cdot \mu_n .\varepsilon_x$

$\varepsilon_n = \dfrac{V_{DS}}{L}$; L is the length of the channel

(Electric field strength depends upon V_{DS}. Channel width 2b depends upon V_{GS}.)

∴ $I_D = 2b\,(x)\, w.\ q.\ N_D \cdot \mu_n \cdot \dfrac{V_{DS}}{L}$

This expression is valid in the linear region only. This is the linear region, It behaves like a resistance, where ohmic resistance depends upon $V_{GS}.V_{DS} / I_D$ is called as the ***ON drain resistance*** or $r_{DS(ON)}$ ON because, JFET is conducting, in the ohmic regions when V_{DS} is small and channel area $A = 2a.w$,

$r_{DS(on)} = L/2\ b(x)\ wq\ N_D\ \mu_n$

$r_{DS(on)}$ will be few Ω to several 100 Ω.

Because, the mobility of electrons is much higher than that for holes, $r_{DS(on)}$ is small for n-channel FETs compared to p-channel FETs.

4.1.6 Pinch off Region

The voltage V_{DS} at which, the drain current I_D tends to level off is called the ***pinch off voltage***. When the value of V_{DS} is large, the electric field that appears along the x-axis, i.e., along the channel ε_x will also be more. When the value of I_D is more, the drain end of the gate i.e., the gate region near the drain is more reverse biased than the source end. Because, the drain is at reverse potential, for n-channel FET, source is n-type, drain is n-type and positive voltage is applied for the drain, gate is also reverse biased. Therefore, the drop across the channel adds to the reverse bias voltage of the gate. Therefore, channel narrows more near the drain region and less near the source region. So, the boundaries of the depletion region are not parallel when V_{DS} is large and hence ε_x is large.

As V_{DS} increases, ε_x and I_D increase, whereas the channel width $b(x)$ narrows (∵ depletion region increases). Therefore the current density $J = I_D / 2b(x)\, w$ increases. If complete pinch off were to take place b = 0. But this cannot happen, because if b were to be 0, $J \rightarrow \infty$ which cannot physically happen.

Mobility μ is a function of electric field intensity μ remains constant for low electric fields when $\varepsilon_x < 10^3$V/cm.

$\mu \;\alpha\; \dfrac{1}{\sqrt{\varepsilon_x}}$ when ε_x is 10^3 to 10^4 V/cm.(for moderate fields).

$\mu \;\alpha\; \dfrac{1}{\varepsilon_x}$ for very high field strength > 10^4 V/cm.

$I_D = 2bwe\ N_D\ \mu_n\ \varepsilon_x.$

As V_{DS} increases, ε_x increases but μ_n decreases, b almost remains constant. Therefore I_D remains constant in the pinch off region.

Pinch off voltage (V_{PO} or V_P) is the voltage at which the drain current I_D levels off. When V_{GS} = 0V, if V_{GS} = – 1,V, the voltage at which the current I_D levels off decreases, (This is not pinch off voltage). If V_{GS} is large – 5, V_{DS} at which the current levels off may be zero. So the relationship between V_{DS}, V_{PD} and V_{GS} is

$V_{DS} = V_{PD} + V_{GS}$ V_{DS} is positive n-channel FET.

V_{PO} = P indicates pinch off voltage, zero indicate the voltage when V_{GS} = 0V.

V_{DS} is the voltage at which the current I_D levels off.(see Fig. 4.6).

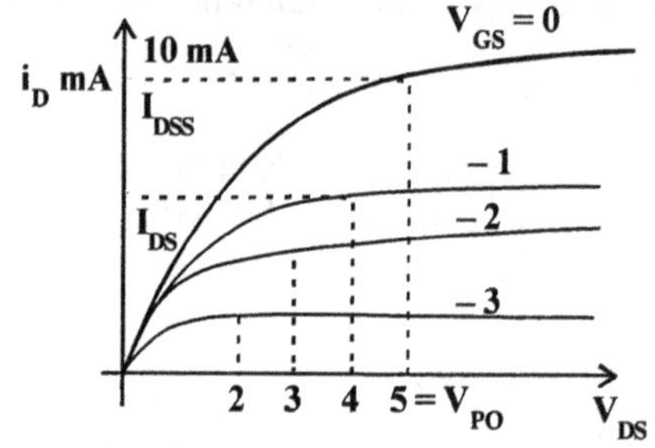

Fig 4.6 JFET drain characteristics.

4.1.7 Expression for Pinch off Voltage (V_P)

Net charge in an alloy junction, must be the same, semiconductor remains neutral.

$$\therefore \quad e.N_A\, w_P = e\, N_D \cdot w_n$$

If $N_A >> N_D$, from the above eq, $w_p << w_n$.
The relation between potential and charge density (ρ) is given by Poisson's equation.

$$\frac{d^2V}{dx^2} = \frac{e.N_D}{\epsilon}$$

We can neglect w_p and assume that the entire barrier potential V_B appears across the increased donor ions. $e.N_D$ = charge density P

$$\text{Integrating} \quad \frac{dv}{dx} = \frac{e.N_D}{\epsilon}.x$$

$$V = \frac{e.N_D}{\epsilon} \cdot \frac{x^2}{2}$$

At the boundary conditions, $x = w_n$

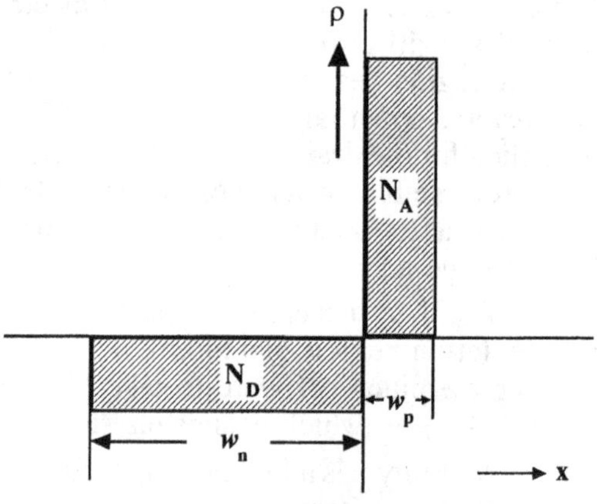

Fig 4.7 Charge density variation

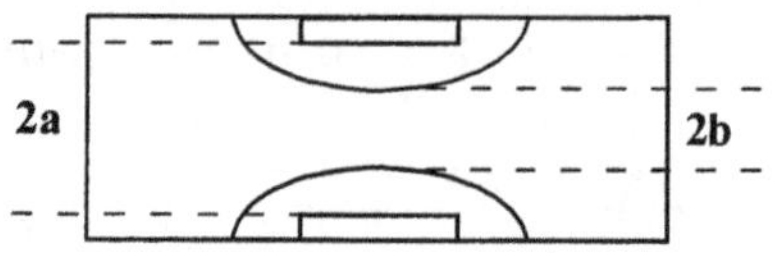

Fig 4.8 Structure

w_n = depletion region width

$$\therefore \quad V = \frac{e.N_D}{2\in} \cdot w_n^2$$

or
$$w_n = \left\{\frac{2\in V}{e.N_D}\right\}^{\frac{1}{2}}$$

This is the expression for the penetration of depletion region into the channel. As the reverse bias potential between Gate and drain increased, the channel width narrows or the depletion region penetrates into the channel. Therefore the general expression for the spacecharge width w_n (x) = w(n).

$$w(n) = \left\{\frac{2\in}{eN_D}\{V_o - V(x)\}\right\}^{1/2}$$

V = Voltage between drain and source V_{DS}

w = Penetration of depletion region into the channel,

Where V = $V_0 - V(x)$; V_0 is the contact potential at x and $V_{(n)}$ is the applied potential across space charge region at x and is negative for an applied potential. This potential (between D and S) is not uniform throughout the channel. The potential is higher near the drain and decreases towards the source. The depletion region is more towards the drain and decrease towards the source.

$$\therefore \quad V' = V_0 - V(x)$$

The actual potential is the difference of V_0 and $V(x)$.

$\in$ = Dielectric constant of channel material

$$w(x) = a - b(x) = \left\{\frac{2\in}{eN_D}(V_0 - V(x))\right\}^{1/2} \quad .....(4.1)$$

$a - b(x)$ = Penetration $w(x)$ of depletion region into channel at a point x along the channel from one side of gate region only.

2a = Total width between the two gate diffusions.

$2b(x)$ = Depletion region width between the two sides of the gate regions.

If $I_D = 0$, $b(x)$ is independent of x and it will be uniform throughout the channel and will be equal to 'b'. V_0 will be much smaller than $V(x)$.

If in equation (1), we substitute b(n) = b = 0, neglect V_0, and solve for V we obtain the pinch off voltage. Because at pinch off, the two depletion regions from both sides meet each other, therefore the spacing between them $b(x) = 0$.

Therefore, w at pinch off, V_0 neglected, b = 0,

$$w = \left\{\frac{2\in}{eN_D}V_P\right\}^{1/2}$$

or
$$w^2 = \frac{2\in}{eN_D} \cdot V_P$$

4.1.8 FET OPERATION

If a p-n junction is reverse biased, the majority carriers will move away from the junction. That is holes on the p-side will move away from the junction leaving negative charge or negative ions on the p-side (because each atom is deprived of a hole or an electron fills the hole. So it becomes negative by charge). Similarly, electrons on the n-side will move away from the junction leaving positive ions near the junction. Thus, there exists space charge on both sides of the junction in a reverse biased p-n junction diode. So, the electric field intensity, the lines of force originate from the positive charge region to the negative charge region. This is the source of voltage drop across the junction. As the reverse bias across the junction increases, the space charge region also increases, or the region of immobile uncovered charges increases (i.e., negative region on p-side and positive region on n-side increases as shown in Fig. 4.9). The conductivity of this region is usually zero or very small. Now in a FET, between gate and source, a reverse bias is applied. Therefore the channel width is controlled by the reverse bias applied between gate and source. Space charge region exists near the gate region on both sides. The space between them is the channel. If the reverse bias is increased, the channel width decreases. Therefore, for a fixed drain to source voltage the drain current will be a function of the reverse biasing voltage across the junction. Drain is at positive potential (for n-type FET). Therefore, electrons tend to move towards drain from the source (1) Because, source is at negative potential, they tend to move towards the drain. But because of the reverse bias applied to the gate, there is depletion region or negative charge region near the gate which restricts the number of electrons reaching the drain. Therefore the drain current also depends upon the reverse bias voltage across the gate junction. The term field effect is used to describe this device because the mechanism of current control is the effect of the extensions, with increase reverse bias of the field associated with region of uncovered charges.

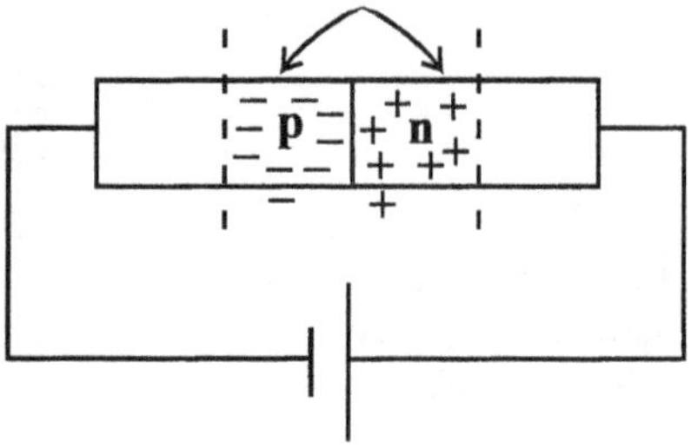

Fig 4.9 Space charge in p-n junction.

The characteristics of n-channel FET between I_D, the drain current and V_{DS} the drain source voltage are as shown in Fig. 4.10, for different values of V_{GS}.

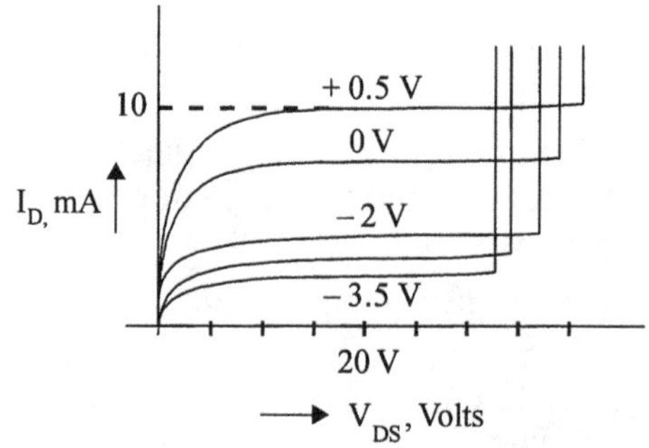

Fig 4.10 Drain charecteristics

To explain these characteristics, suppose $V_{GS} = 0$. When $V_{DS} = 0$, $I_D = 0$, because the channel is entirely open. When a small V_{DS} is applied (source is forward biased or negative voltage is applied to n-type source), the n-type bar acts as a simple semiconductor resistor and so I_D increases linearly with V_{DS}. With increasing current, the ohmic voltage drop between the source and the channel region reverse biases the junction and the conducting portion of the channel begins to constrict. Because, the source gets reverse biased or the negative potential at the n-type source reduces. Because of the ohmic drop, along the length of the channel it self, the constrictions is not uniform, but is more ***pronounced*** at distances farther from the source. i.e., the channel width is narrow at the ***drain*** and wide at the source. Finally, the current will remain constant at a particular value and the corresponding voltage at which the current begins to level off is called as the ***pinch off voltage***. But the channel cannot be closedown completely and there by reducing the value of I_D to zero, because the required reverse bias will not be there.

Now if a gate voltage V_{GS} is applied in a direction to reverse bias the gate source junction, pinch off will occur for smaller values of (V_{DS}), and the maximum drain current will be smaller compared to when $V_{GS} = 0$. If V_{GS} is made + 0.5V, the gate source junction is forward biased. But at this voltage, the gate current will be very small because for Si FET, 0.5V is just equal to or less than the cut in voltage. Therefore, the characteristic for $V_{GS} = +\,0.5V$, I_D value will be comparatively larger, (compared to $V_{GS} = 0V$ or – 0.5V). Pinch off will occur early.

FET characteristics are similar to that of a pentode, in vacuum tubes.

The maximum voltage that can be applied between any two terminals of the FET is the lowest voltage that will cause avalanche breakdown, across the gate junction. From the FET characteristics it can be observed that, as the reverse bias voltage for the gate source junction is increased, avalanche breakdown occurs early or for a lower value of V_{DS}. This is because, the reverse bias gate source voltage adds to the drain voltage because drain though n -type (for n-channel FET), a positive voltage is applied to it. Therefore, the effective voltage across the gate junction is increased.

For n-channel FET, the gate is p-type, source and drain are n-type. The source should be forward-biased, so negative voltage is applied. Positive voltage is applied to the drain. ***Gate source junction should be reverse biased,*** and gate is p-type. Therefore,voltage or negative voltage is applied to the gate. Therefore, n-channel FET is exactly similar to a Vacuum Tube (Triode). Drain is similar to anode (at positive potential), source to cathode and gate to grid, (But the characteristics are similar to pentode).

For p-channel FET, gate is n-type, and positive voltage is applied, drain is at negative potential with respect to source.

Consider n-channel FET. The source and drain are n-type and p-type gate is diffused from both sides of the bar (see Fig. 4.11 below).

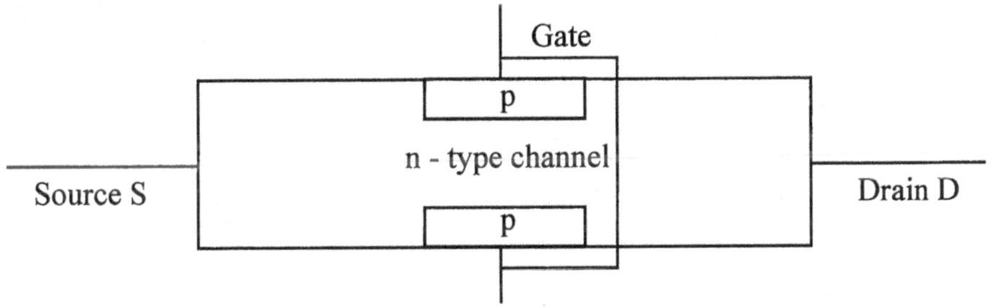

Fig 4.11 Structure of n-channel JFET.

Suppose source and gate are at ground potential and small positive voltage is applied to the drain. Source is n-type. So it is Forward biased. Because drain is at positive potential, electrons from the source will move towards the drain. Negligible current flows between source and gate or gate and drain, since these p-n junctions are reverse biased and so the current is due to minority carriers only. Because gate is heavily doped, the current between S and G or G and D can be neglected. The current flowing from source to drain I_D depends on the potential between source and drain, V_{DS}, resistance of the n-material in the channel between drain to source. This resistance is a function of the doping of the n-material and the channel width; length and thickness. (Fig. 4.12).

If V_{DS} is increased, the reverse bias voltage for the gate drain junction is increased, since, drain is n-type and positive voltage at drain is increased. This problem is similar to that of a reverse biased p-n junction diode.

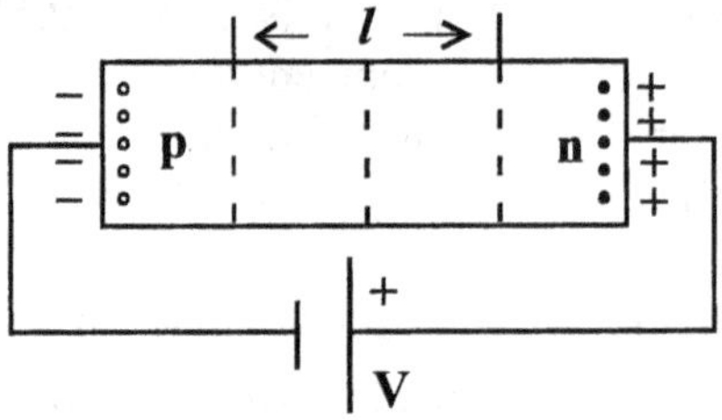

Fig 4.12 Reverse biased G - D junctions

Consider a p-n junction which is reverse biased. The holes on the p-side remain near the negative terminal and electrons on the n-side reach near the positive terminal as shown in Fig.4.12. There are no mobile charges near the junction. So we call this as the depletion region since it is depleted of mobile carriers or charges. As the reverse voltage is increased, the depletion region width '*l*' increases.

This result is directly applicable for JFET. The depletion region extends more towards drain because points close to the drain are at higher positive voltage compared to points close to source. So the depletion region is not uniform $V_{DS} < V_{PO}$. But extends more towards drain than source. V_{P0} is the pinch off voltage.

As V_{DS} is increased, depletion region is increased (shaded portion). So channel width decreases and channel resistance increases. Therefore, the rate at which I_D increases reduces, eventhough there is positive potential for the electrons at the drain and hence they tend to move towards the drain and conventional current flows as shown by the arrow mark, in the Fig.4.13.

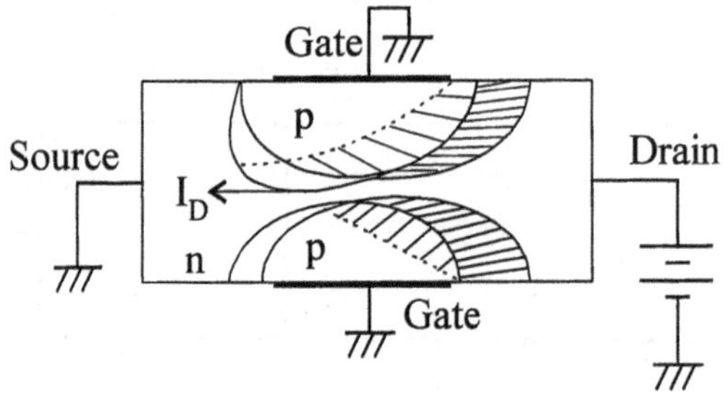

Fig 4.13 Speed of depletion region

When $V_{DS} = V_{PO}$

When V_{DS} is further increased, the depletion region on each side of the channel join together as shown in Fig.4.14. ***The corresponding V_{DS} is called as V_{PD}, the pinch off voltage,*** because it pinches off the channel connection between drain and source.

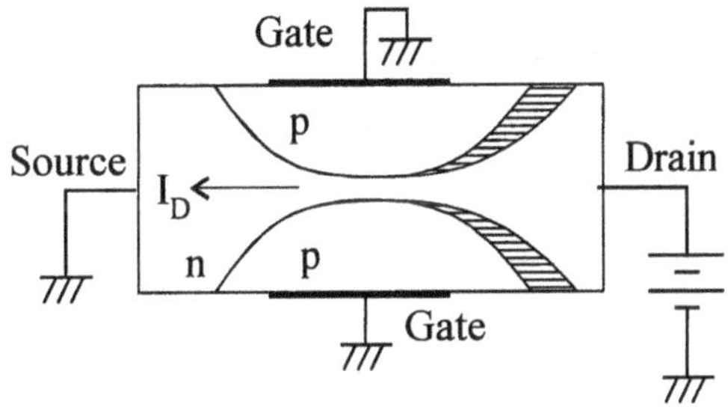

Fig. 4.14

When $V_{DS} > V_{PO}$

If V_{DS} is further increased, the depletion region thickens. So the resistivity of the channel increases. Because drain is at more positive potential, more electrons tend to move towards the drain. Hence I_D should increase. But, because channel resistance increases, I_D decreases. Therefore, the net result is I_D levels off, for any V_{DS} above V_{PO}.

$$I_D = \frac{V_{DS}}{r_{DS}}$$

In the linear region, r_{DS} is almost constant. So as V_{DS} is increased, I_D increases. When the two channels meet, as V_{DS} increases r_{DS} also increases. So I_D remains constant (see Fig.4.15).

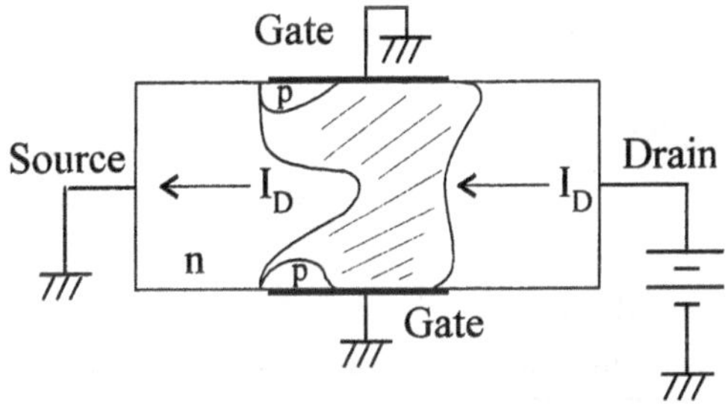

Fig 4.15 Depletion region width variation in JFET.

4.2 JFET VOLT-AMPERE CHARACTERISTICS

Suppose the applied voltage V_{DS} is small. The resulting small drain current I_D will not have appreciable effect on the channel profile. Therefore, the channel cross-section A can be assumed to be constant throughout, its length.

$$\therefore \qquad A = 2b.w,$$

Where 2b is the channel width corresponding to negligible drain current and w is channel dimension perpendicular to the 'b' direction.

$$\therefore \qquad I_D = Ae \,.\, N_D \,\mu_n \in$$

$$A = 2b \,.\, w$$

$$\in = \frac{V_{DS}}{L}$$

$$= 2bwe \,.\, N_D\, \mu_n \,.\, \frac{V_{DS}}{L} \qquad .....(4.2)$$

Where L is the length of the channel. Eliminating 'b' which is unknown,

$$V_{GS} = \left(1-\frac{b}{a}\right)^2 .\, V_P$$

$$\left(1-\frac{b}{a}\right) = \left(\frac{V_{GS}}{V_P}\right)^{1/2}$$

$$\frac{b}{a} = 1 - \left(\frac{V_{GS}}{V_P}\right)^{1/2}$$

$$b = a\left[1-\left(\frac{V_{GS}}{V_P}\right)^{1/2}\right]$$

Substituting, this value in equation (1),

$$\boxed{I_D = \frac{2aweN_D\,\mu_n}{L}\left[1-\left(\frac{V_{GS}}{V_P}\right)^{1/2}\right] V_{DS}}$$

This is the expression for I_D in terms of V_{GS}, V_P and V_{DS} because the value of b is not directly known.

$$V_P = \frac{e.N_D}{2\varepsilon} \,.\, w^2$$

But, $w = a - b\,(x)$

$b = 0$, at pinch off,

$\therefore$ $w = a =$ The spacing between the two gate dopings.

$$\therefore \quad \boxed{|V_P| = \frac{e.N_D}{2\in} \,.\, a^2}$$

V_{DS} controls the width of the depletion region (a – b) because for a given V_{GS}, as V_{DS} increases, the reverse potential between drain and gate increase. Therefore, depletion region width increases.

EXPRESSION FOR V_{GS}

$$V = \frac{e.N_D x^2}{2\in}$$

We can get the expression for V_{GS} if we replace x by (a - b) and V by V_{GS}.

$$\therefore \qquad V_{GS} = \frac{e.N_D(a-b)^2}{2\in}$$

But
$$|V_P| = \frac{e.N_D}{2 \in}.a^2$$

$\therefore$
$$\frac{e.N_D}{2 \in} = \frac{|V_p|}{a^2}$$

Substituting this in expression for V_{Gs},

$$V_{GS} = \frac{V_p}{a^2}(a-b)^2$$

Channel width is controlled by V_{GS}. As V_{GS} increases, channel width reduces.)

Therefore, channel width is controlled by V_{GS} and depletion region width by V_{DS}.

$$\boxed{V_{GS} = \left(1-\frac{b}{a}\right)^2 .V_p}$$

THE ON RESISTANCE r_{ds} (ON)

The drain characteristic is shown in Fig. 4.16.

When V_{DS} is small, the FET behaves like an ohmic resistance whose value is determined by V_{GS}.

The ratio
$$\frac{V_{DS}}{I_D} = r_{ds(ON)}$$

$$I_D = 2bwe\, N_D\, \mu_n \cdot \frac{V_{DS}}{L}$$

$\therefore$
$$\frac{V_{DS}}{I_D} = \frac{L}{2a\,we\, N_D\,\mu_n} = r_{ds(ON)}.$$

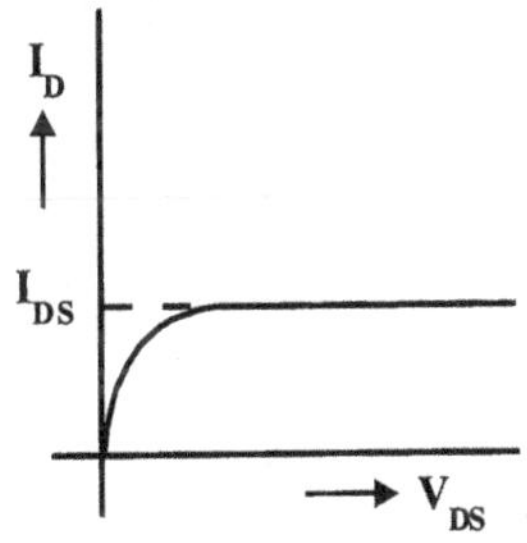

Fig 4.16 Drain characteristics

4.3   TRANSFER CHARACTERISTICS OF FET

$I_{DS} =$ The saturation value of the drain current or the value at which I_D remains constant.

$I_{DSS} =$ The saturation value of the drain current when gate is shorted to source that is $V_{GS} = 0$.

It is found that the transfer characteristics giving the relationship between I_{DS} and V_{GS} can be approximated by parabola.

$$I_{DS} = I_{DSS}\left(1-\frac{V_{GS}}{V_p}\right)^2$$

Transfer characteristic follows the equation, I_D vs V_{GS}

$$I_D = (I_{DSS} + I_{GSS})\left(1-\frac{V_{GS}}{V_p}\right)^2 - I_{GSS}$$

With negligible I_{GSS}, because I_{GS} is the leakage gate current between G and S when D is shorted, to S will be of the order of nano amperes, so it can be neglected.

$$\therefore \qquad I_D = I_{DSS}\left(1 - \frac{V_{GS}}{V_p}\right)^2$$

This V_p is $V_{GS(off)}$ and not V_{DS} because when $V_p = V_{GS(off)}, I_D = 0$.

I_{DSS} = The saturation value of drain current when gate is shorted to source or $V_{GS} = 0$.

The transfer characteristics can be derived from the drain characteristics. (Fig. 4.17).

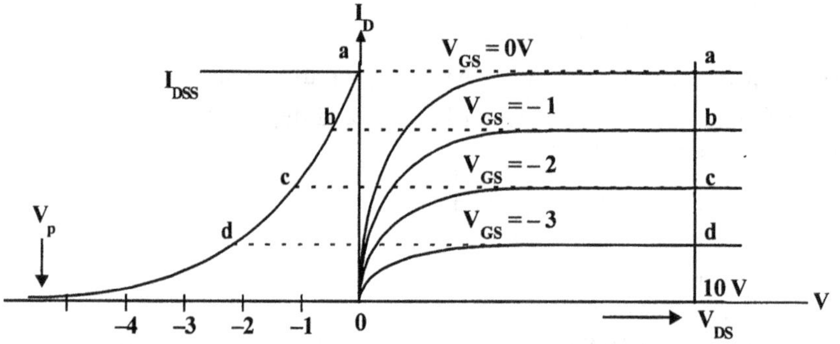

Fig 4.17 Drain and gate characteristics of JFET

To construct transfer characteristics, a constant value of V_{DS} is selected. Normally this is chosen in the saturation region, because the characteristics are flat. The intersection of the vertical line abcd gives a particular value of I_D for different values of V_{GS}. So the transfer characteristics between I_D and V_{GS} can be drawn. (Fig. 4.17).

If the end points V_P and I_{DSS} are known, a transfer characteristics for the device can be constructed.

Here V_p is again called the ***pinch off voltage*** or $V_{GS(off)}$ voltage. This is the voltage between gate and source for which I_D becomes zero. As the reverse potential between G and S increases, depletion region width increases. So channel width decreases. For some value of V_{GS} channel pinches off or I_D practically becomes zero (It cannot be exactly zero but few nano - amps). So this gate source voltage at which I_D becomes zero is also called as pinch off voltage or $V_{GS(off)}$ voltage. This pinch off voltage is different from V_{DS} voltage at which I_D levels off. Since in the latter case I_D is not zero, but channel width becomes zero and channel pinches off. This channel width is made zero either by controlling V_{DS} or V_{GS}. Hence there are two types of pinch of voltages. The specified pinch off voltage V_p for a given FET can be known if it is due to V_{DS} or V_{GS} from the polarity of the voltage V_p. For n-channel FET, V_{GS} is negative (since gate is p-type, G-S junction is reverse biased) V_{DS} is positive. If the specified V_p is –5V, (say) that it is due to $V_{GS(off)}$ because V_P is negative. If V_P is positive, it is due to V_{DS} at which I_{DSS} levels off.

4.3.1 FET Biasing for Zero Drift Current

As Temperature (T) increases, Mobility (μ) decreases at constant Electric Field (ε), therefore I_D increases. As T increases depletion region width decreases. So conductivity σ of the channel increases, I_D also increases (See Fig. 4.18).

I_D decreases by 0.7 % per degree centigrade.

I_D increase at a rate of 2.2 mV/°C change in V_{GS}. Therefore change in V_{GS} = 2.2 mV/°C.

$\therefore$ For zero drift current,

$$0.007\,|I_D| = 0.0022\,g_m$$

or
$$\frac{|I_D|}{g_m} = 0.314\text{ V}$$

or
$$\boxed{|V_p| - |V_{GS}| = 0.63\text{V}}$$

$$I_{DS} = I_{DSS}\left(1 - \frac{V_{GS}}{V_P}\right)^2$$

$$g_m = g_{mo}\left(1 - \frac{V_{GS}}{V_p}\right)$$

$$g_{mo} = \frac{-2I_{DSS}}{V_p}$$

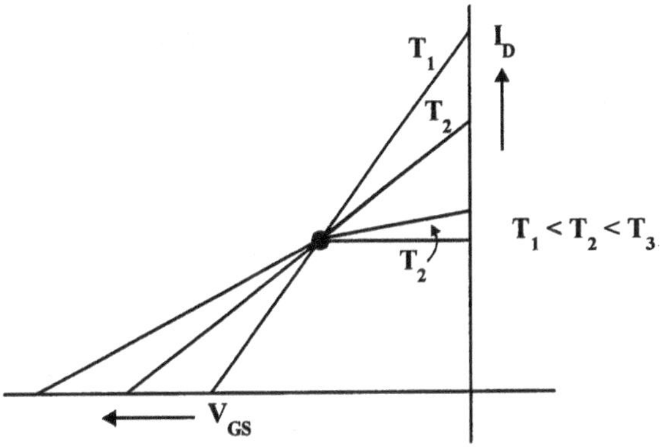

Fig 4.18 Biasing for zero drift current

Problem 4.1

Design the bias circuit for zero drain current drift, for a FET having the following particulars :

$$V_p = -3\text{V}\ ;\ g_{mo} = 1.8\text{ mA/V}$$
$$I_{DSS} = 1.75\text{ mA, if } R_d = 5\text{ k}\Omega$$

(a) Find I_D for zero drift current.

(b) V_{GS}

(c) R_S

(d) Voltage gain with R_s bypassed with a large capacitance,

Solution

For zero drift current,

$$|V_P| - |V_{GS}| = 0.63$$
$$V_p = -3\text{V}$$
$$V_{GS} = V_P - 0.63$$

(a)
$$I_D = I_{DSS}\left(1 - \frac{V_{GS}}{V_p}\right)^2 \simeq 0.08\text{ mA}$$

(b) $\therefore$
$$V_{GS} = -3 - 0.63 = -3.63\text{ V}$$

(c)
$$R_S = \frac{-V_{GS}}{I_D} = \frac{3.63}{I_D} = \frac{3.63}{0.08\text{ mA}} = 43\text{ k}\Omega$$

(d)
$$g_m = g_{mo}\left|\frac{V_{GS} - V_p}{V_p}\right| = 0.378\text{ mA/V}$$

$$A_v = g_m \cdot R_d = 1.89$$

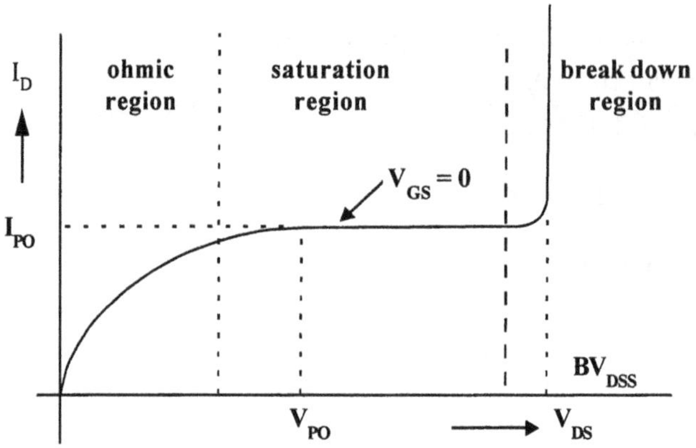

Fig 4.19 Drain characteristics.

The drain current I_D increases rapidly as V_{DS} increases, towards V_{PO}. Above V_{PO}. The current tends to level off at I_{PO} and then rises slowly. When V_{GS} = Breakdown voltage BV_{DSS}, breakdown (avalanche) occurs and the current rises rapidly. Current levels off because as V increases, resistance of the channel also rises. Therefore, I remains constant. As V

increases, R increases, therefore V/R = constant. The region before V_{PO} is called ***ohmic region*** because ohms law is obeyed (see Fig. 4.19). Just as a p-n junction breakdown when reverse voltage is very high FET also break down if V_{DS} is high due to high electric field.

Suppose V_{DS} is fixed and V_{Gs} is varied. As V_{GS} is made negative, p-n junction is reverse biased and depletion region between gate and source increases. This decreases channel width and increases channel resistance. Hence I_D decreases. If gate voltage is made positive, depletion region decreases and hence I_D increases. The p-n junction, between gate and source becomes forward biased and current flows from gate to source. So n-type JFET is usually operated so that $V_{GS} = 0$ or negative.

V_{P0}	:	Pinch off voltage when $V_{GS} = 0$.
		0 indicate $V_{GS} = 0$.
$I_{DSS} (I_{P0})$	:	Saturation value of the drain current when gate is shorted to source $(V_{GS} = 0)$.
I_{DSS}	:	Saturation value of the drain current for $V_{GS} \neq 0. -1, -2$ etc.
BV_{DSS}	:	Breakdown voltage V_{DS}, when gate is shorted to source.
$I_{D(off)}$	:	Drain current when the JFET is in OFF state.
IGSS	:	Gate cut-off current when Drain is shorted to source.
g_{mo}	:	Mutual conductance g_m when $V_{GS} = 0$.

The expression for the pinch off voltage $|V_P| = \dfrac{\epsilon . N_D}{2\epsilon} . a^2$

where

N_D = Donor atom concentration

ϵ = Permitivity $\epsilon_o - \epsilon_r$

a = Effective width of Silicon bar

Problem 4.2

For a p - channel Silicon FET, with $a = 2 \times 10^{-4}$ cm and channel resistivity $\rho = 10\ \Omega$ - cm. Find the pinch off voltage.

Solution

$$V_P = \frac{e.N_A}{2\epsilon}.a^2 \quad \text{(for p - channel FET)}$$

For Si,

$$\epsilon = 12\ e_o\ \mu_p = 500\ \text{cm}^2/\text{v} - \text{sec.}$$

$$\sigma = \frac{1}{\rho} \simeq \rho\mu_p \,.\, e \simeq N_A\ \mu_p \,.\, e$$

or
$$e\,.\,N_A = \frac{1}{\rho\mu_p}$$

$$= \frac{1}{10 \times 500}$$

or
$$e\,.\,N_A = 2 \times 10^{-4}$$

$\therefore$
$$V_p = \frac{2 \times 10^{-4}\left(2 \times 10^{-4}\right)^2}{2 \times 12\left(36\pi \times 10^{11}\right)^{-1}} = 3.78\text{V}$$

CUT OFF

A FET is said to be cut off when the gate to source voltage $|V_{GS}|$ is greater than the pinch off voltage V_P. $|V_{GS}| > |V_p|$. Under these conditions, because for n - channel FET, gate is p - type and source is n - type, there will be some current called ***gate reverse current*** or ***gate cut off current***, designed as I_{GS}, this is when drain is shorted to source. There exists some drain current under these conditions also called $I_{D(off)}$.

I_D (off) and I_{GSS} will be in the range 0.1amps to few nano amps at 25^0C, and increase by a factor of 1000 at a temperature of 150^0C. I_{GSS} is due to the minority carries of the reverse biased gate. Source junction. $I_{D(off)}$ is due to the carriers reaching drain from source because channel resistivity will not become infinity. Because drain is at positive potential for n-channel FET, more electrons will reach the drain.

APPLICATIONS

1. JFETs make good digital and analog switches. off resistance is very high.
2. They are used in special purpose amplifiers such as very high input resistance amplifiers, buffers voltage follower etc.
3. It can be used as a voltage controlled resistance because resistivity of the channel varies with V_{DS}.

4.4 FET SMALL SIGNAL MODEL

The equivalent circuit for FET can be drawn exactly in the same manner as that for a vacuum tube. The drain current I_D is a function of gate voltage V_{GS} and drain voltage (V_{DS})

$$I_D = f(V_{GS}\,.\,V_{DS}) \qquad .....(1)$$

Expanding (1) by Taylor's series,

$$\Delta I_D = \left.\frac{\partial I_D}{\partial V_{GS}}\right|_{V_{DS}} \Delta_{GS} + \left.\frac{\partial I_D}{\partial V_{DS}}\right|_{V_{GS}} . \Delta V_{DS}$$

In the small signal notation the incremental values can be assumed to be a.c quantities.

$$\Delta I_D = I_d ;\ \Delta V_{GS} = V_{gs} ;\ \Delta V_{DS} = V_{ds}$$

$$I_d = g_m . V_{gs} + \frac{1}{r_d} . V_{ds}$$

where, $$g_m = \left.\frac{\partial I_D}{\partial V_{GS}}\right|_{V_{DS}}$$

or $$\left.\frac{\Delta I_D}{\Delta V_{GS}}\right|_{V_{DS}} = \left.\frac{I_d}{V_{gs}}\right|_{V_{DS}}$$

The ***mutual conductance*** or transconductance.

Sometimes it is designated as y_{f_s} or g_{f_s} the second subscript 's' indicating common source and y or g for forwards transadmittance or conductance respectively.

4.4.1 Drain Resistance or Output Resistance (r_d)

Consider the characteristics shown in Fig. 4.20.

$$r_D = \left.\frac{\partial V_{DS}}{\partial I_D}\right|_{V_{GS}} \simeq \left.\frac{\partial V_{DS}}{\partial I_D}\right|_{V_{GS}} = \left.\frac{V_{ds}}{I_d}\right|_{V_{GS}}$$

$$g_D = \frac{1}{r_D}$$

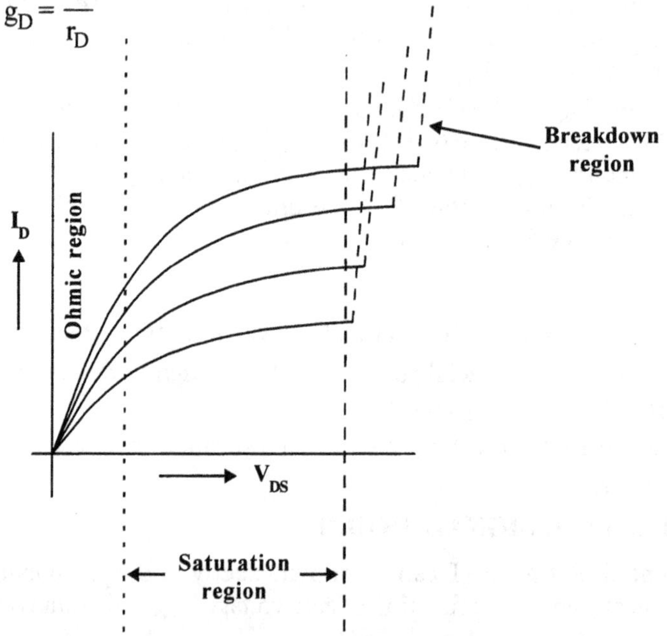

Fig 4.20 Drain characteristics of JFET.

Drain conductance = y_{DS} = output conductance for common source.

g_m of FET is analog to g_m of vacuum tube.

r_D of FET is analog to r_p of vacuum tube.

$$\mu \text{ for FET } = \left.\frac{\partial V_{DS}}{\partial V_{GS}}\right|_{I_D} = \left.\frac{\Delta V_{DS}}{\Delta V_{GS}}\right|_{I_D} = \left.\frac{V_{ds}}{V_{gs}}\right|_{I_D}$$

TYPICAL VALUES

$g_m = 0.1 - 10$ mA/V

$r_d = 0.1 - 1$ MΩ

$\mu = r_d, g_m$

$C_{gs} = 1 - 10$ pf

$C_{ds} = 0.1 - 10$ pf

EQUIVALENT CIRCUIT FOR FET

The expression for drain current I_d in the case of a FET is

$$I_d = g_m V_{gs} + \frac{1}{r_d}.V_{ds}$$

This is similar to the expression I_p of a triode. So the equivalent circuit can be drawn as a current source in parallel with a resistance. Between gate and source the capacitance is C_{gs} and C_{gd} is the barrier capacitance between gate and drain. C_{ds} represents the drain to source capacitance of the channel. (see Fig. 4.21).

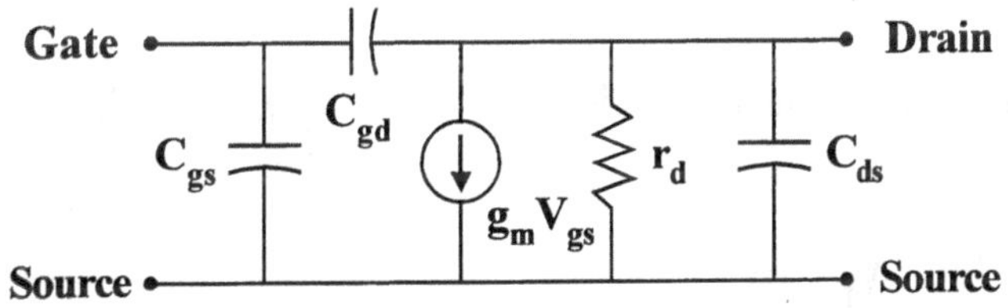

Fig 4.21 JFET equivalent circuit

This is high frequency equivalent circuit because we are considering capacitances. In low frequency circuit, X_C will be very large so C is open circuit and neglected.

FET is a voltage controlled device; by controlling the channel width by voltage, we are controlling the current. If we neglect the capacitances the equivalent circuit of FET can be drawn as given in Fig. 4.22. This is similar of a transistor ***h-parameter*** equivalent circuit.

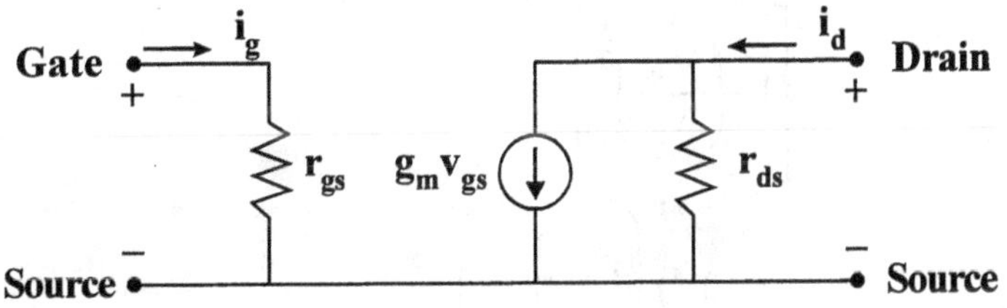

Fig 4.22 Simplified equivalent circuit

Arrow mark is downwards in the current source $g_m V_{gs}$ because the current (conventional current) is flowing from drain to source for n-channel FET (Because electrons are the majority carriers).

Problem 4.3

For a p-channel silicon FET, with $a = 2 \times 10^{-4}$ cm (effective channel width) and channel resistivity $\rho = 10\Omega - \text{cm}$. Find the pinch off voltage.

Solution

$$V_p = \frac{e.N_A}{2\epsilon}.a^2 \text{ for p-channel FET.}$$

For Si, $\epsilon = 12\,\epsilon_0$ $\qquad \mu_p = 500\ \text{cm}^2/\text{v - sec.}$

$$\sigma = \frac{1}{\rho} = p\,\mu_p \,.\, e = N_A\,\mu_p\,.\,e$$

or $$e\,.\,N_A = \frac{1}{\rho\mu_p} = \frac{1}{10\times 500}$$

or $$e\,.\,N_A = 2\times 10^{-4}$$

$\therefore$ $$V_p = \frac{2\times 10^{-4}\left(2\times 10^{-4}\right)^2}{2\times 12\left(36\pi\times 10^{11}\right)^{-1}} = 3.78\text{ V}$$

Problem 4.4

For an n-channel silicon FET, with $a = 3 \times 10^{-4}$ cm and $N_D = 10^{15}$ electrons/cm^3. Find (a) The pinch off voltage. (b) The channel half width for $V_{GS} = \frac{1}{2} V_P$ and $I_D = 0$.

Solution

(a) $$V_p = \frac{e.N_D}{2\epsilon}.a^2$$

$$= \frac{1.6\times 10^{-19}\times 10^{15}\times\left(3\times 10^{-4}\right)^2}{2\times 12\times\left(36\pi\times 10^{11}\right)^{-1}}$$

$$= 6.8\text{V}$$

(b) $$b = ?\ \text{for}\quad V_{GS} = \frac{V_P}{2},$$

$$b = a\left[1-\left(\frac{V_{GS}}{V_P}\right)^{\frac{1}{2}}\right]$$

$$= (3\times 10^{-4})\left(1-\left(\frac{1}{2}\right)^{1/2}\right)$$

$$= 0.87\times 10^{-4}\text{cm}$$

Problem 4.5

Show that for a JFET,

$$g_m = \frac{2}{|V_p|}\sqrt{I_{DSS}.I_{DS}}$$

Solution

$$I_{DS} = I_{DSS}\left(1 - \frac{V_{GS}}{V_p}\right)^2$$

$$g_m = \frac{\partial I_d}{\partial V_{GS}}$$

$$= -\frac{2I_{DSS}}{V_p}\left(1 - \frac{V_{GS}}{V_p}\right)$$

$$= g_m\left(1 - \frac{V_{GS}}{V_p}\right)$$

where g_{mo} is the transistor conductance g_m for $V_{GS} = 0$.

$$g_{mo} = -\frac{2I_{DSS}}{V_P}$$

∴ $$g_m = -\frac{2I_{DSS}}{V_P}\left(1 - \frac{V_{GS}}{V_P}\right)$$

But $$I_{DS} = I_{DSS}\left(1 - \frac{V_{GS}}{V_P}\right)^2$$

or $$\left(1 - \frac{V_{GS}}{V_P}\right) = \sqrt{\frac{I_{DS}}{I_{DSS}}}$$

∴ $$g_m = -\frac{2I_{DSS}}{V_p} \cdot \sqrt{\frac{I_{DS}}{I_{DSS}}}$$

$$= \frac{2}{|V_P|}\sqrt{I_{DSS}.I_{DS}}$$

Problem 4.6

Show that for small values of V_{GS}, compared with V_P, the drain current is $I_D \simeq I_{DSS} + g_{mo} V_{GS}$.

Solution

$$I_D \simeq I_{DSS} + g_{mo} V_{GS}$$

$$I_D = I_{DSS}\left(1 - \frac{V_{GS}}{V_P}\right)^2$$

$$= I_{DSS}\left(1 - \frac{2V_{GS}}{V_P} + \frac{V_{GS}^2}{V_P^2}\right)$$

$$\simeq I_{DSS}\left(1 - 2\frac{V_{GS}}{V_P}\right) V_{GS}$$

$$\simeq I_{DSS} - \left(\frac{2I_{DSS}}{V_P}\right)$$

$$I_D = I_{DSS} + g_{mo} V_{GS} \quad \left[\because g_{mo} = \frac{-2I_{DSS}}{V_P}\right]$$

4.5 FET TREE

The input Z of JFET is much higher compared to a junction transistor. But for MOSFETs, the input Z is much higher compared to JFETs. So these are very widely used in ICs and are fast replacing JFETs (Fig. 4.23).

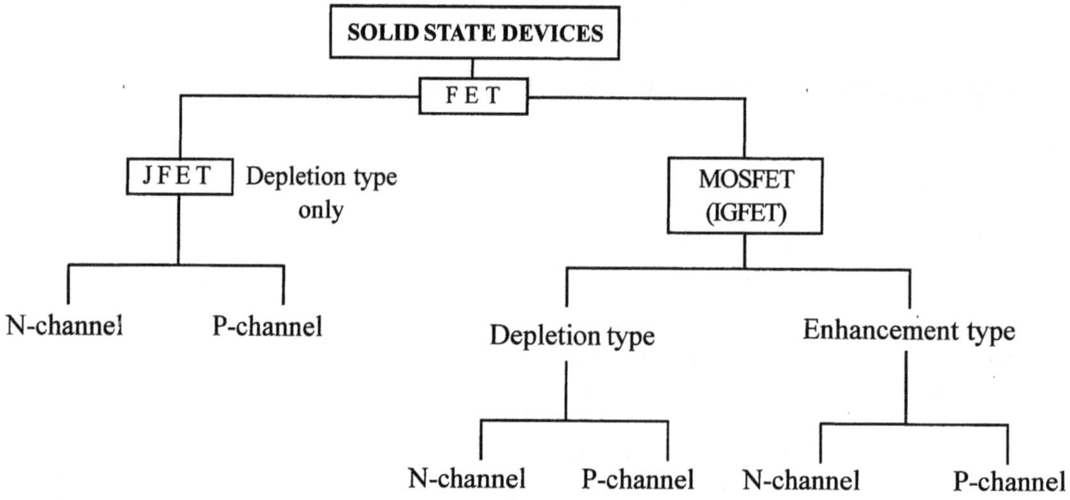

Fig 4.23 FET tree

There are two types of MOSFETs, enhancement mode and depletion mode. These devices, derive their name from Metal Oxide Semiconductor Field Effect Transistor (MOSFET) or MOS transistor (see Fig. 4.23). These are also known as IGFET (Insulated Gate Field Effect Transistor). The gate is a metal and is insulated from the semiconductor (source and drain are semiconductor type) by a thin oxide layer.

In FETs the gate source junctions is a p-n junction. (If the source is n-type, gate will be p-type). But in MOSFETs, there is no p-n junction between gate and channel, but rather a capacitor consisting of metal gate contact, a dielectric of SiO_2 and the semiconductor channel. [two conductors separated by dielectric]. It is this construction which accounts for the very large input resistance of 10^{10} to $10^{15}\Omega$ and is the major difference from the JFET.

The symbols for MOSFETs are,

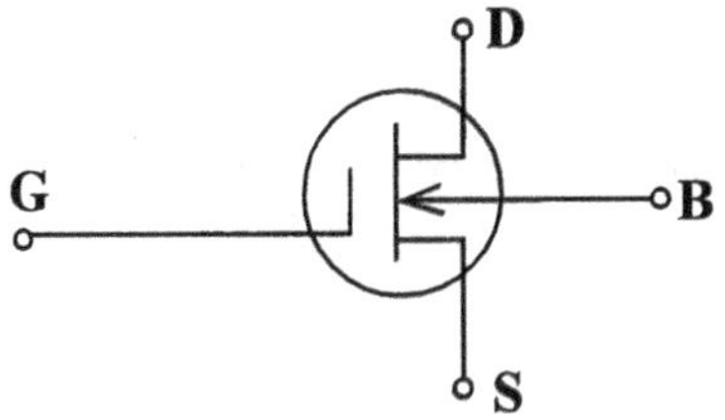

Fig 4.24 n-channel MOSFET, depletion type

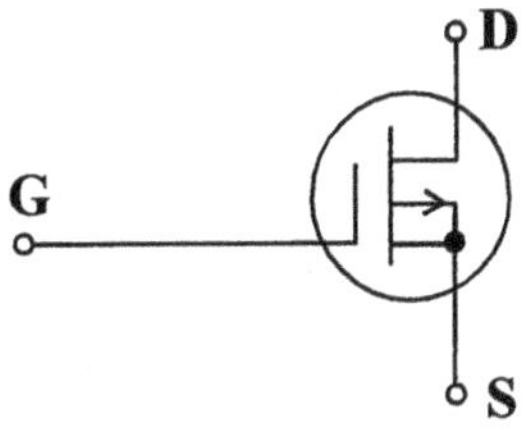

Fig 4.25 p-channel MOSFET, depletion type

Some manufacturers internally connect the bulk to the source. But in some circuits, these two are to be separated (See Fig. 4.24). The symbol, if they are connected together is, given in Fig.4.25.

for p - channel, the arrow will point outwards.

Bulk is the substrate material taken.

Advantages of MOSFETs (over JFETs and other devices)

1. High package density $\simeq 10^5$ components per square cm.
2. High fabrication yield. p - channel Enhancement mode devices require 1 diffusion and 4 photo masking steps.

   But for bipolar devices, 4 diffusions and 8 – 10 photo masking steps are required.
3. Input impedance is very high $Z_n \simeq 10^{14}\Omega$.
4. Inherent memory storage : charge in gate capacitor can be used to hold enhancement mode devices ON.
5. CMOS or NMOS reduces power dissipation, micropower operation. Hence, CMOS ICs are popular.
6. Can be used as passive or active element.

   Active : As a storage device or as an amplifier etc.

   Passive : As a resistance or voltage variable resistance.
7. Self Isolation : Electrical isolation occurs between MOSFETs in ICs since all p-n junctions are operated under zero or reverse bias.

Disadvantage

1. Slow speed switching.
2. Slower than bipolar devices.
3. Stray and gate capacitance limits speed.

4.5.1 Enhancement Type MOSFET

n-channel is induced, between source and drain. So it is called as n-channel MOSFET.

For the MOSFET shown in Fig. 4.26, source and drain are n-type. Gate is A*l* (metal) in between oxide layer is there. The source and drain are separated by 1 mil. Suppose the p substrate is grounded, and a positive voltage is applied at the gate. Because of this an electric field will be directed perpendicularly through the oxide. This field will induce negative charges on the semiconductor side. The negative charge of electrons which are the minority carriers in the p - type substrate form an inversion layer. As the positive voltage on the gate increases, the induced negative charge in the semiconductor increases. The region below the SiO_2 oxide layer has n-type carriers. So the conductivity of the channel between source and drain increases and so current flows from source to drain through the induced channel. Thus, the drain current is enhanced by the positive gate voltage and such a device is called an ***enhancement type MOS***.

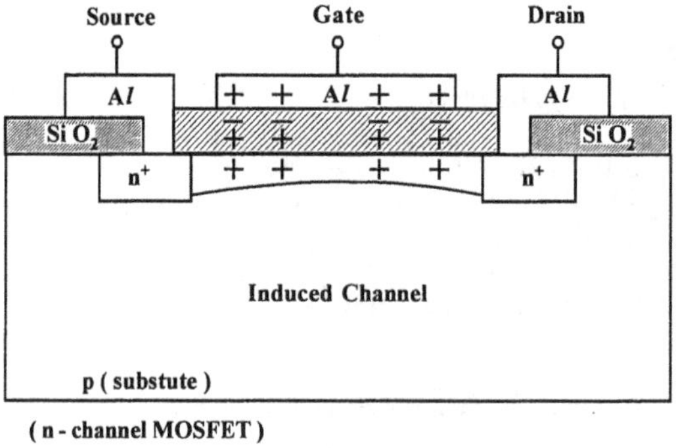

Fig 4.26 n-channel MOSFET

The volt-ampere drain characteristic of an n-channel enhancement mode MOSFET are as shown in Fig. 4.27.

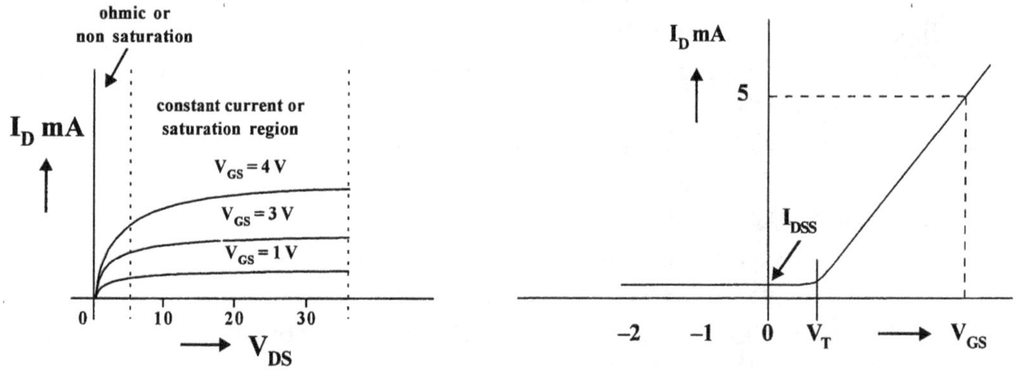

(a) Drain characteristics *(b) Gate characteristics*

Fig 4.27

The current I_D for $V_{GS} \leq 0$ is very small being of the order of few nano amperes. As V_{GS} is made positive, the current I_D increases, slowly first and then much more rapidly with an increase in V_{GS} (Fig. 4.27(b)). Sometimes the manufacturers specify gate, source threshold voltage V_{GST} at which I_D reaches some defined small value $\simeq 10\mu A$. For a voltage $< V_{GST}$, (V_{GS} threshold) I_D is very small. $I_{D(ON)}$ of MOSFET is the maximum value of I_D which remains constant for different values of V_{DS}.

In the ohmic region, the drain characteristic is given by

$$I_D = \frac{\mu C_0 . w}{2L} [2(V_{GS} - V_T) V_{DS} - V^2_{DS}]$$

where, μ = Majority carriers mobility.
C_0 = Gate capacitance per unit area.
L = Channel length.
w = Channel width perpendicular to C

The ohmic region in the I_D vs V_{DS} characteristic of FET or MOSFET is also known as ***triode region***, because like in a triode I_D increases with V_{DS}. The constant current region is known as ***pentode region*** because the current remains constant with V_{DS}.

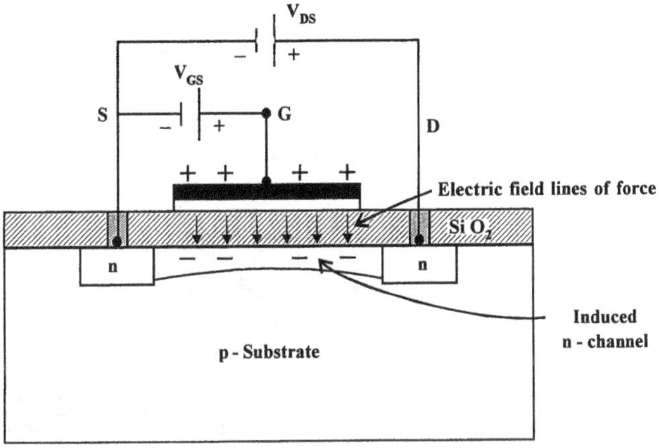

Fig 4.28 MOSFET structure

4.5.2 MOSFET OPERATION

Consider a MOSFET in which source and drain are of n-type (Fig. 4.28). Suppose a negative potential is applied between gate and source. SiO_2 is an insulator. It is sandwiched between two conducting regions the gate (metal) and the S.C p-type substrate. Therefore, equivalent capacitor is formed, with SiO_2 as the dielectric whenever a positive charge is applied to one plate of a capacitor a corresponding negative charge is induced on the opposite plate by the action of the electric field with in the dielectric. A positive potential is applied to the gate. So a negative charge is induced on the opposite plate by the action of electric field within the dielectric, in the p - substrate. This charge results from the minority carriers (electrons) which are attracted towards the area below the gate, in the p - type substrate. The electrons in the p - substrate are attracted towards the lower region of SiO_2 layer because there is positive field acting from the gate through SiO_2 layer, because of the applied positive potential to the gate V_{GS}. As more number of electrons are attracted towards this region, the hole density in the p - substrate (below the SiO_2 layer between n - type source and drain only) decrease. This is true only in the relatively small region of the substrate directly below the gate. An n - type region now extends continuously from source to drain. The n-channel below the gate is said to be induced because it is produced by the process of electric inductor. If the positive gate potential is removed, the induced channel disappears.

If the gate voltage is further increased, greater number of negative charge carriers are attracted towards the induced channel. So as the carriers density increases, the effective channel resistance decreases, because there are large number of free carriers (electrons). So the resistance seen by the V_{DS} depends on the voltage applied to the gate. The higher the gate potential, the lower the channel resistance, and the higher drain current I_D. This process is referred to as enhancement because I_D increases, and the resulting MOSFET is called ***enhancement type MOSFET***. The resistance looking into the gate is high since the oxide is an insulator. The resistance will be of the order of $10^{15}\Omega$ and the capacitance value will also be large. (Fig. 4.29).

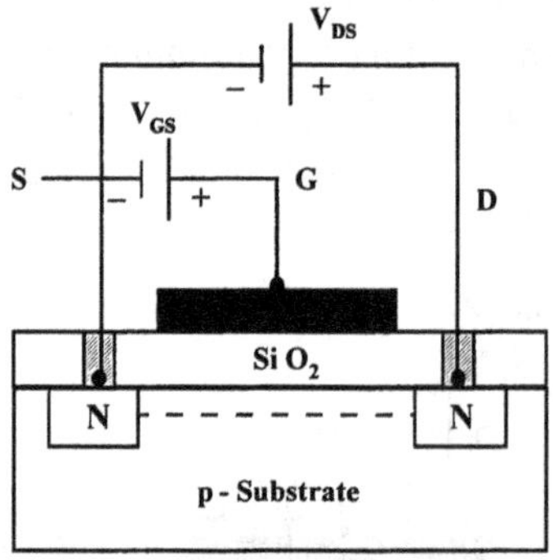

Fig 4.29 MOSFET biasing

Depletion type MOSFET can also be constructed in the same way. These are also known as DE MOSFETs. Here as the gate voltage increases, the channel is depleted of carriers. So channel resistance increases.

Here the region below the gate is doped n - type. If negative voltage is applied to the gate, negative charge on the gate induces equal positive charge i.e., holes. These holes will recombine with the electrons of the n - channel between sources drain since channel resistivity increases. The channel is depleted of carriers. Therefore I_D decreases as V_{GS} increases. (negative voltage) If we apply positive voltage to V_{GS}, then this becomes enhancement type.

4.5.3 MOSFET CHARACTERISTICS

There are two types of MOS FETs,

1. Enhancement type
2. Depletion type.

Depletion type MOSFET can also be used as enhancement type. But enhancement type cannot be used as depletion type. So to distinguish these two, the name is given as depletion type MOSFET which can also be used in enhancement mode.

As V_{GS} is increased in positive values (0, +1, +2 etc) if I_D increases, then it is enhancement mode of operation because I_D is enhanced or increased.

As V_{GS} is increased in negative values (0, –1, –2, etc) if I_D decreases, then it is depletion mode of operation.

Consider n - channel MOSFET, depletion type. Fig. 4.30.

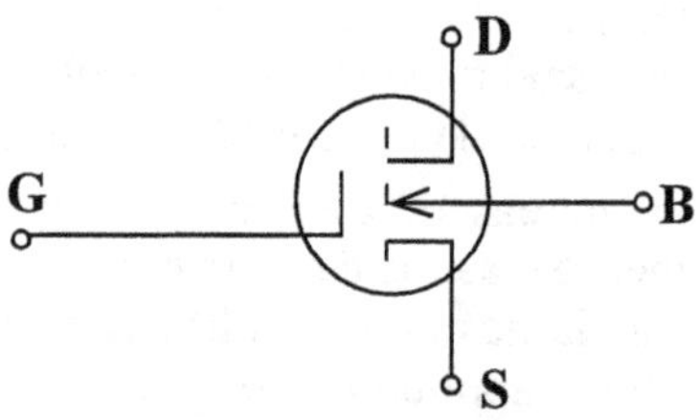

Fig 4.30 n-channel MOSFET

4.5.4 Drain Characteristic : I_D Vs. V_{DS}

When V_{GS} = OV, for a given V_{DS}, significant current flows just like in a JFET. When the gate is made negative (i.e., $V_{GS} = -1V$,) it is as if a negative voltage is applied for one plate of plate capacitor. So a positive charge will be induced below the gate in between the n type source and drain. Because it is semiconductor electrons and holes are induced in it below the gate. The channel between the source and drain will be depleted of majority carriers electrons, because these induced holes will recombine with the electrons. Hence, the free electron concentration in the channel between source and drain decreases or channel resistivity increases. Therefore current decreases as V_{GS} is made more negative i.e., –1, –2, etc. This is the depletion mode of operation, because the channel will be depleted of majority carriers as V_{GS} is made negative (for n-channel MOSFET).

In a FET, there is a p - n junction between gate and source. But in MOSFET, there is no such p - n junction. Therefore, positive voltage positive V_{GS} can also be applied between gate and source. Now a negative charge is induced in the channel, thereby increase free electron

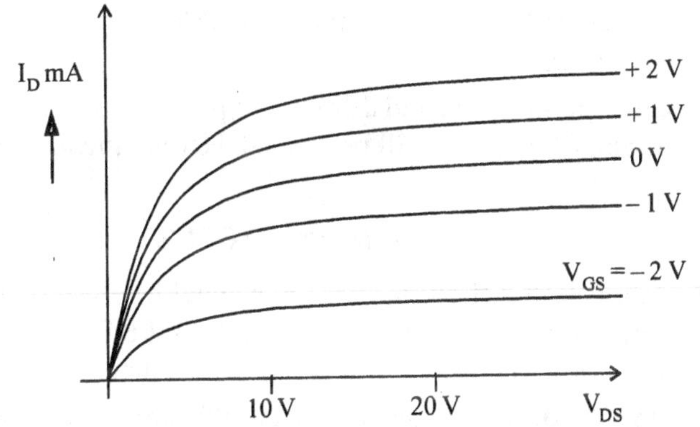

Fig 4.31 Drain characteristics (N MOS)

concentration. So channel conductivity increases and hence I_D increases. Thus, the current is enhanced. So, the device can be used both in enhancement mode and depletion mode.

In the transfer characteristics as (Fig. 4.32) V_{GS} increases, V_D increases. Similarly in the depletion mode as V_{GS} is increases in negative values, I_D decreases

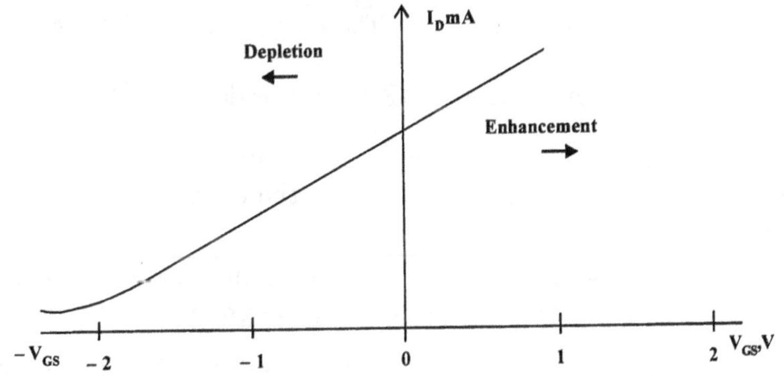

Fig 4.32 Gate characteristics

A JFET is a depletion type, because as V_{GS} is made positive (+1, +2 etc for n-channel FET) I_D increases. So it can be used in enhancement type. But there is no other type of JFET which is used only in enhancement mode and no depletion mode, hence no distinction is made.

4.5.5 MOSFET Gate Protection

The SiO_2 layer of gate is extremely thin, it will be easily damaged by excessive voltage. If the gate is left open circuited, the electric field will be large enough (due to accumulation of charge) to cause punch through in the SiO_2 layer or the dielectric. To prevent this damage, some MOS devices are fabricated with zener diode between gate and substrate. When the potential at the gate is large, the zener will conduct and the potential at the gate will be limited to the zener breakdown voltage. When the potential at the gate is not large, the zener is open circuited and has no effect on the device.

If the body (bulk) of MOS transistor (MOSFET) is p - type Silicon, and if two 'n' regions separated by the channel length are diffused into the substrate to form source and drain n-channel enhancement device (designated as NMOS) is obtained. In NMOS, the induced mobile channel surface charges would be ***electrons***.

Similarly, if we take n -type substrate and diffuse two p - regions separated by the channel length to form source and drain, PMOSFET will be formed. Here in the induced mobile channel surface charges would be ***holes***.

4.5.6 Comparison of p-channel and n-channel MOSFETs

Initially there were some fabrication difficulties with n-channel MOSFETs. But in 1974, these difficulties were overcame and mass productions of n-channel MOSFETs began. Thus NMOSFETs have replaced PMOSs and PMOSFETs have almost become obsolete.

The hole mobility in Silicon at normal fields is 500 cm^2/v.sec. Electron mobility = 1300cm^2/v-Sec. Therefore p-channel ON resistance will be twice that of n-channel MOSFET ON resistance (ON resistance means the resistance of the device when I_D is maximum for a given V_{DS}) ON resistance depends upon 'μ' of carriers because $\sigma = \frac{1}{\rho} = ne\mu_n$ or $pe\mu_p$.

If the ON resistance of a p-channel device were to be reduced or to make equal to that of n - channel device, at the same values of I_D and V_{DS} etc, then the p-channel device must have more than twice the area of the n-channel device. Therefore n - channel devices will be smaller or packing density of n-channel devices is more ($R = \frac{\rho l}{A}$ .R is decreased by increase in A).

The second advantage of the NMOS devices is fast switching. The operating speed is limited by the internal RC time constant of the device. The capacitance is proportional to the junction cross sections.

The third advantage is NMOS devices are TTL compatible since the applied gate voltage and drain supply are positive for an n - channel enhancement MOS.

(Because in n-channel MOS, source and drain are n-type. So drain is made positive. For enhancement type the gate which is Al metal is made positive).

4.5.7 ADVANTAGES OF NMOS OVER PMOS

1. NMOS devices are fast switching devices since electron mobility is less than holes.
2. NMOS devices are TTL compatible since V_{GS} and V_D to be applied for NMOS devices are positive
3. Packing density of NMOS devices is more
4. The ON resistance is less because conductivity of NMOS devices is more since μ of electrons is greater than that of holes.

4.5.8 THE DEPLETION MOSFET

In enhancement type MOSFET, a channel is not diffused. It is of the same type (p-type or n-type) as the bulk or substrate. But if a channel is diffused between source and drain with the same type of impurity as used for the source and drain diffusion depletion type MOSFET will be formed. (Fig. 4.33).

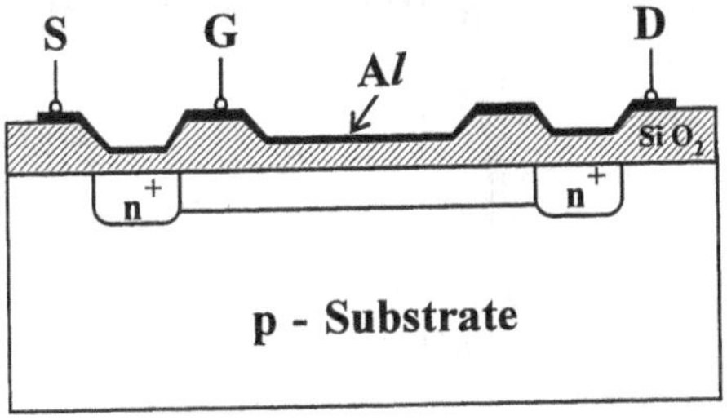

Fig 4.33 Depletion MOSFET.

The conductivity of the channel in the case of depletion type is much less compared to enhancement type. The characteristics of depletion type MOSFETs are exactly similar to that of JFET.

When V_{DS} is positive and $V_{GS} = 0$, large drain current denoted as I_{DSS} (Drain to source current saturation value) flows. If V_{GS} is made negative, positive charges are induced in the channel through SiO_2 of the gate capacitor. But the current in MOSFET is due to majority carriers. So the induced positive charges in the channel reduces the resultant current I_{DS}. As V_{GS} is made more negative, more positive charges are induced in the channel. Therefore its conductivity further decreases and hence I_D decreases as V_{GS} is made more negative (for n-channel depletion type). The current I_D decreases because, the electrons from the source recombine with the induced positive charges. So the number of electrons reaching the drain reduces and hence I_D decreases. So because of the recombination of the majority carriers, with the induced charges in the channel, the majority carriers will be depleted. Hence, this type of MOSFET is knows as ***depletion type MOSFET*** (Fig. 4.33). JFET and depletion MOSFET have identical characteristics.

A MOSFET of depletion type can also be used as enhancement type. In the case of n-channel depletion type (source and drain are n-type), if we apply positive voltage to the gate-source junction, negative charges are induced in the channel. So, the majority carriers (electrons in the source) are more and hence I_D will be very large. Thus, depletion type MOSFET can also be used as enhancement type by applying positive voltage to the gate (for n-channel type). The drain characteristics of DMOSFET are shown in Fig. 4.34 and Gate characteristics are shown in Fig. 4.35.

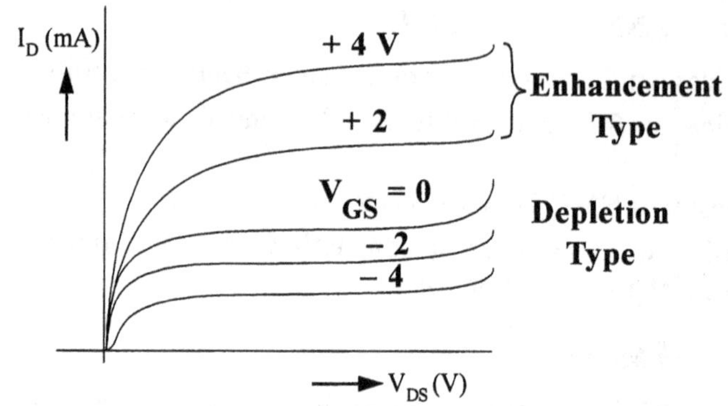

Fig 4.34 Drain characteristics of DMOSFET.

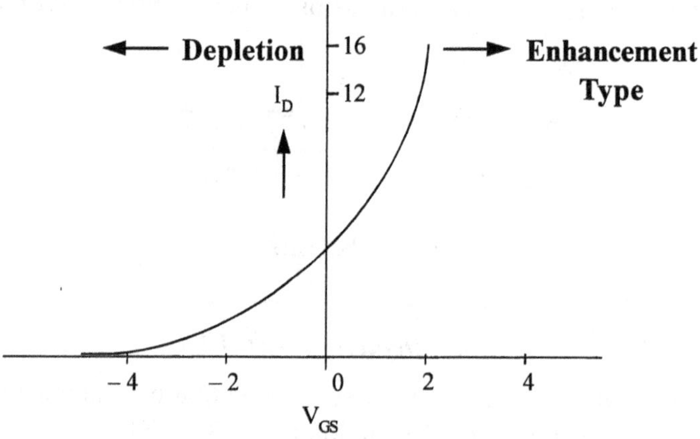

Fig 4.35 Gate characteristics of depletion type MOSFET.

Problem 4.7

A JFET is to be connected in a circuit with drain load resistor of 4.7 kΩ, and a supply voltage of V_{DD} = 30V. V_D is to be approximately 20V, and is to remain constant within ± 1V. Design a suitable self bias circuit (Fig. 4.36).

Solution

$$V_D = V_{DD} - I_D \cdot R_L$$

$$I_D = \frac{V_{DD} - V_D}{R_L}$$

$$= \frac{30\text{V} - 20\text{V}}{4.7\text{k}\Omega} = 2.1 \text{ mA}$$

for V_D to be constant, it should be within ± 1V,

$$\Delta I_D = \frac{\pm V}{R_L}$$

$$= \frac{\pm 1\text{V}}{4.7\text{k}\Omega} \simeq \pm 0.2\text{mA}$$

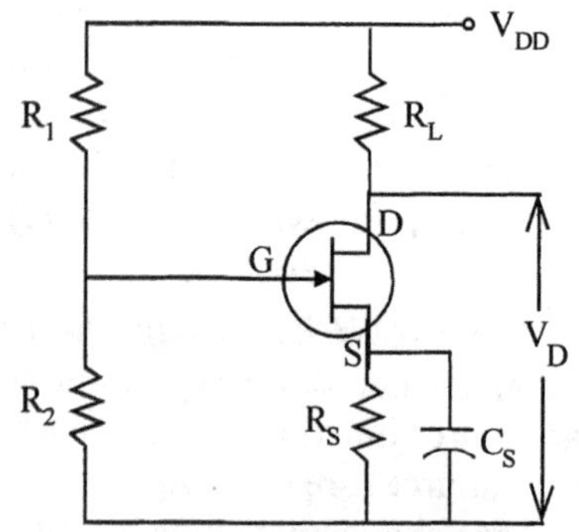

Fig 4.36 For Problem 4.7.

$$I_D = (2.1 \pm 0.2mA)$$
$$I_D(min) = (2.1 - 0.2) = 1.9mA$$
$$I_D\,(max) = (2.1 + 0.2) = 2.3mA$$

Indicate the points A and B on the maximum and minimum transfer characteristics of the FET. Join these two points and extend it till it cuts at point C. (Fig. 4.37).

The reciprocal of the slope of the line gives R_S

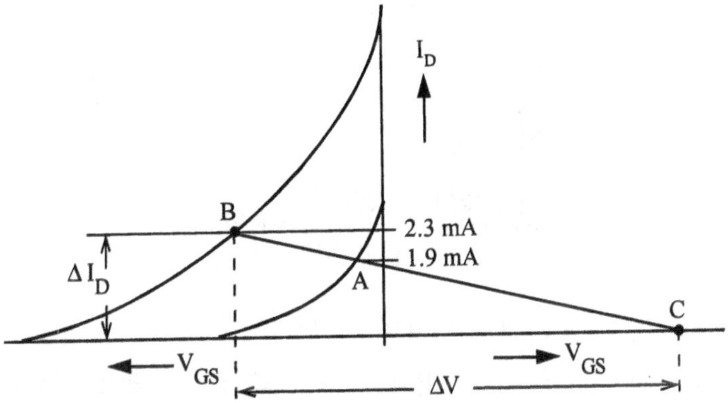

Fig 4.37 For Problem 4.7

$$R_S = \frac{\Delta V}{\Delta I} = \frac{10V}{2.5mA} = 4k\Omega$$

The bias line intersects the horizontal axis at $V_G = 7V$

∴ An external bias of 7V is required.

$$V_G = \frac{R_2}{R_1 + R_2}.V_{DD}$$

$$\therefore \quad \frac{R_2}{R_1 + R_2} = \frac{V_G}{V_{DD}} = \frac{7V}{30v}$$

$$\therefore \quad \frac{R_2}{R_1} = \frac{7}{23}$$

R_2 and R_1 should be large to avoid overloading input signals. If $R_2 = 700$ kΩ,

$$R_1 = 2.3\ M\Omega$$

Problem 4.8

(a) For the common source amplifier, calculate the value of the voltage gain, given

$r_d = 100$ kΩ, $g_m = 300\ \mu$, $R_L = 10k\Omega$, $R_o = 9.09k\Omega$

Solution

$$A_V = \frac{-g_m.r_d.R_L}{rd + R_L} = \frac{-3000 \times 10^{-6} \times 100 \times 10^3 \times 10 \times 10^3}{(100 \times 10^3) + (10 \times 10^3)} = -27.3$$

(b) If C_{DS} = 3pf, determine the output impedance at a signal frequency of 1MHz.

Solution

$$X_C = \frac{1}{2\pi f\ C_{DS}}$$

f = 1 MHz,

$$X_C = \frac{1}{2\pi \times 1 \times 10^6 \times 3 \times 10^{-12}} = 53\text{k}\Omega$$

$$|Z_0| = \frac{R_0 . X_C}{\sqrt{R_0^2 + X_C^2}}$$

$$|Z_0| = \frac{9.09\text{k}\Omega \times 53\text{k}\Omega}{\sqrt{(9.09\text{k}\Omega)^2 + (53\text{k}\Omega)^2}} = 8.96\text{k}\Omega$$

CMOS STRUCTURE (COMPLEMENTARY MOS)

This is evolved because of circuits using both PMOS and NMOS devices. It is the most sophisticated technology. CMOS devices incorporate p-channel MOSFET and n-channel MOSFET. The advantages that we get with this complementary use of transistor are :

1. Low power consumption.
2. High speed.

But the disadvantages are

1. Fabrication is more difficult. More number of oxidation and diffusion steps are involved.
2. Fabrication density is less. Less number of devices per unit area, because the device chip area is more.

MNOS STRUCTURE

In this process, we will have Silicon semiconductor, SiO_2 over it and then layer of Silicon Nitride (Si_3N_4) and a metal layer. So we get the ***Metal, Nitride,*** (Si_3N_4) Oxide (SiO_2) and Semiconductor structure. Hence, the name MNOS. So the dielectrics between metal (Al) and semi-conductor (Si) is a sandwich of SiO_2 and Si_3N_4, whereas in ordinary MOSFETs we have only SiO_2. (Fig. 4.38).

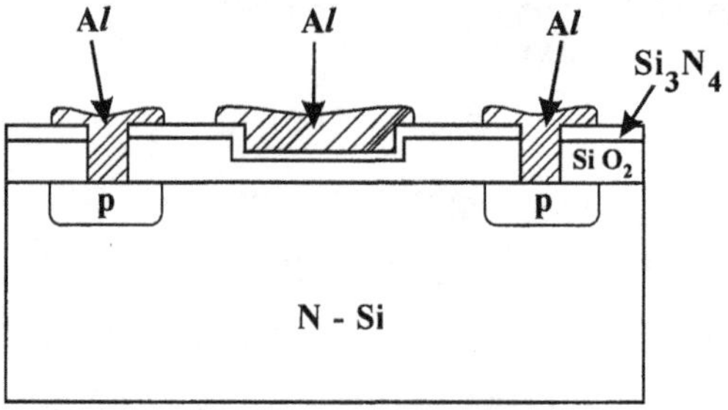

Fig 4.38 MNOS FET.

The advantages are :

1. Low threshold voltage V_T. (V_T is the voltage of the drain, (V_D) beyond which only the increase in drain current I_D will be significant)
2. Capacitance per unit area of the device is more compared to MOS structure. Because the dielectric constant will have a different value since because C is more, charge storage is more. It is used as a memory device.

SOS - MOS Structure

This is ***Silicon On Saphire*** MOS structure. The substrate used (Fig.4.39) is the silicon crystal grown on Saphire subtrate. For such a device, the parasitic capacitance will be very low.

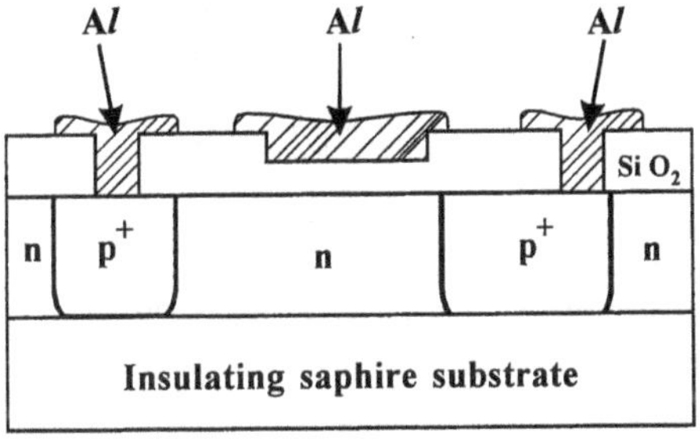

Fig 4.39 SOS – MOS structure.

Voltage Breakdown in JFETs

There is a p-n junction between gate and source and gate and drain. Just as in the case of a p-n junction diode if the voltage applied to the p-n junction of G and S, or G and D junction or D and S junctions increase, avalanche breakdown will occur. (Fig. 4.40).

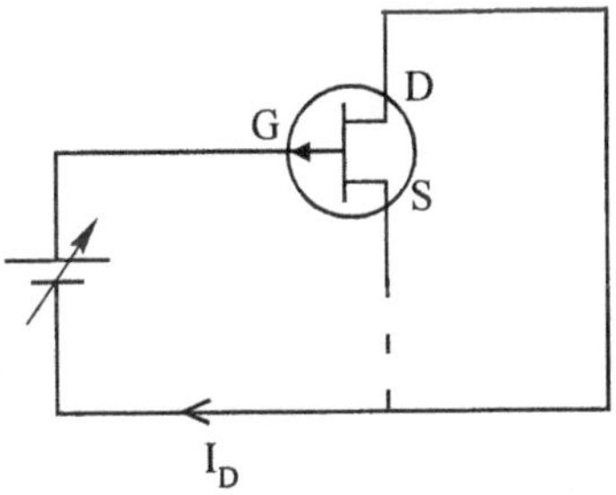

Fig 4.40 JFET circuit.

BV_{DGO} :

This is the value of V_{DG} that will cause junction breakdown, when the source is left open i.e., $I_S = 0$

BV_{GSS} :

This is the value of V_{GS} that will cause breakdown when drain is shorted to source. When drain is shorted to source, there will not be much change in the voltage which causes breakdown.

$$\text{So, } BV_{GSS} = -BV_{DGO}$$

For a p-channel FET, Gate is n-type, S and D are p - type. A positive voltage is applied to G and negative voltage to drain. Therefore with respect to D, V_{DG} is negative voltage. V_{GS} is positive voltage because G is at positive potential with respect to source. Since $BV_{GSS} = -BV_{DGO}$ gate junction breakdown can also result by the application of large V_{DS}.

$$V_{DS} = V_{DG} + V_{GS}$$

$\therefore$ More V_{DS} that can be applied to a FET = Max. V_{DG} + Max. V_{GS}

BV_{DS} X = Breakdown voltage V_{DS} for a given value of V_{GS}.

X indicates a specific value of V_{GS}

$$BV_{DSX} = BV_{DGO} + V_{GSX}$$

Problem 4.9

Determine the values of V_{GS}, I_D, and V_{DS} for the circuit shown in using the following data.

$$I_{DSS} = 5mA;\ V_P = -5\ V;\ R_S = 5k\Omega;\ R_L = 2k\Omega;\ V_{DD} = 10V$$

Solution

The circuit is shown in Fig. 4.41.

The gate is at DC ground potential, since no DC voltage is applied between gate and Source.

$\therefore$ $V_{GS} + I_S \cdot R_S = 0$

But $I_S \simeq I_D$, I_G is negligible

$\therefore$ $V_{GS} + I_D R_S = 0$

or $V_{GS} = -I_D \cdot R_S$

$= -5000\ I_D$.

V_{GS} is unknown and I_D is unknown.

$$I_D = I_{DSS}\left(1 - \frac{V_{GS}}{V_P}\right)^2$$

$$= 5 \times 10^{-3}\left(1 - \frac{V_{GS}}{-5}\right)^2$$

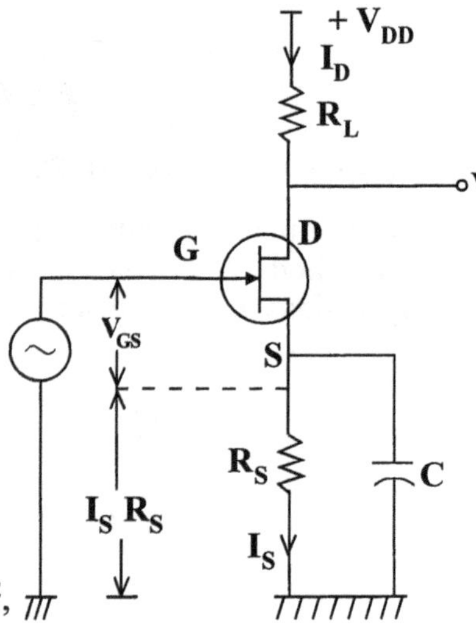

Fig 4.41 For Problem 4.9

But $I_D = -\dfrac{V_{GS}}{5000}$

$V_{GS} = (-5000)(5 \times 10^{-3})(1 + 0.2V_{GS})^2$,

$V_{GS}^2 + 11\ V_{GS} + 25 = 0$

$V_{GS} = -3.2V$ and $V_{GS} = -7.8V$.

The pinch off voltage V_P for which I_D becomes zero is given as 5V therefore V_{GS} cannot be 7.8 V. Therefore $V_{GS} = -3.2V$.

$\therefore$ $I_D = (5 \times 10^{-3})(1 - 0.64)^2 = 0.65$ mA

$I_D \cdot R_L = 2000\ I_D = 1.3V$

$I_D \cdot R_S \simeq 5000\ I_D = 3.25V$

$\therefore$ $V_{DS} = 10 - 1.3 - 3.25 = 5.45V$.

The FET must be conducting.

If $V_{GS} = -7.8V$, the FET in cut off. Therefore $V_p = -5V$. Therefore V_{GS} is chosen as $-3.2V$.

4.6 COMMON SOURCE AMPLIFIER (C.S)

Capacitance and DC voltages are shorted. FET is replaced by its small signal model so the equivalent circuit is as shown in Fig. 4.42.

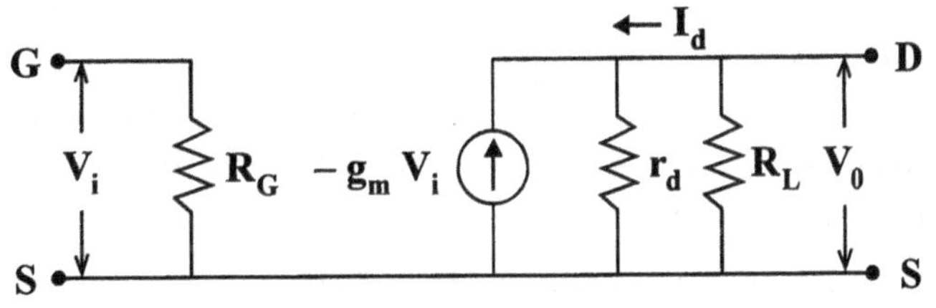

Fig 4.42 C.S amplifier.

I_d = AC drain current

r_d = drain resistance

$$V_0 = I_d \times (r_d \,||\, R_L)$$

$$= I_d \times \left[\frac{r_d . R_L}{r_d + R_L}\right]$$

But $I_d = - g_m V_i$ (The value of the current source)

$$V_0 = - g_m . V_i \left\{\frac{r_d \times R_L}{r_d + R_L}\right\}$$

∴ Voltage = Av

$$= \frac{V_0}{V_i}$$

$$= \frac{-g_m . r_d . R_L}{rd + R_L}$$

But $r_d >> R_L$

∴ $r_d + R_L \simeq r_d$

∴ $Av \simeq \frac{-g_m . r_d . R_L}{r_d}$

$$Av \simeq - g_m . R_L$$

For a FET, Y_{fs} = forward transfer admittance in C.S. configuration

$= g_m$

Y_{OS} = Out put admittance in C.S configuration

$= \frac{1}{r_d}$

At low frequencies, $Z_0 = r_d \,||\, R_L$

$\simeq R_L$ ∵ r_d is large.

At high frequencies r_d and R_L are shunted by C_{ds}.

$$Z_0 = R_0 \parallel X_{ds} \qquad X_{ds} = \frac{1}{2\pi f C_{ds}}$$

At low frequencies, input impedance

$$Z_i = R_G \,.\parallel R_{GS}$$

But R_{GS} is a very large value. Therefore $Z_i \simeq R_G$.

At high frequencies, the input capacitance shunting R_G becomes effective. The actual capacitance presented to an input signal is amplified by the Miller effect.

$\therefore \qquad C_{in} = C_{gs} + (1 + Av) \,.\, C_{gd}$

where $\qquad A_v$ = circuit voltage gain $g_m . (R_L \parallel rd)]$

$$C_{in} = C_{gs} + [1 + g_m (R_L \parallel r_d)]\, C_{gd}$$

4.7 COMMON DRAIN AMPLIFIERS (C.D)

This is also known as ***source follower***. This is FET equivalent of emitter follower or common collectors transistor amplifier circuit. (Fig. 4.43).

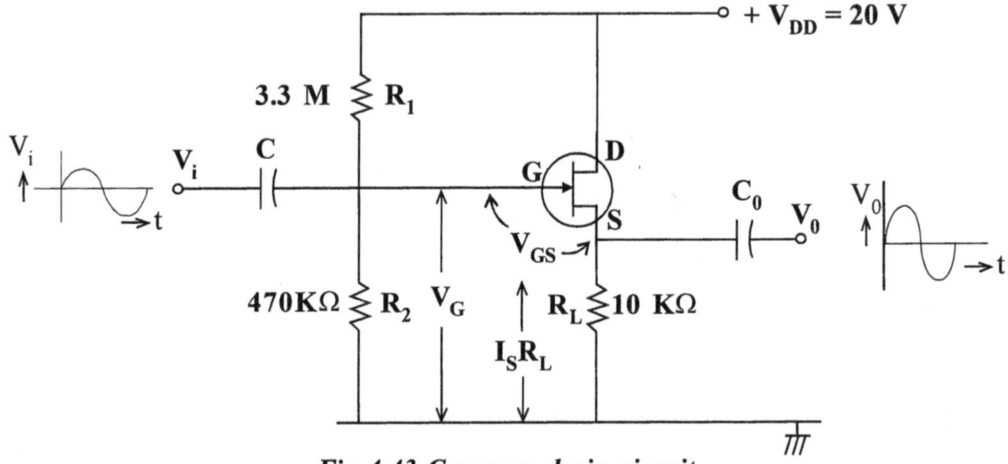

Fig 4.43 Common drain circuit.

V_G is not the same as V_{GS}. S is not at ground partial. The equivalent circuit is shown in Fig. 4.44

$$V_G = V_{GS} + V_{RL}$$
$$= V_{GS} + I_D \,.\, R_L$$

If the input increases by 1V, output also increases by IV. So, there is no phase-shift and voltage gain = 1.

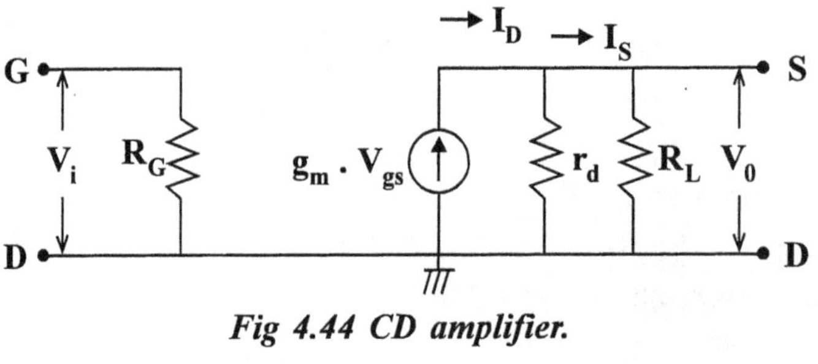

Fig 4.44 CD amplifier.

$$V_0 = I_D \times (r_d \parallel R_L)$$

$$= I_D \cdot \left\{ \frac{r_d \times R_L}{r_d + R_L} \right\}$$

$$I_D = g_m \cdot V_{gs}$$

$$V_{GS} = (V_i - V_0)$$

$$I_D = g_m (V_i - V_0) \quad \text{from the circuit}$$

$$\therefore \quad V_0 = g_m (V_i - V_0) \cdot \frac{r_d \times R_L}{r_d + R_L}$$

Solving for V_0

$$V_0 (r_d + R_L) = g_m \cdot V_i \cdot r_d \cdot R_L - g_m \cdot V_0 r_d \cdot R_L$$

$$V_0(rd + R_L + g_m rd \cdot R_L) = g_m \cdot V_i \cdot r_d \cdot R_L$$

$$V_0 = g_m \cdot V_i \cdot \left\{ \frac{r_d . R_L}{r_d + R_L + g_m r_d . R_L} \right\}$$

$$\therefore \text{ Voltage gain} \quad A_v = \frac{V_d}{V_i}$$

$$\boxed{A_V = g_m \cdot \frac{r_d . R_L}{r_d + R_L + g_m r_d R_L}}$$

If $g_m \cdot r_d \cdot R_L >> (r_d + R_L)$, $A_v \simeq 1$.

Z_0 :

$$V_0 = g_m \cdot V_i \cdot \frac{r_d . R_L}{r_d + R_L + g_m r_d . R_L}$$

$g_m V_i$ is output current proportional to V_i, and,

$$Z_0 = \frac{(r_d . R_L)}{r_d + R_L + g_m r_d R_L}$$

$$\therefore \quad Z_0 = \frac{r_d . R_L}{r_d + R_L(1 + g_m r_d)}$$

$$= \frac{[r_d / 1 + g_m r_d] R_L}{[(r_d / 1 + g_m r_d)] + R_L}$$

$$= R_L \parallel \frac{r_d}{1 + g_m r_d}$$

$$Z_0 \simeq R_L \parallel \frac{1}{g_m}$$

Z_i :

$Z_i = R_1 \parallel R_2$

C_{gd} is in parallel with Z_i

4.8 COMMON GATE FET AMPLIFIER CIRCUIT (C.G)

The circuit is shown in Fig. 4.45.

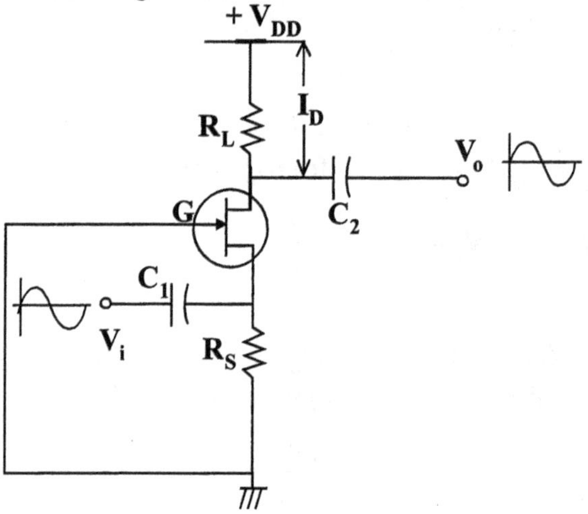

Fig 4.45 Common gate circuit.

The equivalent circuit is shown in Fig. 4.46.

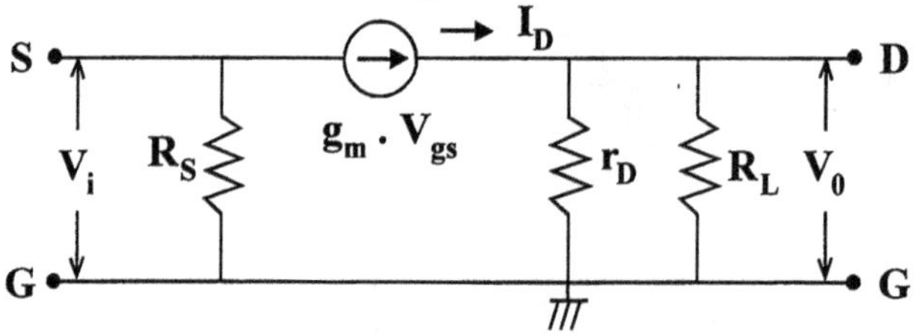

Fig 4.46 C.G configuration.

$$V_0 = I_D \times (r_d \parallel R_L)$$

$$= I_D \cdot \left\{ \frac{r_d . R_L}{r_d + R_L} \right\}$$

$$I_D = g_m \cdot V_{gs} = g_m . V_i$$

$$V_0 = g_m \cdot V_i \times \frac{r_d . R_L}{r_d + R_L}$$

$$\therefore \quad \text{voltage gain} = \boxed{A_v = \frac{V_0}{V_i} = \frac{g_m . r_d . R_L}{r_d + R_L}}$$

$$Z_0 = R_2 \parallel r_d \simeq R_L$$

$$I_D = g_m \cdot V_{gs} = g_m \cdot V_i$$

$$\therefore \quad Z_i = \frac{V_i}{I_D} = \frac{1}{g_m}.$$

This is the input impedance of the device.

The circuit input impedance is, $\frac{1}{g_m} \parallel R_S$.

4.9 COMPARISON OF FET AND BJT CHARACTERISTICS

From construction point of view triode is similar to BJT (Transistor).

Cathode E ; B → Grid; Anode → collector.

But from characteristics point of view pentode is similar to a BJT. (Fig. 4.47).

1. In FET and pentode current is due to one type of carriers only. In pentode it is only due to electrons. But in BJT it is due to both electron and holes.
2. In FET pinch off occurs and current remains constant. But no such channel closing in Transistor and Pentode.
3. As T increases, I_{CO} increases, so I in a BJT increases. But in FET, as temperature increases, 'μ' decreases. So I_D decreases. In pentode, also as temperature increases, I increases. But it is less sensitive compared to a transistor or FET.

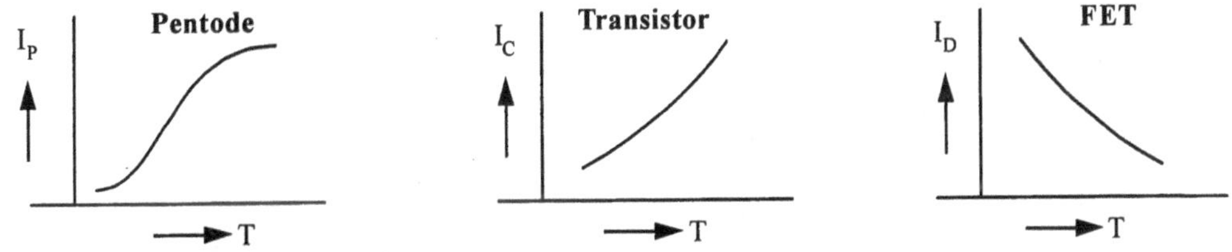

Fig 4.47

4. As I_B decreases I_C decreases in a transistor.
   As V_{GS} decreases, I_D decreases in FET.
   As $V_{control\ grid}$ increases I_p decreases in pentode.
5. FET and pentode are voltage dependent devices. Transistor is current dependent.
6. r_D of FET is very large. MΩ, r_C of transistor is less (in kΩ).
7. Breakdown occurs in FET and BJT for small voltages 50, 60V etc. But in pentode it is much higher.
8. The characteristics are shown in Fig. 4.48 and 4.49.

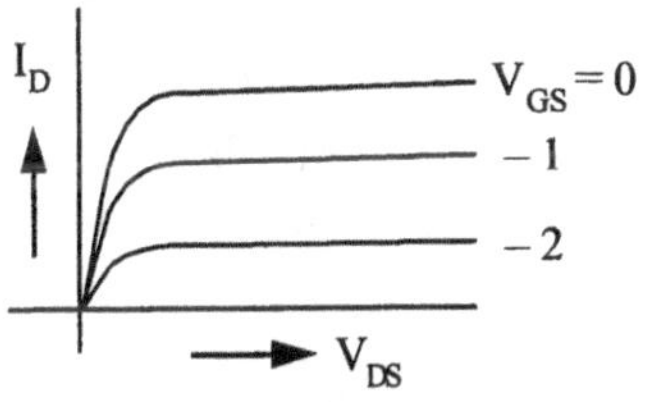

Fig 4.48

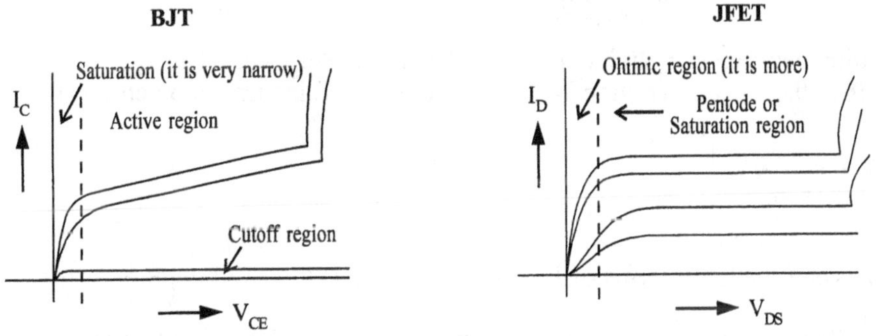

Fig 4.49

SUMMARY

- The relation between Emitter Efficiency γ, Transportation Factor β^* and Current Gain α is,

$$\alpha = \beta^* \times \gamma$$

- Transistor is an acronym for the words Transfer Resistor. As the input side in forward biased and output side is reverse biased, there is transfer of resistance from a lower value on input side to a higher value on the output side.
- Transitor can be used as an amplifiers, when operated in the Active Region. It is also used as a Switch, when operated in the cut-off and saturation regions.
- The three configurations of Transistor are Common Emitter, Common Base and Common Collector.
- The proper name for this device being referred as transistor is Bipolar Junction Transistor (BJT).
- The three regions of the output characteristics of a transistor are
  1. Active Region
  2. Saturation Region
  3. Cut-off Region
- JFET is UNIPOLAR Device (Unipolar only one type of carriers either holes or electrons)
- JFET device has Higher input resistance compared to BJT and Lower input resistance compared to MOSFET.
- The disadvantage of JFET amplifier circuits is Smaller Gain - Bandwidth product compared to BJT amplifier circuits.
- $r_{ds}\ (ON) = \left.\dfrac{\partial V_{DS}}{\partial I_D}\right|_{V_{GS} = K}$
- $g_m = \left.\dfrac{\partial I_D}{\partial V_{GS}}\right|_{V_{DS} = K}$
- $\mu = \text{Amplification factor} = \left.\dfrac{\partial V_{DS}}{\partial V_{GS}}\right|_{I_D = K}$
- Relation between μ, r_d and g_m for a JFET is $\mu = r_d \times g_m$
- Width of the depletion region W_n for n-channel JFET, interms of pinch off voltage V_P is $W = \left\{\dfrac{2 \in}{eN_D} V_P\right\}^{1/2}$
- Expression for I_{DS} in terms of I_{DSS}, V_{GS} and V_P is $I_D = I_{DSS}\left\{1 - \dfrac{V_{GS}}{V_P}\right\}^2$
- I_{DSS} = Satuaration value of Drain current when gate is shorted to source.

- For zero drift current in the case of JFET, $0.007|I_D| = 0.0022\ g_m$
- For zero drift current, in the case of JFETs, $|V_p| - |V_{GS}| = 0.63V$
- Expression for g_m for JFET interms of g_{mO} is, $g_m = \left(1 - \frac{V_{GS}}{V_P}\right)$
- The two types of n-channel or p-channel MOSFETs are
    1. Depletion type
    2. Enchancement type
- In JFET terminology, $B\ V_{DGO}$ parameter means the value of breakdown voltage V_{DG} at breakdown when the source is left open i.e., $I_S = 0$.
- The parameters $B\ V_{GSS}$ stands for breakdown voltage V_{GS} when drain is shorted to source.
- At high frequencies, the expressions for input capacitance C_{in} of JFET shunting R_G is, $C_{in} = C_{gs} + [1 + g_m (R_L \parallel r_d)]\ C_{gd}$
- The purpose of swamping resistor r_s connected in series with source resistance R_S in common source (C.S) JFET amplifier is to reduce distortion.
- Simplified expression for voltage gian A_V in the case of common Drain JFET amplifier is $A_V = \dfrac{R_S}{R_S + \dfrac{1}{g_m}}$
- The distortion caused due to the non linear transfer curve of JFET device, in amplifier circuits is known as Square Law Distorion.

OBJECTIVE TYPE QUESTIONS

1. For identical construction, ... types of bipolar junction transistors (BJTs) have faster switching times.
2. The arrow mark in the symbols of BJTs indicates ....
3. For NPN Bipolar Junction Transistor Emitter Efficiency γ = ...
4. For PNP Transistor (BJT), the transportation factor β^* = ...
5. Relation between α, β^* and γ for a transistor (BJT). α = ...
6. The expression for α in terms of β for a BJT is ....
7. The expression for β in terms of α for a BJT is ....
8. Expression for I_C in terms of β, I_B and I_{CBO} is ....
9. Expression for I_{CEO} in terms of I_{CBO} and α is ....
10. β' of a BJT is defined as ....
11. β' of a transistor (BJT) is ....
12. Relation between β' and β is ....
13. β' is synonymous to .... and β is same as ....
14. β is .... (for DC Currents).
15. β' is .... (for AC Currents).

16. Base width modulation is ....
17. IGFET is the other name for .... device.
18. In JFET recombination noise is less because it is .... device.
19. The disadvantage of JFET amplifier circuit is ....
20. The D, G, S terminals of JFET are similar to .... terminals of BJT respectively.
21. The voltage V_{DS} at which I_D tends to level off, in JFET is called ....
22. The voltage V_{GS} at which I_D becomes zero in the transfer characteristic of JFET is called ....
23. The range of $r_{DS\,(ON)}$ for JFET is ....
24. For low electric fields E_x of the order of 10^3V/cm, $\mu \alpha$ ....
25. Expression for V_{GS} in terms of V_P, ...
26. Expression for I_{DX} in terms of I_{DSS} in I_{DS} ....
27. I_{DSS} in defined as ....
28. JFET can be used as .... resistor.
29. Relation between μ, r_d and g_m for JFET is ....
30. The square law device is ....
31. The MOSFET that can be used in both enhancement mode and depletion mode is ....

ESSAY TYPE QUESTIONS

1. With the help of a neat graph qualitatively explain the Potential distribution through a transistor (BJT).
2. Explain about the different current components in a transistor.
3. Derive the relation between β and β'.
4. Derive the relation between α, β^* and γ
5. Differentiate between the terms h_{FE} and h_{fe}. Derive the relationship between them.
6. Qualitatively explain the input and output.
7. Explain the V-I characteristics in Common Emitter Configuration.
8. Describe the V-I characteristics of a transistor in Common Collector Configuration and Explain.
9. Draw the Eber-Moll Model of a transistor for NPN transistor and explain the same.
10. Compare the input and output characteristics of BJT in the three configurations, critically.
11. Compare BJT, JFET and MOSFET devices in all respects.
12. Derive the expression for the width of Depletion region 'W' in the case of p-channel JFET.
13. Obtain the expression for the pinch off voltage V_P in the case of n-channel JFET
14. Deduce the condition for JFET biasing for zero drift current.

15. Draw the structure and explain the static Drain and Gate characteristics of n-channel JFET. Repeat the same for p-channel JFET.
16. Draw the structure of p-channel MOSFET and Qualitatively explain the static Drain and Gate characteristics of the device.
17. What are the applications of JFET and MOSFET devices?
18. Give the constructional features of UJT.
19. Qualitatively explain the static V-I characteristics of UJT.
20. What is the significance of negative resistance region? Explain how UJT exhibits this characteristics ?

Specifications of a Bipolar Junction Transistor (BJT)

Sl.No.	Parameter	Symbol	Typical value	Units
1.	Collector-Emitter Voltage	V_{CE}	40	V
2.	Collector-Base Voltage	V_{CB}	60	V
3.	Emitter-Base Voltage	V_{EB}	6	V
4.	Collector-Current	I_C	500	mA
5.	Total Power Dissipation	P_D	0.4	W
6.	Operating Temperature	T_{op}	25-80	°C
7.	Collector- Emitter Break down voltage	BV_{CEO}	40	V
8.	Emitter-Base Break down voltage	BV_{EBO}	50	V
9.	Current - Gain Bandwidth product	f_T	250	MHz
10.	Output Capacitance	C_0	8	p_f
11.	Input Capacitance	C_{in}	20	p_f
12.	Small Signal Current gain	h_{fe}	100	-
13.	Output admittance	h_{oe}	10	m℧
14.	Voltage feedback ratio	h_{re}	8×10^{-4}	-
15.	Input impedance	h_{ie}	3	kΩ
16.	Delay time	t_d	10	nSec.
17.	Rise time	t_r	20	nSec.
18.	Fall time	t_f	50	nSec.
19.	Storage time	t_s	200	nSec.
20.	Noise Figure	NF	4	dB

Nomanclature used for Transistors (BJTs)

BC107	B	:	Silicon Device
(NPN Transistor)	C	:	Audio Amplifier application
	107	:	Type Number (No Significance)
AF 114	A	:	Germanium Device
	F	:	High Frequency Applications
	114	:	Type No.
2N 2222	2	:	Two Junction semiconductor device
(NPN Transistor)			
AC 128		:	Germanium Audio Frequency range BJT
BC 147		:	Silicon Audio Frequency range Transistor
2N 3866		:	NPN RF Power Transistor, 5W, 30V 0.4A.
2 N6253		:	NPN High power, 115W, 45V, 15A.
2 N 1481		:	NPN 5W, 40V, 1.5A

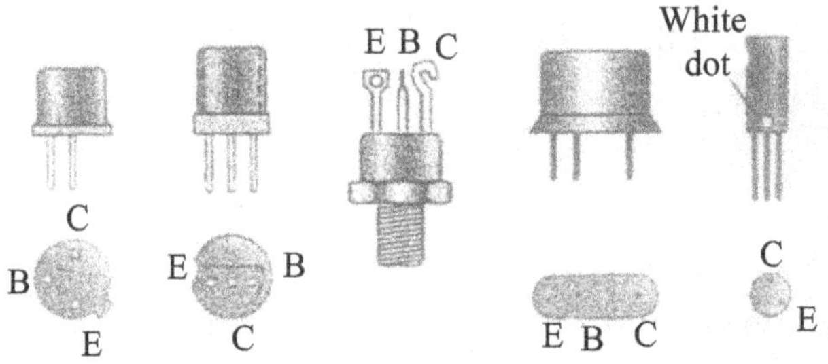

Fig. 4.38 Transistor terminal identification

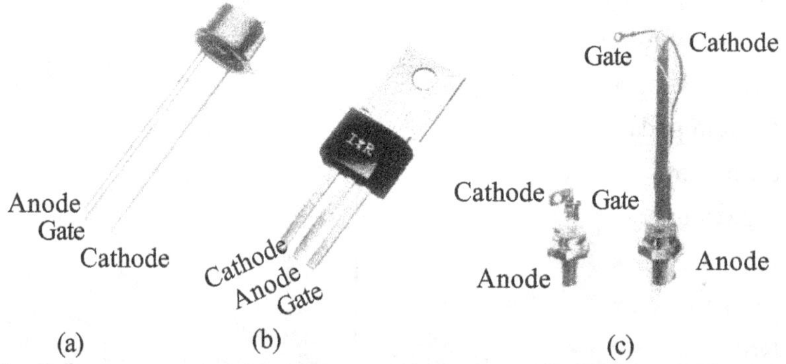

Fig. 4.39 SCR case construction and terminal identification [(a) courtesy general electric company; (b) and (c) courtesy international rectifier corporation]

Type Numbers of JFETs

2 N 4869	:	N - channel Silicon JFET
BF W10	:	N - channel Silicon JFET
BF W 11	:	N - channel Silicon JFET

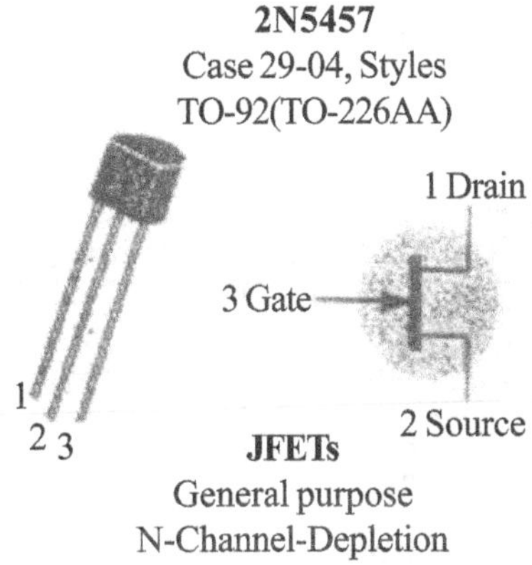

Fig. 4.40 Identification of JFET leads

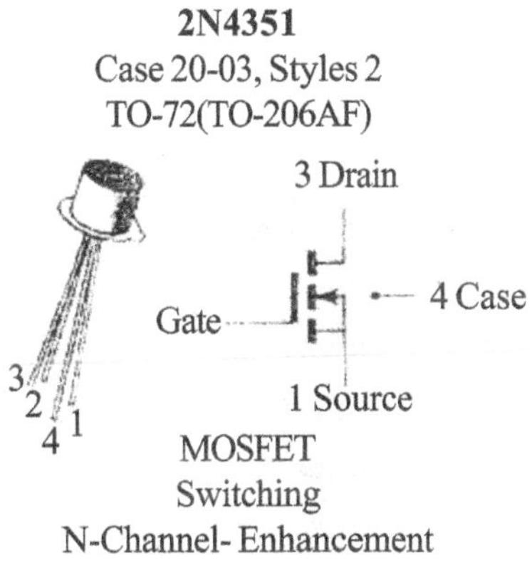

Fig 4.41 Identification of MOSFET leads

Specifications of JFETs (Typical Values)

Sl.No.	Parameter	Symbol	Typical Value	Units
1	Drain-Source voltage	V_{DS}	25	V
2.	Drain-Gate voltage	V_{DG}	25	V
3.	Reverse Gate - Source voltage	V_{GSR}	-25	V
4.	Gate Current	I_G	10	mA
5.	Junction temperature range	Tj	125	°C
6.	Gate source Breakdown voltage	$V_{(BR)GSS}$	-25	V

Type No. 2N5457 n- channel JFET
Type No. 2 N 4351 n-channel Enhancement type MOSFET
MOSFET

Sl.No.	Parameter	Symbol	Typical Value	Units
1.	Drain-Source voltage	V_{DS}	25	V
2.	Drain-Gate voltage	V_{DG}	30	V
3.	Gate-source voltage	V_{GS}	30	V
4.	Drain current	I_D	30	mA
5.	Zero gate voltage drain current	I_{DSS}	10	nA
6.	Gate threshold voltage	$V_{GS\ (th)}$	1	V

UJT

Sl.No.	Parameter	Symbol	Typical Value	Units
1.	Power Dissipation	P_D	300	mw
2.	RMS Emitter current	I_E (rms)	50	mA
3.	Emitter reverse voltage	$V_{E\ R}$	30	V
4.	Inter base voltage	V_B are (inter)	30	V
5.	Intrinsic stand off ratio	η	0.56	-
6.	Peak Emitter current	I_p	0.14	μA
7.	Valley point current	I_V	4	mA
8.	Emitter reverse current	I_{EO}	1	μA

Specifications of Light Emitting diode (LED)

Sl.No.	Parameter	Symbol	Typical Value	Units
1.	Average forward Current	I_F	20	mA
2.	Power Dissipation	PD	120	mw
3.	Peak forward Current	i f (peak)	60	mA
4.	Axial luminous Intensity	I_V	1	mcd milli candelas
5.	Peak wavelength	λ Peak	600	nm
6.	Luminous efficiency	nv	140	*l*m/w

ANSWERS TO OBJECTIVE TYPE QUESTIONS

1. NPN
2. the direction of conventional current when the E-B Junction is forward biased.
3. $\frac{I_{PE}}{I_E}$
4. $\frac{I_{pC}}{I_{pE}}$
5. $\alpha = \beta^* \gamma$
6. $\alpha = \frac{\beta}{(1+\beta)}$
7. $\beta = \frac{\alpha}{(1-\alpha)}$
8. $I_C = \beta I_B + (\beta + 1) I_{CBO}$
9. $I_{CEO} = \frac{I_{CBO}}{(1-\alpha)}$
10. $\beta' = \left.\frac{\partial I_C}{\partial I_B}\right|_{V_{CE}=K}$
11. Small Signal Common Emitter Forward Current Gain
12. $\beta' = \frac{\beta}{\left[1-(I_{CBO}+I_B)\frac{\partial h_{FE}}{\partial I_C}\right]}$
13. h_{fe} and h_{FE}
14. Large Signal Current Gain
15. Small Signal Current Gain
16. Change in base width with the change in the base voltage V_{BE}
17. MOSFET
18. UNIPOLAR
19. gain band width product is less.
20. C, B, E
21. Pinch off Voltage

22. Pinch off Voltage
23. few tens of ohms to few hundred ohms.
24. $\dfrac{1}{\sqrt{E_x}}$
25. $V_{GS} = \left(1 - \dfrac{b}{a}\right)^2 . V_P$
26. $I_{DS} = I_{DSS}\left(1 - \dfrac{V_{GS}}{V_P}\right)^2$
27. the saturation value of the drain current when gate is shorted to source. i.e., $V_{GS} = 0$
28. voltage variable
29. $\mu = r_d * g_m$
30. JFET
31. Depletion MOSFET (DMOSFET)

5 Feedback Amplifiers

In this Chapter,

- The concept of Feedback is introduced.
- Effect of negative feedback on amplifier characteristics is explained. Necessary equations are derived.
- Voltage series and shunt amplifiers, current series and shunt amplifier circuits are given.

5.1 FEEDBACK AMPLIFIERS

Any system whether it is electrical, mechanical, hydraulic or pneumatic may be considered to have at least one input and one output. If the system is to perform smoothly, we must be able to measure or control output. If the input is 10mV, gain of the amplifier is 100, output will be 1V. If the input deviates to 9mV or 11mV, output will be 0.9 V or 1.1V. So there is no control over the output. But by introducing feedback between the output and input, there can be control over the output. If the input is increased, it can be made to increase by having a link between the output and input. By providing feedback, the input can be made to depend on output.

One example is, the temperature of a furnace. Suppose, inside the furnance, the temperature should be limited to 1000^0C. If power is supplied on, continuously, the furnace may get over heated. Therefore we must have a thermocouple, which can measure the temperature. When the output of the thermocouple, reaches a value corresponding to 1000^0C, a relay should operate which will switch off the power supply to the furnace. Then after sometime, the temperature may come down below 1000^0c. Then again another relay should operate, to switch on the power. Thus the thermocouple and relay system provides the feedback between input and output.

Another example is traffic light. If the timings of red, green and yellow lights are fixed, on one side of the road even if very few vehicles are there, the green light will be on for sometime. On the other hand, if the traffic is very heavy on the other side of the road, still if the green lamp glows for the same period, the traffic will not be cleared. So in the ideal case, the timings of the red and green lamps must be proportional to the traffic on the road. If a traffic policeman is placed, he provides the feedback.

Another example is, our human mind and eyes. If we go to a library, our eyes will search for the book which we need and indicates to the mind. We take the book, which we need. If our eyes are closed, we can't choose the book we need. So eyes will provide the feedback.

Basic definitions

Ideally an amplifier should reproduce the input signal, with change in magnitude and with or without change in phase. But some of the short comings of the amplifier circuit are

1. Change in the value of the gain due to variation in supplying voltage, temperature or due to components.
2. Distortion in wave-form due to non linearities in the operating characters of the amplifying device.
3. The amplifier may introduce noise (undesired signals)

The above drawbacks can be minimizing if we introduce feedback

5.2 CLASSIFICATION OF AMPLIFIERS

Amplifiers can be classified broadly as,

1. Voltage amplifiers.
2. Current amplifiers.
3. Transconductance amplifiers.
4. Transresistance amplifiers.

This classification is with respect to the input and output impedances relative to the load and source impedances.

5.2.1 VOLTAGE AMPLIFIER

This circuit is a 2-port network and it represents an amplifier (Fig 5.1). Suppose $R_i >> Rs$, drop across Rs is very small.

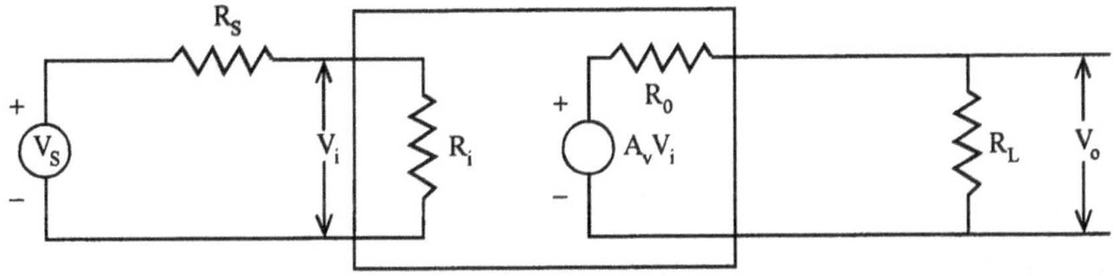

Fig 5.1 Equivalent circuit of voltage amplifiers.

$\therefore \qquad V_i \simeq V_S.$

Similarly, if $R_l >> R_o$, $V_o \simeq A_V . V_i$.

But $V_i \sim V_S$.

$\therefore \qquad V_o \simeq A_V . V_S.$

$\therefore$ Output voltage is proportional to input voltage.

The constant of proportionality A_V doesn't depend on the impedances. (Source or load). Such a circuit is called as ***Voltage Amplifier***.

Therefore, for ideal voltage amplifier

$$R_i = \infty .$$

$$R_o = 0.$$

$$A_V = \frac{V_o}{V_i}$$

with $\qquad R_L = \infty .$

A_V represents the open circuit voltage gain. For ideal voltage amplifier, output voltage is proportional to input voltage and the constant of proportionality is independent of R_S or R_L.

5.2.2 CURRENT AMPLIFIER

An ideal current amplifier is one which gives output current proportional to input current and the proportionality factor is independent of R_S and R_L.

The equivalent circuit of current amplifier is shown in Fig. 5.2.

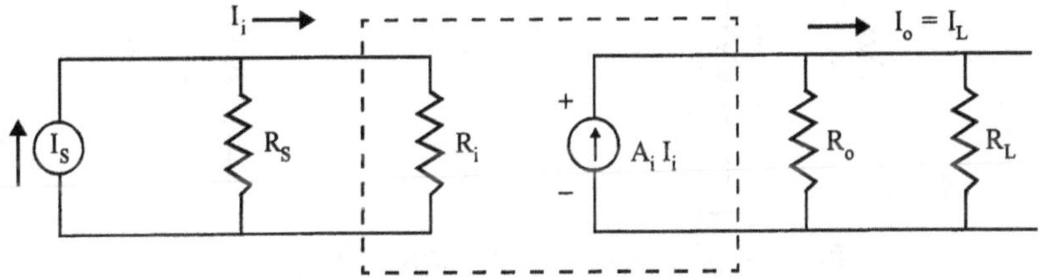

Fig 5.2 Current amplifier.

For ideal Current Amplifier,

$$R_i = 0$$

$$R_o = \infty.$$

If $R_i = 0, I_S \sim I_i$.

$\therefore$ $R_o = \infty$,

$$I_L = I_o = A_i I_i = A_i I_S.$$

$\therefore$ $A_I = \dfrac{I_L}{I_i}$ with $R_L = 0$.

$\therefore$ A_I represents the short circuit current amplification.

5.2.3 Transconductance Amplifier

Ideal Transconductance amplifier supplies output current which is proportional to input voltage independently of the magnitude of R_S and R_L.

Ideal Transconductance amplifier will have

$$R_i = \infty.$$

$$R_o = \infty.$$

In the equivalent circuit, on the input side, it is the Thevenins' equivalent circuit. A voltage source comes in series with resistance. On the output side, it is Norton's equivalent circuit with a current source in parallel with resistance. (Fig. 5.3).

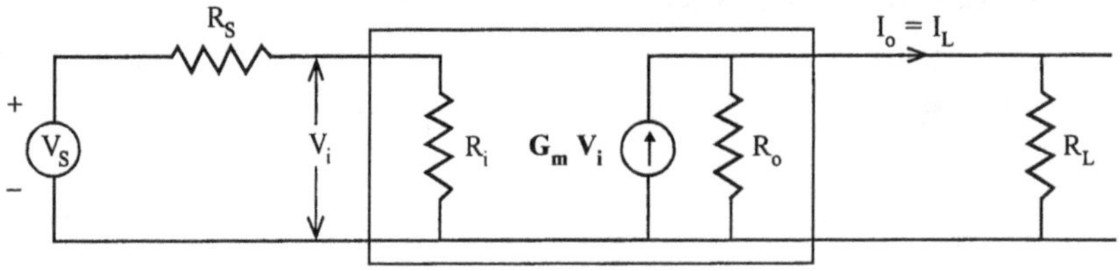

Fig 5.3 Equivalent circuit of transconductance amplifier.

5.2.4 Transresistance Amplifier

It gives output voltage V_0 proportional to I_s, independent of R_s α R_L. For ***ideal amplifiers***

$$R_i = 0, R_o = 0$$

Equivalent circuit **(Fig. 5.4)**

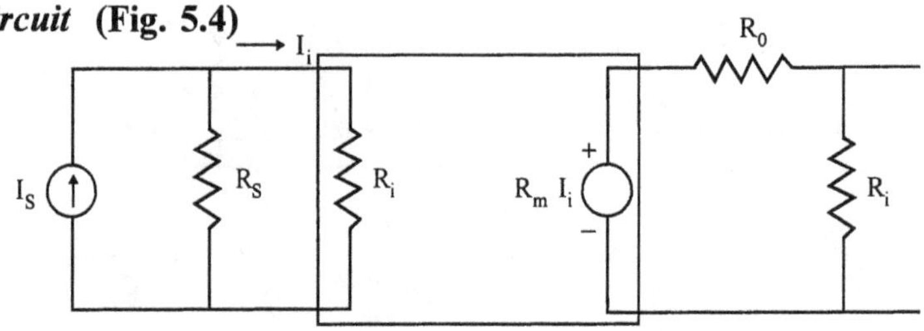

Fig 5.4 Transresistance amplifier.

Norton equivalent circuit on the R_i input side.

Thevenins' equivalent circuit on the output side.

5.3 FEEDBACK CONCEPT

A sampling network samples the output voltage or current and this signal is applied to the input through a feedback two port network. The block diagram representation is as shown in Fig. 5.5.

GENERALIZED BLOCK SCHEMATIC

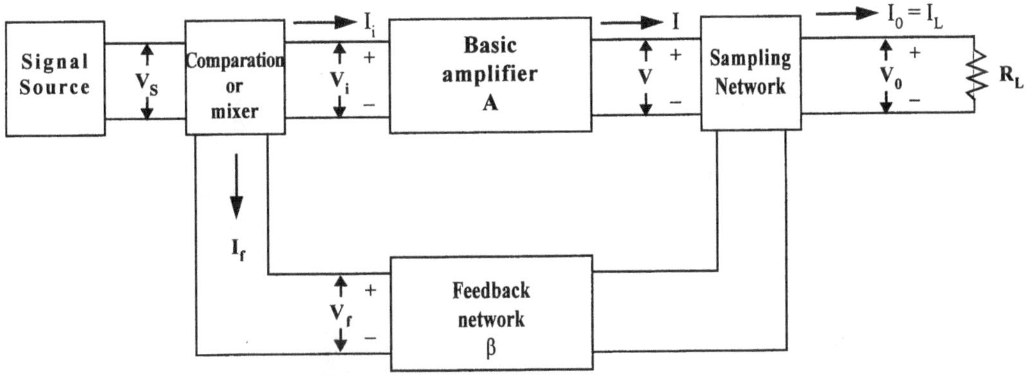

Fig 5.5 Block diagram of feedback network.

Signal Source

It can be a voltage source V_S or a current source I_S

FEEDBACK NETWORK

It is a passive two port network. It may contain resistors, capacitors or inductors. But usually a resistance is used as the feedback element. Here the output current is sampled and feedback. The feedback network is connected in series with the output. This is called as ***Current Sampling or Loop Sampling***.

A voltage feedback is distinguished in this way from current feedback. For voltage feedback, the feedback element (resistor) will be in parallel with the output. For current feedback the element will be in series.

COMPARATOR OR MIXER NETWORK

This is usually a differential amplifier. It has two inputs and gives a single output which is the difference of the two inputs.

V = Output voltage of the basic amplifier *before sampling* [see the block diagram of feedback]

V_i = Input voltage to the basic amplifier

A_V = Voltage amplification = V/V_i

A_I = Current amplification = I/I_i

G_M = Transconductance of basic amplifier = I/V_i

R_M = Transresistance = V/I_i.

All these four quantities, A_v, A_I, G_M and R_M represent the transfer gains (Though G_M and R_M are not actually gains) of the basic amplifier without feedback. So the symbol 'A' is used to represent these quantities.

A_f is used to represent the ratio of the output to the input **with feedback**. This is called as the transfer gain of the amplifier with feedback.

$\therefore$

$$A_{Vf} = \frac{V_o}{V_s}$$

$$A_{If} = \frac{I_o}{I_s}$$

$$G_{Mf} = \frac{I_o}{V_s}$$

$$R_{Mf} = \frac{V_o}{I_s}$$

Feedback amplifiers are classified as shown below.

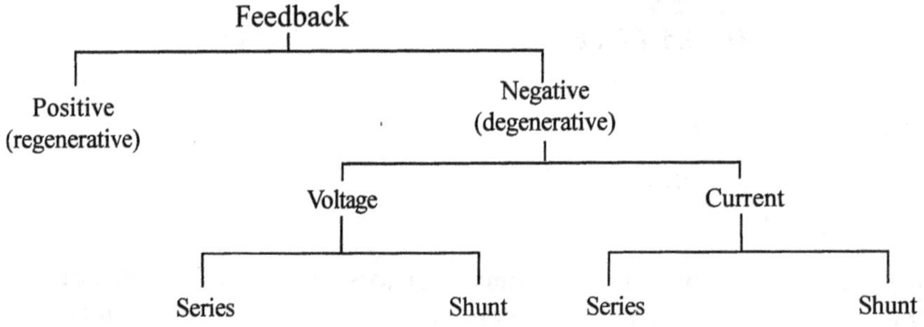

$$A = \text{voltage gain} = \frac{V_o}{V_i}$$

Feedback Factor, $\quad \beta = \dfrac{V_f}{V_o}$ (This β is different from β used in BJTs)

$$(\beta)(A)(-1) = -\beta A$$

Loop gain / Return Ratio

-1 is due to phase-shift of $180°$ between input and output in Common Emitter Amplifier. Since

$$\text{Sin}(180°) = -1.$$

Return difference $D = 1-$ loop gain negative sign is because it is the difference

$$D = 1 - (-\beta A)$$

$$\boxed{D = 1 + \beta.A.}$$

Types of Feedback

How to determine the type of feedback ? Whether current or voltage ? If the feedback signal is proportional to voltage, it is **Voltage Feedback.**

If the feedback signal is proportional to current, it is **Current Feedback.**

Conditions to be satisfied

1. Input signal is transmitted to the output through amplifier A and not through feedback network β.
2. The feedback signal is transmitted to the input through feedback network and not through amplifier.
3. The reverse transmission factor β is independent of R_s and R_L.

5.4 TYPES OF FEEDBACK

Feedback means a portion of the output of the amplifier circuit is sent back or given back or feedback at the input terminals. By this mechanism the characteristics of the amplifier circuit can be changed. Hence feedback is employed in circuits.

There are two types of feedback.

1. Positive Feedback
2. Negative Feedback

Negative feedback is also called as ***degenerative feedback***. Because in negative feedback, the feedback signal opposes the input signal. So it is called as degenerative feedback. But there are many advantages with negative feedback.

Advantages of Negative Feedback

1. Input impedance can be increased.
2. Output impedance can be decreased.
3. Transfer gain A_f can be stabilized against variations in ***h-parameter*** of the transistor with temperature etc.
   i.e. stability is improved.
4. Bandwidth is increased.
5. Linearity of operation is improved.
6. Distortion is reduced.
7. Noise reduces.

5.5 EFFECT OF NEGATIVE FEEDBACK ON TRANSFER GAIN

Consider the black schematic shown in Fig. 5.6.

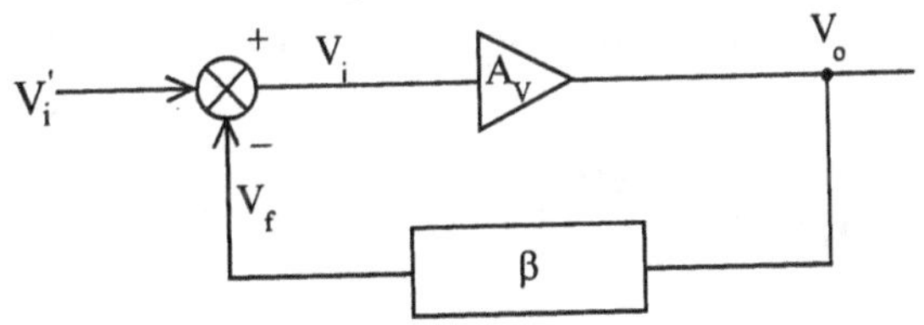

Fig 5.6 Block schematic for negative feedback

$$A_V = \frac{V_o}{V_i}$$

$$A_V' = V_o / V_i'$$

$$V_i' = V_i - \beta V_o$$

$$V_o = A_v (V_i - \beta V_o)$$
$$V_o = A_v . V_i - A_v . \beta V_o$$
$$V_o (1 + \beta A_v) = A_v V_i$$
$$\therefore \quad A_{vf} = \frac{V_o}{V_i} = \frac{A_v}{1+\beta A_v}$$

5.5.1 Reduction in Gain

$$A_V' = \frac{A_v}{1-\beta A_v}$$

(for positive feedback)

A_v = Voltage gain without feedback. (Open loop gain).

If the feedback is negative, β is negative.

$$\therefore \quad A'_V = \frac{A_v}{1-(-\beta . A_v)}$$

For negative feedback $A'_V = \dfrac{A_v}{1+\beta A_v}$

Denominator is > 1. $\therefore \quad A_V' < A_V$

$\therefore$ There is reduction in gain.

5.5.2 Increase in Bandwidth

If, f_H is upper cutoff frequency.

f_L is lower cut off frequency.

f is any frequency.

Expression for A_v (voltage gain at any frequency f) is,

$$A_v = \frac{A_v(\text{mid})}{1+j.\dfrac{f}{f_H}} \qquad ..... (5.1)$$

A_v (mid) = Mid frequency gain

$$A_v = \frac{A_v(\text{mid})}{1+\beta A_v} \text{ for negative feedback.} \qquad .....(5.2)$$

Substituting equation (5.1) in (5.2) for A_V,

$$A_v = \frac{A_v(\text{mid})/1+j.\dfrac{f}{f_H}}{1+\beta\left[\dfrac{A_v \text{mid}}{1+jf/f_H}\right]} \quad \text{(for negative Feedback, } \beta \text{ is } V_e\text{)}$$

Simplifying,
$$A_V = \frac{\dfrac{A_V(\text{mid})}{f_H + j.f}\left[1 + j\dfrac{f}{f_H}\right]}{1 + j.f + \beta\,[A_V\ \text{mid}]}$$

$$= \frac{A_V(\text{mid})}{f_H + j.f + \beta.A_{V\text{mid}} \cdot f_H}$$

$$A_v' = \frac{A_v(\text{mid}) / 1 + \beta_v A_v(\text{mid})}{1 + \dfrac{jf}{f_H\left[1 + \beta_v A_v(\text{mid})\right]}} \quad .....(5.3)$$

Equation (5.3) can be written as,

$$A_v' = \frac{A_V'(\text{mid})}{1 + \dfrac{jf}{f_H'}}$$

Where
$$A_V'{}_{(\text{mid})} = \frac{A_v(\text{mid})}{1 + \beta_v A_v(\text{mid})}$$

and
$$f_H' = f_H\,(1 + \beta_v A_{V\,(\text{mid})})$$

$\because$ β is negative for negative feedback, $f_H' > f_H$.

$\therefore$ ***Negative feedback, increases bandwidth.***

Similarly,
$$f_L' = \frac{f_L}{1 + \beta_v A_{v(\text{mid})}} \qquad A_v = \frac{A_{v(\text{mid})}}{1 - j\left(\dfrac{f_L}{f}\right)}$$

or
$$A_v' = \frac{A_{V\,(\text{mid})} / 1 + \beta A_{V(\text{mid})}}{1 + j\dfrac{f_L}{f(1 + \beta.A_V)}}$$

$$A_v' = \frac{A_V}{1 + j\dfrac{f_L'}{f}}$$

5.5.3 Reduction in Distortion

Suppose, the amplifier, in addition to voltage amplification is also producing distortion D.

$$V_o = A_V \,.\, V_i + D.$$

Where, $V_i = V_i' - \beta_V.\, V_o$ (for negative feedback)

$V_i' = V_i - \beta\, V_o$ and β is negative for negative feedback,

$\therefore$ $V_o' = A_v\,[V_i' - \beta_v V_o] + D$

$$V_o\left[1+A_V\ \beta_V\right]=V_i'.A_V+D$$

$$\therefore \qquad V_o=\frac{V_i'.A_v}{\left(1+A_V.\beta_V\right)}+\frac{D}{\left(1+\beta_V A_V\right)}$$

For negative feedback, β is negative therefore denominator is > 1

∴ The distortion in the output is reduced.

$$=\frac{D}{1+\beta_v A_v}\ \text{is} < D$$

The physical explanation is, suppose the input is pure sinusoidal wave. There is distortion in the output as shown, in Fig. 5.7.

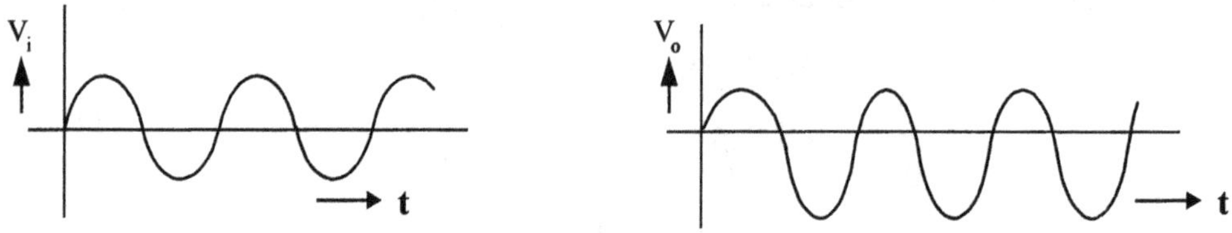

(a) Input signal

(b) Output with distortion (without negative feedback)

Fig 5.7

Now, if a part of the distorted output is fed back to the input, so as to oppose the input, the feedback signal and input will be out of phase.

So the new input V_i will have distortion introduced init, because of mixing of distorted V_f with pure V_i. So the distortion in the output will be reduced because, the distortion introduced in the input cancels the distortion produced by the amplifier. Because these two distortions are out of phase. The feedback signal cancels the distortion produced by the amplifier, Therefore these two are out of phase.

5.5.4 Feedback to Improve Sensitivity

Suppose an amplifier of gain A_1 is required. Build an amplifier of gain $A_2 = DA_1$ in which D is large. Feedback is now introduced to divide the gain by D. Sensitivity is improved by the same factor D, because both gain and instability are divided by D. The stability will be improved by the same factor.

5.5.5 Frequency Distortion

$$A_f=\frac{A}{1+\beta A}$$

If the feedback network does not contain reactive elements. The overall gain is not a function of frequency. So frequency duration is less. If β depends upon frequency, with negative feedback, Q factor will be high.

5.5.6 Band Width

Consider the Fig. 5.8 and 5.9.

It increases with negative Feedback (as shown in Fig. 5.8)

$$f_1' < f_1,\ f_2' > f_2$$

$$BW' = (f_1' - f_2') : BW = (f_2 - f_1)$$

$$f_1' = \frac{f}{1+\beta A_m} \quad f_2' = f_2\ (1+\beta A_m) \qquad BW' > BW$$

$$BW = f_2 - f_1 \simeq f_2$$

$$(BW)_f = f_2' - f_1 \sim f_2'$$

$$\therefore \quad (BW)_f = (1+\beta A_m)\ BW$$

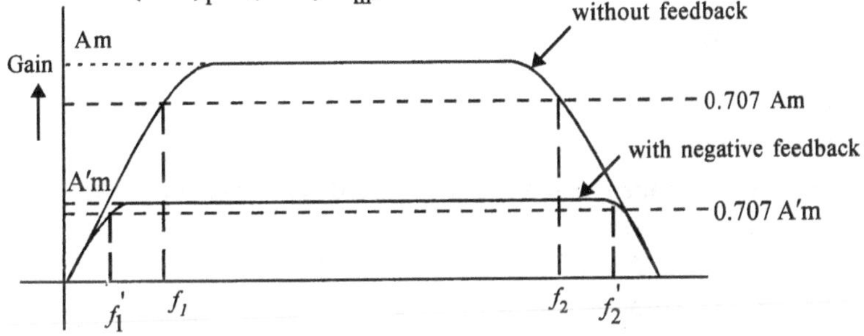

Fig 5.8 Frequency response with and without negative feedback

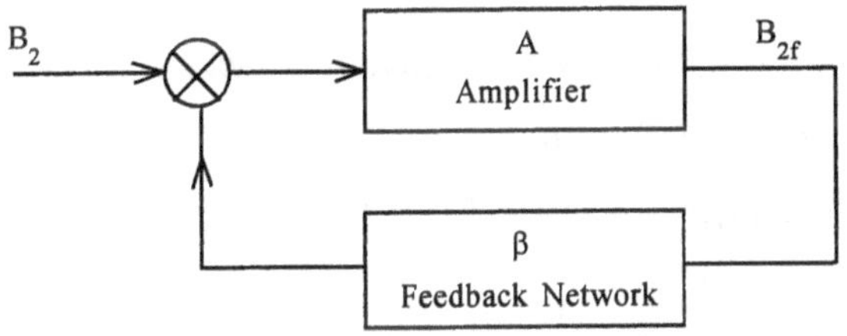

Fig 5.9 Feedback network

5.5.7 Sensitivity of Transistor Gain

Due to aging, temperature effect etc., on circuit capacitance, transistor or FET, stability of the amplifier will be affected

Fractional change in amplification with feedback divided by the fractional change without feedback is called ***Sensitivity*** of Transistor.

$$= \frac{\left|\dfrac{dA_f}{A_f}\right|}{\left|\dfrac{dA}{A}\right|}$$

$$A_f = \frac{A}{1+\beta A}$$

Differentiating, $$\frac{dA_f}{A_f} = \frac{1}{|1+\beta A|}\left|\frac{dA}{A}\right|$$

$\therefore$ Sensitivity, $$= \frac{1}{1+\beta A}$$

Reciprocal of sensitivity is ***Densitivity*** $\boxed{D = (1 + \beta A).}$

5.5.8 Reduction of Nonlinear Distortion

Suppose the input signal contains second harmonic and its value is B_2 before feedback. Because of feedback. B_{2f} appears at the output. So positive βB_{2f} is fed to the input. It is amplified to $-A\beta B_{2f}$.

$\therefore$ Output with two terms $B_2 - A\beta B_2 f$.

$\therefore$ $$B_2 - A\beta B_2 f = B_{2f}.$$

or $$B_{2f} = \frac{B_2}{1+\beta A} \quad B_{2f} < B_2$$

So it is reduced.

5.5.9 Reduction of Noise

Let N be noise constant without feedback and N_F with feedback. N_F is fed to the input and its value is βN_F. It is amplified to $-\beta A N_F$.

$\therefore$ $$N_F = N - \beta A N_F$$

$$N_F(1 + \beta A) = N$$

as $$N_F = \frac{N}{1+\beta A}$$

$\therefore$ $N_F < N$. Noise is reduced with negative feedback.

5.6 TRANSFER GAIN WITH FEEDBACK

Consider the generalized feedback amplifier shown in Fig.5.10.

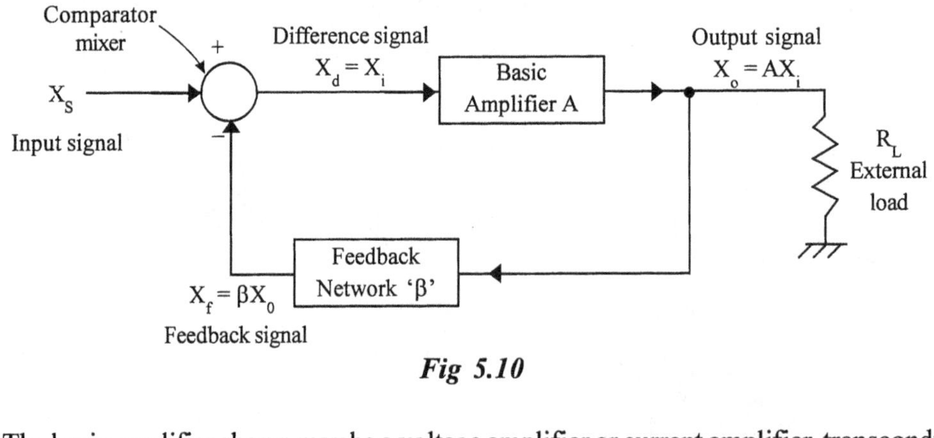

Fig 5.10

The basic amplifier shown may be a voltage amplifier or current amplifier, transconductance or transresistance amplifier.

The four different types of feedback amplifiers are,

1. Voltage series feedback.
2. Voltage shunt feedback.
3. Current series feedback.
4. Current shunt feedback.

X_s = Input signal.
X_0 = Output signal.
X_f = feedback signal.
X_d = Difference signal. [Difference between the input signal and the feedback signal].
β = Reverse transmission factor or feedback factor = X_f / X_o.

The feedback network can be a simple resistor. The mixer can be a difference amplifier. The output of the mixer is the difference between the input signal and the feedback signal.

$$X_d = X_s - X_f = X_i.$$

X_d is also called as error or comparison signal.

$$\beta = \frac{X_f}{X_o}$$

It is often a positive or negative real number.

This β should not be confused with the β of a transistor $\frac{I_C}{I_B}$

Transfer gain $$A = \frac{X_o}{X_i} = \frac{X_o}{X_d}$$

$$X_i = X_d \qquad .....(5.4)$$

Gain with feedback $A_f = \frac{X_o}{X_s}$

But $$X_d = X_s - X_f = X_i \qquad .....(5.5)$$

$$\beta = \frac{X_f}{X_o} \qquad .....(5.6)$$

From (5.5), $X_d = X_S - X_f$.

Substitute (5.5) in (5.4).

$$\therefore \qquad A = \frac{X_o}{X_s - X_f}$$

Dividing Numerator and Denominator by X_s,

$$A = \frac{X_o/X_s}{1-\frac{X_f}{X_s}} = \frac{X_o/X_s}{1-\frac{X_f}{X_o}.\frac{X_o}{X_s}}.$$

We know that $X_o / X_s = A_f$; $\frac{X_f}{X_o} = \beta$.

$$\therefore \quad A = \frac{A_f}{(1 - \beta.A_f)}$$

or $\quad A - \beta.\ A_f\ .\ A = A_f.$

$$A = A_f(\ 1 + \beta A)$$

$$\therefore \quad \boxed{A_f = \frac{A}{1+\beta A}}$$

A_f = gain with feedback.

A = transfer gain without feedback.

If $|A_f| < |A|$ the feedback is called as negative or degenerative, feedback

If $|A_f| > |A|$ the feedback is called as positive or regenerative, feedback

5.6.1 Loop Gain

In the block diagram of the feedback amplifier, the signal X_d which is the output of the comparator passes through the amplifier with gain A. So it is multiplied by A. Then it passes through a feedback network and hence gets multiplied by β, and in the comparator it gets multiplied by –1. In the process we have started from the input and after passing through the amplifier and feedback network, completed the loop. So the total product $A\beta$.

In order that ***series feedback*** is most effective, the circuit should be driven from a ***constant voltage*** source whose internal resistance R_s is small compared to R_i of the amplifier. If R_s is very large, compared with R_i, V_i will be modified not by V_f but because of the drop across R_s itself. So effect of V_f will not be there. Therefore for series feedback, the voltage source should have less resistance. (See Fig.5.11).

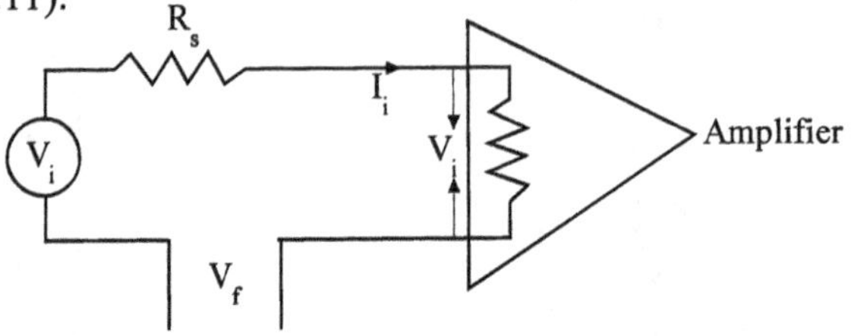

Fig 5.11 Series feedback

In order that ***shunt feedback*** is most effective, the amplifier should be driven from a constant current source whose resistance R_s is very high ($R_s >> R_i$). (Fig. 5.12).

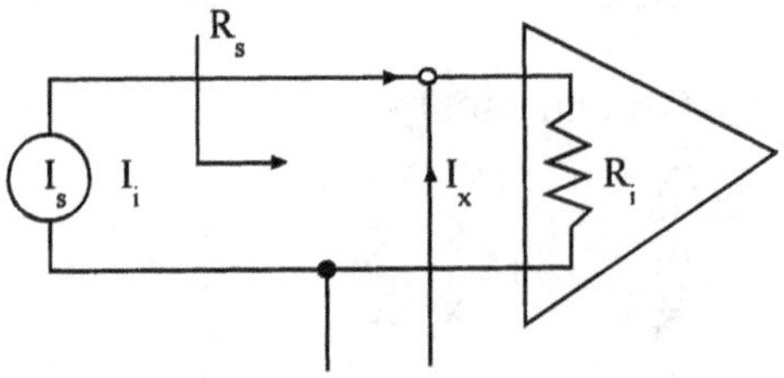

Fig 5.12 Current shunt feedback

If the resistance of the source is very small, the feedback current will pass through the source and not through R_i. So the change in I_i will be nominal. Therefore the source resistance should be large and hence a current source should be used.

If the feedback current is same as the output current, then it is series derived feedback. (Fig. 5.13).

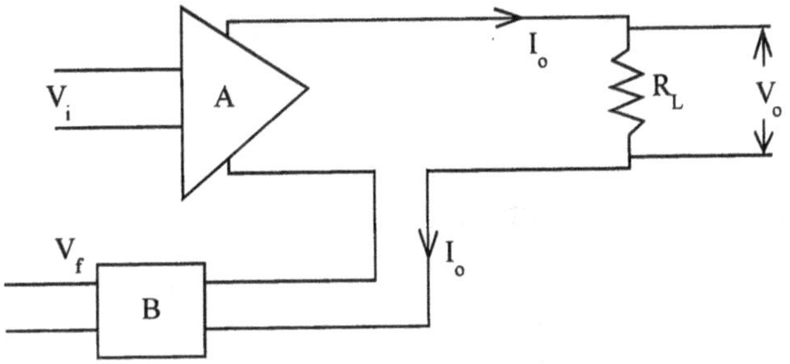

Fig 5.13 Series derived feedback

When the feedback is shunt derived (Fig. 5.14), output voltage is simultaneously present across R_L and across the input to the feedback. So in this case V_f is proportional to V_o.

An amplifier with shunt derived negative Feedback increases the output resistance. When the feedback is series derived, I_o remains constant, so R_o increase.

Similarly if the feedback signal is ***shunt fed*** it reduces the input resistance.

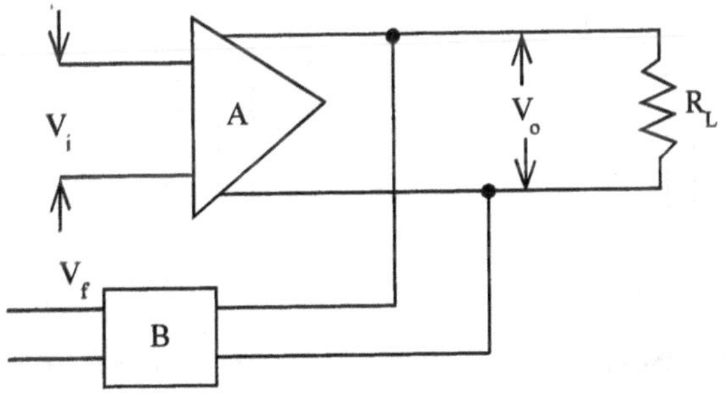

Fig. 5.14 Shunt derived feedback

If it is ***series fed,*** it increases the output resistance. Therefore I_o remains constant, even through V_o increase, so R_0 increases.

Return Ratio

βA = Product of feedback factor β and amplification factor A is called as ***Return Ratio.***

Return Difference (D)

The difference between unity (1) and return ratio is called as ***Return difference.***

$$D = 1-(-\beta A) = 1+\beta A.$$

Amount of feedback introduced is expressed in decibels

$$= N \log\left(\frac{A'}{A}\right) \quad N = 20 \log\left|\frac{A_f}{A}\right| = 20 \log\left|\frac{1}{1+A\beta}\right|$$

If feedback is negative N will be negative because $A' < A$.

5.7 CLASSIFACTION OF FEEDBACK AMPLIFIERS

There are four types of feedback,

1. Voltage series feedback.
2. Voltage shunt feedback.
3. Current shunt feedback.
4. Current series feedback.

Voltage Series Feedback

Feedback signal is taken across R_L.proportional to V_o. So it is voltage feedback. V_f is coming in series with V_i. So it is Voltage series feedback. (See Fig. 5.15(a)). The other configurations are shown in Figs. 5.15 (b), (c) and (d).

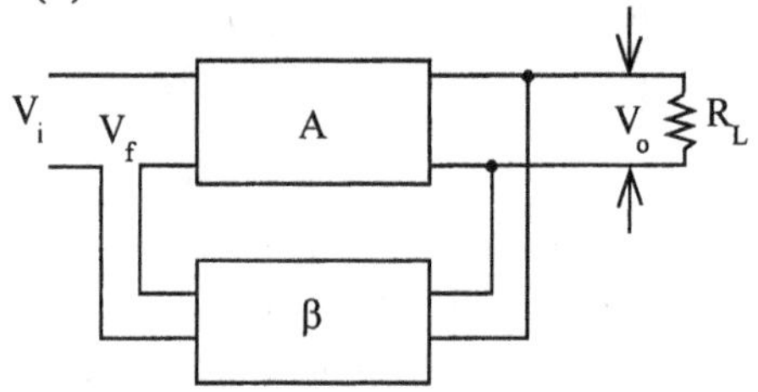

(a) Schematic for voltage series feedback

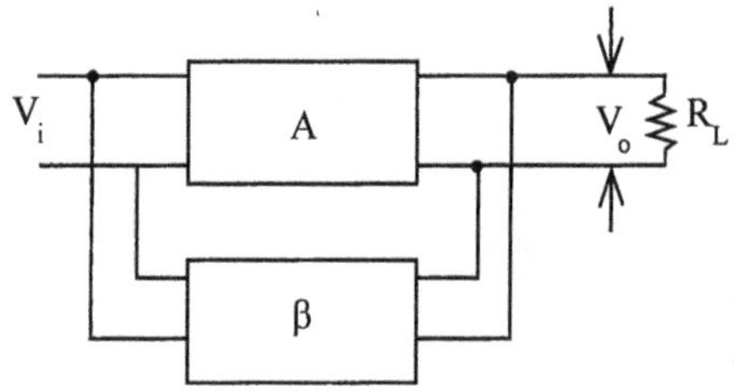

(b) Voltage shunt feedback

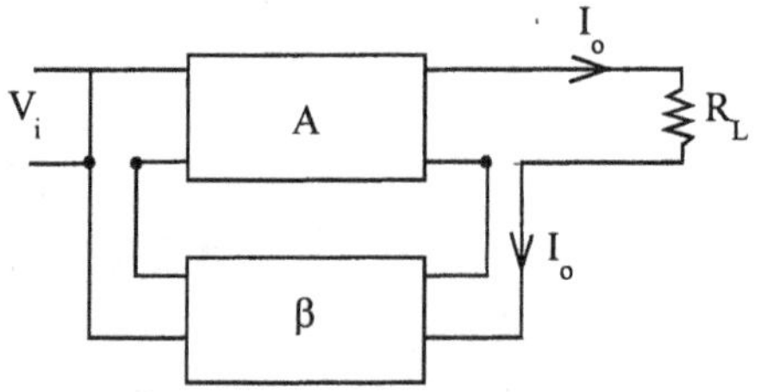

(c) Current shunt feedback

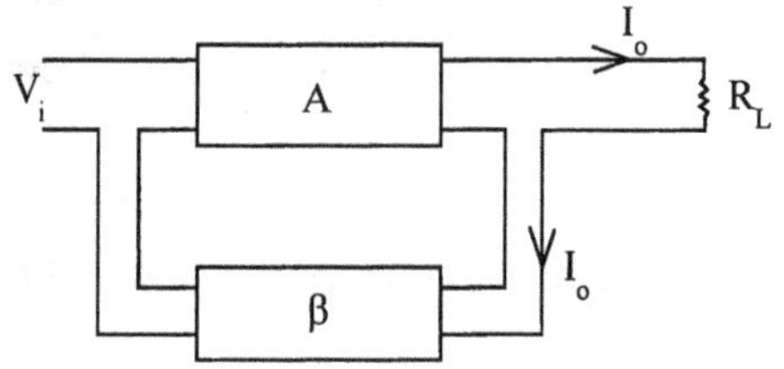

(d) Current series feedback

Fig 5.15

Improvement of Stability with Feedback

Stability means, the stability of the voltage gain. The voltage gain must have a stable value, with frequency. Let the change in A_v is represented by S.

$$\frac{dA_V}{A_v} = S\left(\frac{dA_V'}{A_V'}\right) = S = \frac{\left(\frac{d\,A_V'}{A_v'}\right)}{\left(\frac{d\,A_V}{A_V}\right)}$$

$$A_V' = \frac{V_0}{V_S} \text{ (for negative feedback)}$$

$$A_V' = \frac{A_V}{1 + A_V \beta_1}$$

Differentiating with respect to A_V, $\frac{dA_V}{A_V} = \frac{\left(1+\beta.A_V\right) - A_V\left(+\beta\right)}{\left(1+\beta.A_V\right)^2} = \frac{1}{\left(1+A_V\beta\right)^2}$

Dividing by A_V on both sides, of $\frac{dA_V'}{dA_V} = \frac{1}{\left(1+\beta A_V\right)^2}$

$$\frac{dA_V'}{A_V}\frac{1}{dA_V} = \frac{1}{\left(1+\beta A_V\right)} \cdot \frac{1}{A_V'}$$

But $A_V' = \frac{A_V'}{1+\beta A_V}$ and $\frac{dA_V'}{A_V'} = S'$ $\frac{dA_V}{A_V} = S$

$$S = \frac{dA_V'}{dA_V} = \frac{1}{\left(1+\beta A_V\right)^2}$$

$$\boxed{S = \frac{dA_V}{\left(1+\beta A_V\right)}}$$

For negative Feedback, β is negative $\therefore$ denominator > 1.

$\therefore$ $S' < S$

i.e., variation in A_v, or % change in A_v is less with -negative feedback.

$\therefore$ ***Stability*** is good.

5.8 EFFECT OF FEEDBACK ON INPUT RESISTANCE

5.8.1 INPUT RESISTANCE WITH SHUNT FEEDBACK

With feedback $R_{if} = \frac{V_i}{I_s}$

$$I_f = \beta_i\, I_o. \quad \left[\because \beta_i = \frac{I_f}{I_o}\right]$$

$$I_s = I_i + \beta\, I_o$$

$$R_i^{'} = \frac{V_i}{I_i + \beta_i I_o}$$

$$I_o = A_i . I_i$$

$$R_i^{'} = \frac{V_i}{I_i + \beta_i A_i I_i}$$

$$R_i = \frac{V_i}{I_i(1 + \beta_i A_i)}$$

$$\boxed{R_i^{'} = \frac{R_i}{(1 + \beta_i A_i)}}$$

∴ ***Input resistance decreases with shunt feedback.***

If the feedback signal is taken across R_L, it is α V_o or so it is ***Voltage feedback.***

If the feedback signal is taken in series with the output terminals, feedback signal is proportional to I_o. So it is ***current feedback.***

If the feedback signal is in series with the input, it is ***series feedback.***

If the feedback signal is in shunt with the input, it is ***shunt feedback.***

EXPRESSION FOR R_I WITH CURRENT SHUNT FEEDBACK

Consider Fig. 5.16.

A_i = Shunt circuit current gain of the BJT

A_I = practical current gain I_o / I_i

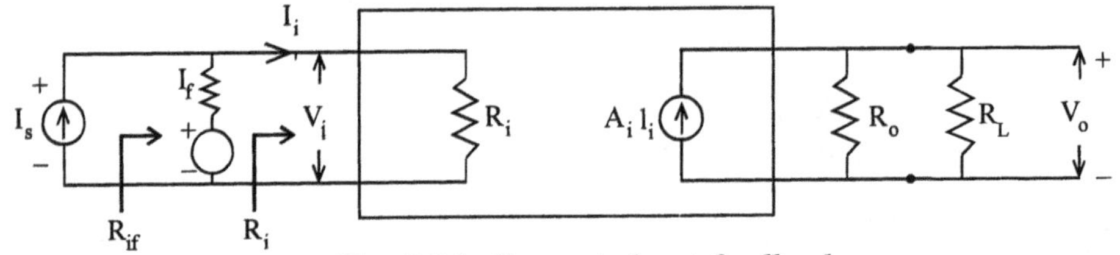

Fig. 5.16 Current shunt feedback

A_i represents the Shunt circuit current gain taking R_s into accounts,

$$I_S = I_i + I_f = I_i + \beta I_o$$

and $$I_o = \frac{A_i R_o I_i}{R_o + R_L} = A_I I_i \quad \therefore \text{ Let } A_I = \frac{A_i R_o}{R_o + R_L}$$

$$A_I = \frac{I_o}{I_i} = \frac{A_i R_o}{R_o + R_i} = (I_i + I_f)$$

$$I_S = I_i + \beta . I_0 = I_i \left(1 + \beta . \frac{I_o}{I_i}\right) = I_i (1 + \beta . A_I)$$

$$R_{if} = V_i \mid I_s \qquad R_i = V_i \mid I_i$$

$$\boxed{R_{if} = \frac{V_i}{(1 + \beta A_I) I_i} = \frac{R_i}{1 + \beta . A_I}}$$

for shunt feedback the input resistance decreases.

5.8.2 Input Impedance with Series Feedback

Consider the schematic shown in Fig. 5.17.

$$V_i' = V_i + \beta V_o \text{ (in general case).}$$

For negative feedback, β is negative.

$$V_i = \frac{V_i^1}{(1+\beta A)}$$

$$\frac{V_i}{I_i} = \frac{V_i'}{I_i(1+\beta A)}$$

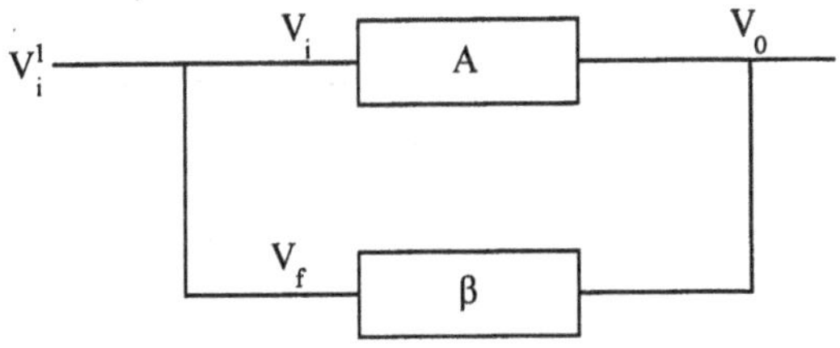

Fig 5.17 Feedback network.

In general, R_i increases,

$$V_i' = V_i + \beta\, V_o$$

$$V_i' = V_i + \beta.A.\, V_i \qquad \because \quad V_0 = A.\, V_i$$

$$V_i' = V_i\,(1+\beta A)$$

But $\quad V_i' = I_i\, R_i$

$\therefore \quad V_i' = (1+\beta A)\, R_i\, I_i$

$$\frac{V_i'}{I_i} = \text{Input Z seen by the source} = R_i\,(1+\beta A)$$

$\therefore \quad R_{if} = R_i\,(1+\beta A)$

Expression for R_I with Voltage Series Feedback

In this circuit A_v represents the open circuit voltage gain taking R_s into account. (Fig. 5.18).

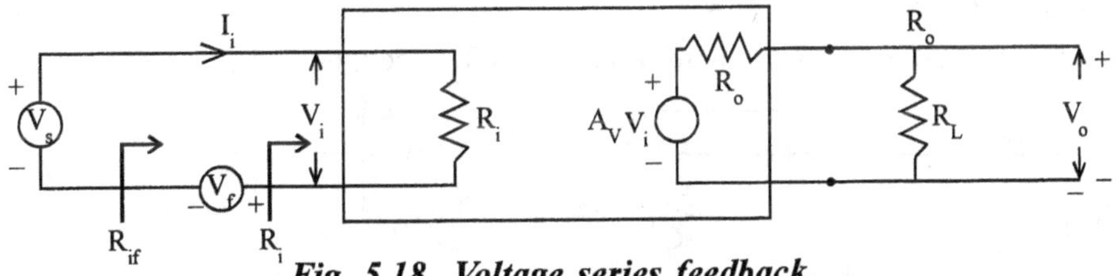

Fig. 5.18 Voltage series feedback.

$$V_i = V_S - V_f$$

or $\quad V_s = V_i + V_f$

$$R_0' = R_{if} = V_S/I_i.$$

$$V_f = I_i\, R_i + V_f = I_i\, R_i + \beta V_0$$

$$V_0 = \frac{A_v V_i R_L}{R_o + R_L} = (A_V\, I_i\, R_i)$$

Let, $\quad \frac{A_v.R_L}{R_o + R_L} = A_V,\ V_i = I_i.\, R_i.$

$$A_V = \frac{V_o}{V_i} = \frac{A_v R_L}{R_o + R_L}$$

$$V_S = I_i R_i + \beta . V_0$$

$$\boxed{\frac{V_s}{I_i} = R_i + \frac{\beta . V_o}{I_i} = R_i\left(1 + \frac{\beta . V_o}{I_i \times R_i}\right)}$$

$$R_{if} = R_i\left(1 + \beta . \frac{V_o}{V_i}\right) = R_i(1 + \beta . A_V)$$

A_V = voltage gain, without feedback.

for series feedback input resistance increases.

5.9 EFFECT OF NEGATIVE FEEDBACK ON R_o

Voltage feedback (series or shunt) R_o decreases.

Current feedback (series or shunt) R_o increases.

Series feedback (voltage or current) R_i increases.

Shunt feedback (voltage or current) R_i decreases.

OUTPUT RESISTANCE

Negative feedback tends to decrease the input resistances. Feeding the voltage back to the input in a degenerative manner is to cause lesser increase in V_0. Hence the output voltage tends to remain constant as R_L changes because output resistance with feedback $R_{of} << R_L$.

Negative feedback, which samples the output current will tend to hold the output current constant. Hence an output current source is created ($R_{of} >> R_L$). So this type of connection increases output resistance.

For voltage sampling $R_{of} < R_0$. For current feedback $R_{of} > R_o$.

5.9.1 VOLTAGE SERIES FEEDBACK

Expression for R_{of} looking into output terminals with R_L disconnected, (Fig. 5.19).

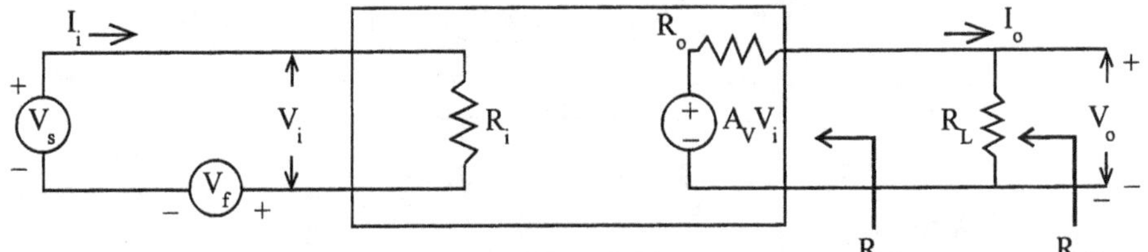

Fig. 5.19

R_0 is determined by impressing voltage 'V' at the output terminals or messing 'I', with input R_{of}' terminals shorted.

Disconnect R_L. To find R_{of}, remove external signal (set $V_s = 0$, or $I_s = 0$)

Let $R_L = \infty$.

Impress a voltage V across the output terminals and calculate the current I delivered by V. Then, $R_{0f} = V|I$.

$$I = \frac{V_0 - A_V\, V_i}{R_0} = \frac{V_0 + \beta A_v V}{R_o}$$

$\because$ V_0 = output voltage.

$V_i = -\beta_V$

Because with $V_S = 0,\ V_i = -V_f = -\beta_V$

Hence, $R_{of} = \frac{V_0}{I} = \frac{R_o}{1+\beta A_v}$

This expression is excluding R_L. If we consider R_L also R_{of} is in parallel with R_L.

$$R_{of}' = \frac{R_{of}.R_L}{R_{of} + R_L}$$ Substitute the1 value of R_{of}

$$R_{of}' = \frac{\frac{R_o}{1+\beta A_V} \times R_L}{\frac{R_0}{1+\beta A_V} + R_L} = \frac{R_o R_L}{R_o + R_L + \beta.A_v R_L}$$

5.9.2 Current Shunt Feedback

Consider the circuit schematic shown in Fig. 5.20.

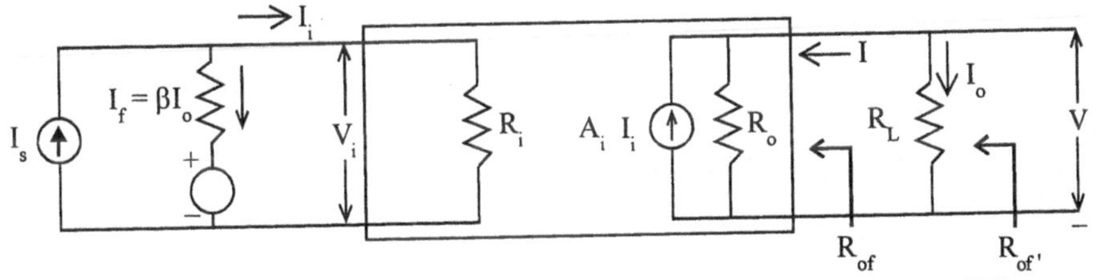

Fig. 5.20 Block schematic for current shunt feedback amplifier.

$\therefore$ $I_o = -I$

A_i = Shunt circuit current gain

A_I = Practical current gain $\left(\frac{I_o}{I_i}\right)$

$$I = \frac{V}{R_o} - A_i\, I_i$$

With, $I_S = 0,\ I_i = -I_f = -\beta I = +\beta I,$

$$I = \frac{V}{R_o} - \beta A_i\, I \text{ or } I\,(1+\beta A_i) = \frac{V}{R_o}$$

$$R_{of} = \frac{V}{I} = R_o\,(1+\beta A_i)$$

A_i = short circuit current gain.

R_{of}' includes R_L as part of the amplifier.

$$R'_{of} = \frac{R_{of}.R_L}{R_{of} + R_L}$$

$$= \frac{R_o(1+\beta A_i)R_L}{R_o(1+\beta A_i)+R_L}$$

$$= \frac{R_oR_L}{R_o + R_L} \times \frac{1+\beta A_i}{1+\dfrac{\beta A_i R_o}{R_o + R_L}}$$

$$A_I = \frac{I_o}{I_i} = \frac{A_i R_o}{R_o + R_L}$$

$$R'_o = \frac{R_o.R_L}{R_o + R_L}$$

$\therefore$ $$R'_{of} = R'_o \frac{1+\beta A_i}{1+\beta A_I}$$

If, $R_L = \infty$,

$A_I = 0,\ R'_o = R_o$

$\therefore$ $$\boxed{R'_{of} = R_o\,(1+\beta\, A_i)}$$

Problem 5.1

The following information is available for the generalized feedback network. Open loop voltage amplification (A_v) = – 100. Input voltage to the system (V_i') = 1mV. Determine the closed loop voltage amplification, the output voltage, feedback voltage, input voltage to the amplifier, and type of feed back for (a) $\beta = 0.01$, (b) $\beta = -0.005$ (c) $\beta = 0$ (d) $\beta = 0.01$.

Solution

$$A'_v = \text{closed loop voltage amplification} = \frac{A_v}{1-\beta_v A_v}$$

Sign must be considered,

$$A'_V = \frac{-100}{1-0.01(-100)} = -50.$$

$\because$ It is positive feedback, A_v' is less negative $\therefore$ there is increase in gain.

V_o Output voltage $= V_i A_V$

$= -50 \times 10^{-3}$ mV

$= -50$ mV.

V_x Feedback voltage $= \beta V_o = 0.01\,(-50 \times 10^{-3})$

$= -0.5$ m V.

$V_i = V'_i + \beta_v . V_o$

$= 10^{-3} + (-0.5 \times 10^{-3})$

$= 0.5$ m V.

This is negative feedback because, $V_i < V_i'$.

$$\therefore \qquad |A_V'| < |A_v|.$$

Problem 5.2

In the above problem determine the % variation in A_v^1 resulting from 100 % increase in A_v when $\beta_v = 0.01$.

When $A_v = -100$, $A_v^1 = -50$.

Solution

If A_v increases by 100 %, Then new value of $A_v = -200$.

$$A_V' = \frac{A_v}{1-\beta_v A_v}$$

$$A_V' = \frac{A_v}{1-\beta_v A_v}$$

Now, $$A_V' = \frac{-200}{1-0.01(-200)} = -66.7.$$

Change in A_V' is $-66.7 - 50 = 16.7$

$$\therefore \qquad \text{Variation} = \frac{16.7}{50} \times 100 \quad = 33.3\%$$

Problem 5.3

An amplifier with open loop voltage gain $A_v = 1000 \pm 100$ is available. It is necessary to have an amplifier where voltage gain varies by not more than $\pm$ 0.1 %

(a) Find the reverse transmission factor β of the feedback network used

(b) Find the gain with feedback.

Solution

(a) $$\frac{dA_f}{A_f} = \frac{1}{1+\beta A} \cdot \frac{dA}{A}$$

or $$\frac{0.1}{100} = \frac{1}{(1+\beta A)} \cdot \frac{100}{1000}$$

$\therefore \quad (1+\beta A) = 100$ or $\beta A = (100-1) = 99$.

Hence, $$\beta = \frac{99}{1000} = 0.099$$

(b) $$A_f = \frac{A}{1+\beta A} = \frac{1000}{1+99} = 10$$

Problem 5.4

A gain variation of + 10 % is expected for an amplifier with closed loop gain of 100. How can this variation be reduced to + 1 %

Solution

$$S' = \frac{S}{1-\beta A_v}.$$

for positive feedback,

$$S' = \frac{S}{1-\beta A_v}$$

$$S' = 0.01, \quad S = 0.1$$

for negative Feedback, $(1+\beta A_v) = \dfrac{0.1}{0.01} = 10 = \dfrac{S}{S'}$

$$A_V' = 100 = \frac{A_v}{1+\beta A_v} = \frac{A_v}{10}$$

$\therefore$ $$A_V' = 100 \times 10 = 1000.$$

$$1 + \beta A_v = 10\,; \quad \therefore \quad \beta . A_v = 10 - 1 = 9$$

$\therefore$ $$\beta = \frac{9}{A_v} = \frac{9}{1000} = 0.009$$

$\therefore$ By providing negative feedback, with $\beta = 0.009$, we can improve the stability to 1%.

Problem 5.5

An amplifier with $A_v = -500$, produces 5% harmonic distortion at full output. What value of β is required to reduce the distortion to 0.1% ? What is the overall gain?

Solution

$$D' = \frac{D}{1+\beta A_v} \quad \text{(for negative feedback)}$$

$$D = 5; \quad D' = 0.1$$

$$N' = \frac{N}{(1+\beta A_v)}$$

$$0.1 = \frac{5}{1+\beta\,[500]}$$

or $$\beta = \frac{4.9}{50} = 0.098.$$

$$A_V' = \frac{-A_v}{1+\beta A_v} = -\frac{500}{50} = -10.$$

$$\frac{D'}{D} = \frac{A_v'}{A_v}.$$

$$\therefore \qquad A_V' = A_V\left(\frac{0.1}{5}\right) = -10$$

Problem 5.6

An amplifier has voltage gain of 10,000 with ground plate supply of 150 V and voltage gain of 8000 at reduced rate supply of 130 V. On application of negative feedback. The voltage gain at normal plate supply is reduced by a factor of 80. Calculate the 1) voltage gain of the amplifier with feedback for two values of plate supply voltage. 2) Percentage reduction in voltage gain with reduction in plate supply voltage for both condition, with and without feedback.

Solution

$$A = \frac{V_o}{V_s}.\ \text{When } V_s = 150\text{ V and } A = 10{,}000.$$

$$\therefore \qquad V_o = 10{,}000 \times 150 = 15 \times 10^5\text{ V. (at normal supply voltages)}$$

$$\beta = \frac{V_f}{V_o}.\ A_{fb} = \frac{A}{1+\beta A}$$

But $$A_{fb} = \frac{10{,}000}{80}$$

$\because$ it is reduced by a factor of 80 with feedback. $\left(\frac{1}{80}^{Th} \text{ of } 10{,}000\right)$

$$\therefore \qquad \frac{10{,}000}{80} = \frac{10{,}000}{1+\beta \times 10000}$$

$$1 + \beta\,(10{,}000) = 80$$

$$\therefore \qquad \beta = +\frac{79}{10^4}$$

$$= 0.0079$$

V_o when $V_s = 130$ V is gain $\times$ $V_s = 8000 \times 130 =$ **104×10^4 V**

Voltage gain of amplifier with feedback when $V_s = 150$V,

$$A_{fb_1} = \frac{10{,}000}{80} = 125.$$

$$V_s = 130\text{ V.}$$

$$A_{fb_2} = \frac{8000}{1+0.0079 \times 8000} = 124.7$$

$$\%\ \text{stability of gain without feedback} = \frac{10{,}000 - 8000}{10{,}000} \times 100 = 20\%$$

$$\%\ \text{stability of gain with feedback} = \frac{125 - 124.7}{125} \times 100 = 0.24\%$$

$\therefore$ With negative feedback stability is improved.

5.10 ANALYSIS OF FEEDBACK AMPLIFIERS

When an amplifier circuit is given, separate the basic amplifier block and the feedback network. Determine the gain of the basic amplifier A. Determine the feedback factor β. Knowing A and β of the feedback amplifier, the characteristics of the amplifiers R_i, R_o, noise figure, A_f etc., can be determined. (Fig. 5.21).

Complete analysis of the feedback amplifier is done by the following steps :

Step I : First identify whether the feedback signal X_f is a voltage feedback signal or current feedback signal. If the feedback signal X_f is applied in shunt with the external signal, ***it is shunt feedback***. If the feedback signal is applied in series, it is ***series feedback***.

Then determine whether the sampled signal X_o is a voltage signal or current signal. If the sampled signal X_o is taken between the output node and ground, it is voltage feedback.

If the sampled signal is taken from the output loop, it is current feedback.

Step II :

2. Draw the basic amplifier without feedback
3. Replace the active device (BJT or FET) by proper model (hybrid – π equivalent circuit or ***h-parameter*** model)
4. Indicate V_f and V_0 in the circuit.
5. Calculate $\beta = V_f / V_0$
6. Calculate 'A' of basic amplifier.
7. From A and β, calculate $A_f = \dfrac{A}{1+\beta A}$; R_{if} and R_{of}.

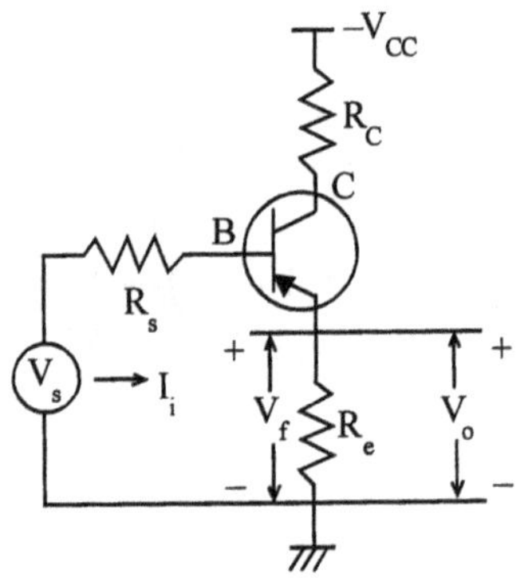

Fig. 5.21 Circuit for voltage series feedback.

5.10.1 VOLTAGE SERIES FEEDBACK

Amplifier circuit is shown in Fig. 5.21. Simplified circuits as shown in Fig. 5.22 and block schematic in Fig. 5.23.

This is emitter follower circuit because output is taken across R_e. The feedback signal is also across R_e. So $V_f = V_o$. because feedback signal is proportional to output voltage. If V_o increases V_f also increase and if V_0, V_f also decreases because V_f a V_o. So it is

Voltage feedback. Now the drop across R_e, i.e. V_f opposes the input voltage. It reverse biases the feedback in V_f coming in series with V_{BE} and it opposes it. So it is negative ***series voltage feedback***.

In the current series feedback, V_f is the voltage across R_e but, output is taken across R_c or R_L and not Re. So in that case $V_f \propto I_o$ or I_c or not V_o. But in this case, it is emitter follower circuit. Output is taken across R_e and that itself is the feedback signal V_f.

Let us draw the base amplifier without feedback.

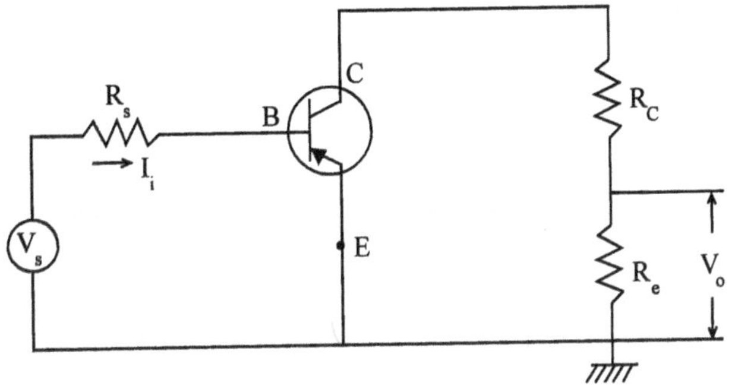

Fig. 5.22 Simplified circuit.

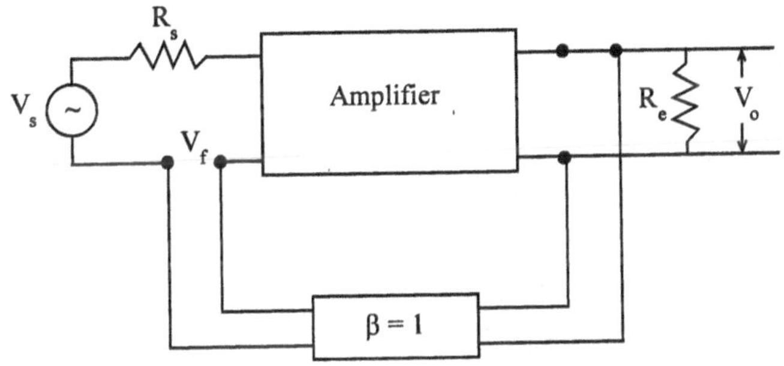

Fig. 5.23 Block schematic.

EQUIVALENT CIRCUIT

Consider the circuit shown in Fig. 5.24.

Now replace the transistor by its low frequency h-parameter equivalent circuit.

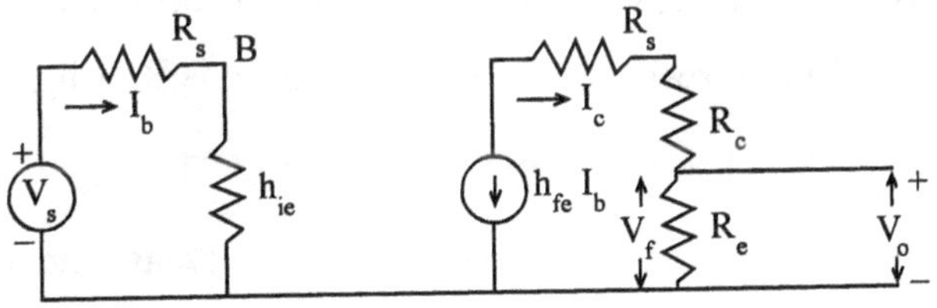

Fig. 5.24 Equivalent circuit

$$V_o = V_f$$

$$\therefore \qquad \boxed{\beta = \frac{V_o}{V_f} = 1.}$$

V_S is considered as part of the amplifier. So $V_i = V_S$.

$$A_v = \frac{V_o}{V_i} \,.\, V_o = h_{fe} \cdot I_b \cdot R_e$$

$I_C \simeq I_e$; $I_C = h_{fe} \cdot I_b$ ($\because$ It is voltage source, R_s in services with h_{ie})

$V_i = (R_s + h_{ie})\, I_i$

$I_i = I_b$

$$A_v = \frac{h_{fe}\, I_b\, R_e}{I_b (R_s + h_{ie})}$$

$$\therefore \quad A_v = \frac{h_{fe} . R_e}{R_s + h_{ie}} = \text{Voltage gain without feedback}$$

A'_V : Voltage gain with feedback.

$$A'_V = \frac{A_v}{1 + \beta A_v}$$

$\beta = 1$

$$A_V = \frac{h_{fe} . R_e}{R_s + h_{ie}}$$

$$= \frac{\dfrac{h_{fe} \cdot R_e}{R_S + h_{ie}}}{1 + \dfrac{1.h_{fe} \cdot R_e}{R_S + h_{ie}}}$$

$$A'_V = \frac{h_{fe} . R_e}{R_s + h_{ie} + h_{fe} R_e}$$

$h_{fe}\, R_e >> (R_s + h_{ie})$

$\therefore$ $A_{Vf} \cong 1.$ or <1, which is true for emitter follower.

R'_i = Input resistance without feedback is $R_s + h_{ie}$.

R'_i = Input resistance with feedback

$R'_i = R_i (1 + \beta A_v)$ (voltage Feedback increase input resistance)

$\beta = 1$,

$$A_V = \frac{h_{fe} . R_e}{R_s + h_{ie}}$$

$$\therefore \quad R'_i = (R_s + h_{ie}) \left(1 + \frac{h_{fe} . R_e . 1}{R_s + h_{ie}}\right)$$

$$R_i' = \frac{(R_S + h_{ie})(R_S + h_{ie} + h_{fe}R_e)}{(R_S + h_{ie})}$$

$$\boxed{R_i' = R_s + h_{ie} + h_{fe} R_e}$$

R_i' : Output resistance with feedback

Output resistance of the circuit is R_e. Because output taken across R_e and not R_c and ground. Load resistance is also R_e.

∴ Considering load resistance,

$$R_0' = \frac{R_o'}{1+\beta A_v}$$

$$R_0' = \frac{R_e}{1+\frac{1.h_{fe}.R_e}{R_s + h_{ie}}}$$

Problem 5.7

Calculate A_{Vf}, R_{of} and R_{if} for the voltage series feedback for the circuit shown in Fig. 5.25.

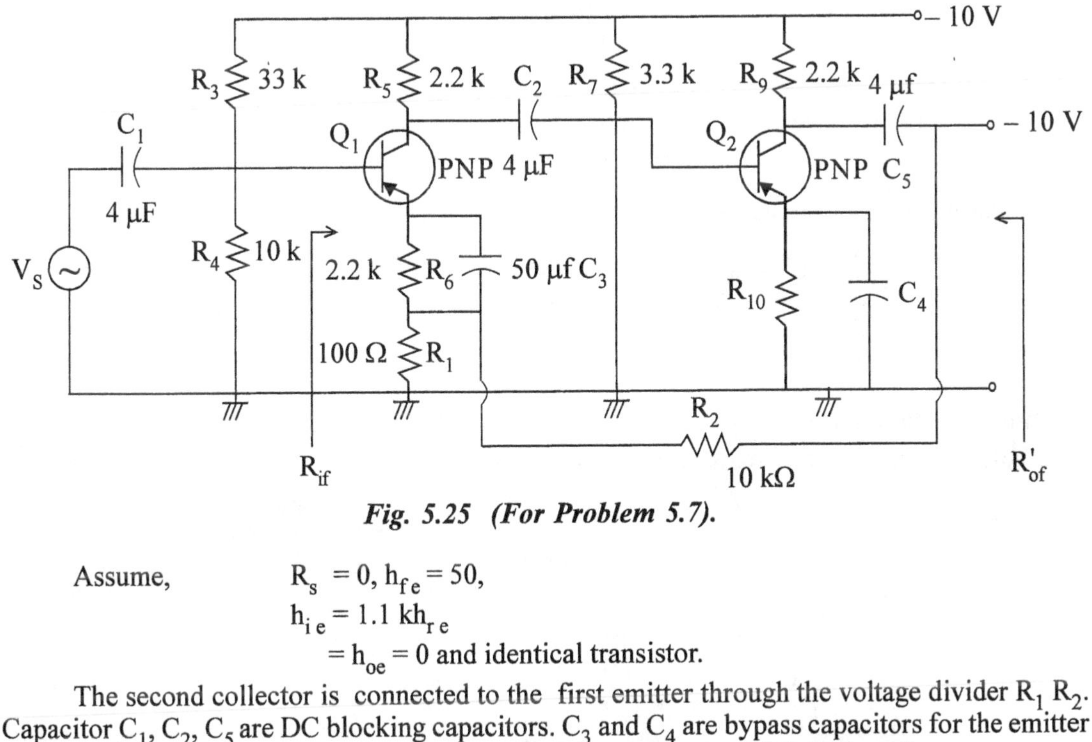

Fig. 5.25 (For Problem 5.7).

Assume, $R_s = 0$, $h_{fe} = 50$,
$h_{ie} = 1.1$ kh_{re}
$= h_{oe} = 0$ and identical transistor.

The second collector is connected to the first emitter through the voltage divider R_1 R_2. Capacitor C_1, C_2, C_5 are DC blocking capacitors. C_3 and C_4 are bypass capacitors for the emitter resistors. All these capacitances represent negligible reactance at high frequencies.

Solution

Voltage gain without feedback $A_V = A_{V_1} \times A_{V_2}$.

Load resistance, $R_{L_1}^{'}$

R_5 in parallel with R_7 ($\because$ at high frequency X_{C_2} is negligible), in parallel with

R_8, and h_{ie2} $R_{L_1}^{'} = 2.2 \parallel 33 \parallel 4.3 \parallel 1.1 \text{ k}\Omega = 980\Omega.$

$$R_L^{'} = R_5 \parallel R_7 \parallel R_8 \parallel h_{ie_2}$$

Effective load resistance $R_{L2}^{'}$ of transistor Q_2 is its collector resistance R_9 in parallel with

$(R_1 + R_2)$; $R_{L_2}^{'} = R_9 \parallel (R_1 + R_2)$

$\therefore$ $R_{L_2}^{'} = 2.2 \text{ k}\Omega \parallel^{le} \text{with } 10.1\text{k}\Omega \simeq 2 \text{ k}\Omega$

Effective emitter impedance R_e of Q_1 is R_1 parallel with R_6.

$R_{e_1} = R_1 \parallel^{le} R_6$ $\qquad \because$ C_3 is short for AC

$R_e = 0.1 \parallel^{le} \text{with } 2.2 \text{ k}\Omega = 0.098 \text{ k}\Omega = 98\Omega$

The voltage gain A_{V_1} of Q_1 for a common emitter transistor with emitter resistance

is $\quad A_{V_1} = \dfrac{V_1}{V_i} = \dfrac{-h_{fe} R_L^{'}}{h_{ie} + (1 + h_{fe}) R_e}$

For Q_1 emitter is not at GROUND potential. So for A_{V_1} this formula must be used.

$$= \frac{-50 \times 0.980}{1.1 + (51) \times 0.098} \simeq -7.78$$

Voltage gain A_{v2} of transistor Q_2,

$$A_{v_2} = A_I \cdot \frac{R_L}{R_i} = -\frac{h_{fe} R_L}{h_{ie}}$$

$$A_{v_2} = -\frac{h_{fe} \cdot R_{L_2}}{h_{ie}} = -\frac{50 \times 2}{1.1} = -91$$

For Q_2 emitter is bypassed. So $R_e = 0$. $\therefore$ For A_v this formula is used.

Voltage gain A_v of the two stages is cascade without feedback

$$A_v = \frac{V_o}{V_i} = A_{v1} \times A_{v2} = 7.78 \times 91 \simeq 728$$

$$\beta = \frac{R_1}{R_1 + R_2} = \frac{100}{100 + 10{,}000} = \frac{100}{10100} = \frac{1}{101} \simeq 0.01$$

$A_v \times \beta = 728 \times 0.01 = 7.28$

$D = 1 + \beta A_v = 8.28$.

$$A_{vf} = \frac{A}{1 + \beta A} = \frac{A_v}{D} = \frac{728}{8.28} \simeq 90$$

$$A_{vf} = \frac{A}{1 + \beta A} = \frac{A_v}{D} = \frac{728}{8.28} \simeq 90$$

Input resistance without external feedback

$$R_i = h_{ie} + (1+h_{fe})\, R_e = 1.1 + (1+50)\, 0.098 \simeq 6.1K\,\Omega$$

$$R_i' = R_{if} = R_i(1+\beta A) = 6.1\ (1+7.28) = 50.5 \approx 51$$

Output resistance without feedback $R' = R'_{L_2} = 2\ k\Omega$

Output resistance without feedback $R'_{0f} = \frac{R_o}{1+\beta A} = \frac{2}{8.28} = 0.24\ K\,\Omega$

Equivalent Circuit

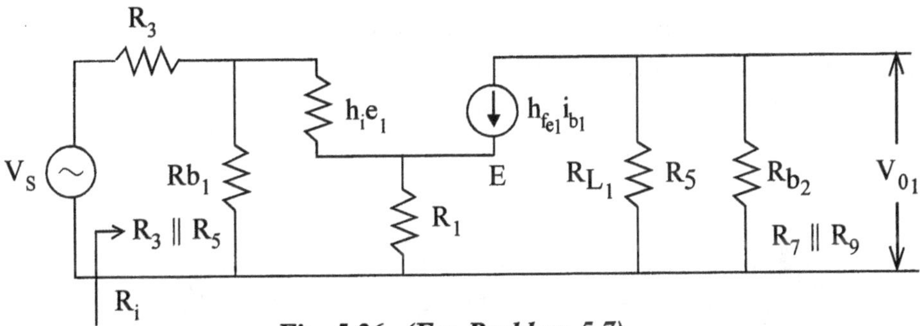

Fig. 5.26 (For Problem 5.7).

BJT will not behave like a fixed resistor with h_{ie}, value . So circuit analysis is not simply parallel or series combination.

$$A_V = \frac{V_o}{V_i} = \frac{R_L}{R_s + r_{bb'}}$$

$$R_o = \frac{V_o}{R_L} = \frac{V_i}{R_s + r_{bb}'}$$

$$A_I = \frac{I_o}{I_i} = \frac{R_s}{R_s + r_{bb}'}$$

5.10.2 CURRENT SHUNT FEEDBACK

The circuit is shown in Fig. 5.27.

Amplifier circuit parameters in terms of transistor h-parameters.

$$h_{ie} = R_i$$
$$h_{fe} = \beta$$
$$h_{oe} = R_0$$

The circuit shows two transistors Q_1 and Q_2 in cascade(See Fig.5.30). Feedback is provided from the emitter of Q_2 to the base of Q_1. This is negative feedback because, V_{i_2} the input voltage to Q_2 is >> V_{i1}.V_{i2} is out of phase with V_{i1}. $V_{i\,2}$ >> V_{i1} because Q_1 is in Common emitter configuration. A_V is large. Also, V_{i_2} is 180^0 out of phase with V_{i1}. Q_2 is emitter follower because emitter is not at ground potential. Voltage is taken across R_e. [This voltage follows the collector voltage. So it is emitter follower]. So $A_v < 1. \simeq 0.99$

$\therefore$ V_{e2} is slightly less than V_{i2}, and there is no phase shift. [because emitter follows action]

$\therefore$ V_{e2} is in phase with V_{i2}.

V_{i2} is out of phase with V_{i1}.

$\therefore$ V_{e2} is out of phase with V_{i1} (180^0). So it is negative feedback

$V_{i2} >> V_{i1}$. $V_{e2} \simeq V_{i2}$.

$V_{e2} >> V_{i1}$.

If the input signal V_s increases, I_s' the input current from source also increases.If I_s' increases,

I_f also increases ($\because$ V_{e2} increases as I_s' increases)

$I_i = I_s' - I_f$. (I_i is the base current for the transistor Q_1. So it is negative feedback)

This is current shunt feedback because,

$$I_f = \frac{V_{e_2} - V_i}{R'}$$

But $$V_{i1} << V_{e2}.$$

$$I_f \simeq + \frac{V_{e_2}}{R'}.$$

I_0 = collector current of Q_2. $\qquad V_{e2} = (I_0 - I_f)\, R_e$

$\simeq$ emitter current of Q_2.

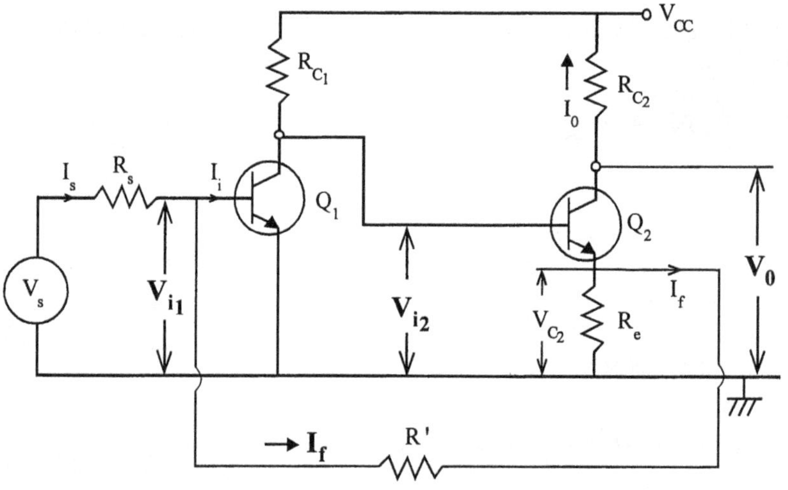

Fig. 5.27 Current shunt feedback.

Dividing by R.

$$\frac{V_{e2}}{R'} = I_f = \frac{(I_o - I_f)R_e}{R'}.$$

$$\therefore \quad I_f = + \frac{(I_o - I_f)R_e}{R'};$$

$$I_f + \frac{I_f.R_e}{R'} = \frac{I_o.R_e}{R'}$$

$$I_f\left(1+\frac{R_e}{R'}\right) = \frac{I_o.R_e}{R'}$$

$$I_f = \frac{R'.I_0.R_e}{(R'+R_e)R'} = \frac{R_e}{(R'+R_e)} I_0$$

$$\boxed{\beta = \frac{I_f}{I_o} = \frac{R_e}{R'+R_e}}$$

$I_f \alpha I_0$

$\therefore$ This is current feedback

A_I' : Current gain with feedback.

$$A_I' = \frac{I_o}{I_s'}$$

$I_S' = I_i + I_f$ which is small μA $I_i = I_b$

$\therefore \quad I_S' \simeq I_f$

$\therefore \quad A_I' = \frac{I_o}{I_f} = \frac{1}{\beta} \qquad \because \beta = \frac{I_f}{I_o}$

$\therefore \quad \boxed{A_I' = \frac{R'+R_e}{R_e}}$

A_V' : Voltage gain with feedback.

$\frac{V_o}{V_s}$ (with feedback)

$V_0 = I_0 . R_{c2}$

$V_s = I_s . R_s.$

$$\frac{V_o}{V_s} = \frac{I_o.R_{c2}}{I_sR_s} \qquad I_S \simeq I_f$$

$$A_V' \simeq \frac{I_o}{I_f}.\frac{R_{c2}}{R_s} \qquad \frac{I_0}{I_f} = \frac{1}{\beta}$$

$$= \frac{1}{\beta}.\frac{R_{c2}}{R_s} \qquad \beta = \frac{R_e}{R'+R_e}$$

$$\boxed{A_V' = \frac{R'+R_e}{R_e}.\frac{R_{c2}}{R_s}}$$

We shall determine R_i', R_0' taking numerical values. The actual circuit, without feedback, considering R' can be drawn as shown in Fig. 5.28 To represent R' on the input side, with output terminals open circuited ($\therefore$ it is emitter follower configuration, output is taken across emitter and ground). When E_2 is left open (To make $I_o = 0$). R' is in series with R_{e2} and thus total resistance is between base B_1, and ground. Hence on the input side, R' is in series with R_{e2}.

To find the output resistance, the base of Q_1 is shorted to ground because one terminal of R' is connected to (input shorted looking the high output terminals) B_1 is at ground. R' or R_{e2} are in parallel.

$\therefore$ The circuit is as shown below, (Fig. 5.28)

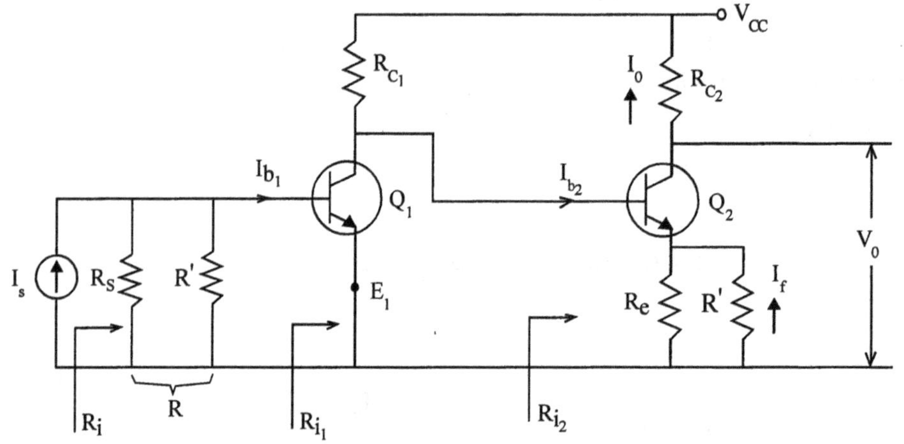

Fig. 5.28 Current shunt feedback without R'.

Because it is shunt feedback, we have shown this as a current source with R_s in parallel with I_S.

Problem 5.8

To find A_V', R_i', R_0'. A_I' for the circuit shown in Fig. 5.31.

$$R_{c1} = 3\text{ k}\Omega, \qquad R_{c2} = 500\Omega; \quad R_{e2} = 50\Omega$$
$$R' = R_s = 1.2\text{ k}\Omega \quad h_{fe} = 50.$$
$$h_{ie} = 1.1\text{ k}\Omega \qquad h_{re} = h_{re} = 0.$$

Solution

In this problem, output is taken at the collector 2 and not emitter 2. So the formulae derived earlier can be used directly.

$$A_I = \frac{-I_{c_2}}{I_s} = \frac{-I_{c2}}{I_{b2}} \cdot \frac{I_{b2}}{I_{c1}} \cdot \frac{I_{c1}}{I_{b1}} \cdot \frac{I_{b1}}{I_s}$$

(Multiplying and dividing by I_{b2}, I_{c1} and I_{b1})

$$\frac{-I_{c_2}}{I_{b_2}} = -h_{fe} = -50; \text{ (Emitter follower)}$$

$$\frac{I_{c_1}}{I_{b_1}} = + h_{fe} = +50 \text{ (common emitter configuration)}$$

$I_{b2} \# I_{c1}$,

∵ The current gets divided in the ratio of the resistances, current gets divided depending upon R_{c1} and R_{i_2} of transistor Q_2.

$$\frac{I_{b2}}{I_{c1}} = \frac{-R_{c_1}}{R_{c_1} + R_{i_2}}$$

$$= \frac{-3}{3+3.55} = -0.457.$$

$$R_{i2} = h_{ie} + (1+h_{fe})(R_{e2} \parallel R').$$

(For emitter follower configuration this is the input resistance).

$$= 1.1 + (50+1)\left[\frac{0.05 \times 1.20}{1.25}\right] \quad (R_{e\,2} = 0.05 \text{ k}\Omega)$$

$$= 3.55 \text{ k}\Omega$$

Let $R = R_S$ in parallel with $(R' + R_{e2})$

$$= \frac{1.2 \times 1.25}{1.2 + 1.25} = 0.612 \text{ k}\Omega$$

$$= \frac{I_{b_1}}{I_S} = \frac{R}{R + h_{ie}}$$

$$= \frac{0.61}{0.61 + 1.1} = 0.358\ \Omega$$

∴ $A_I = (-50)(-0.457)(50)(0.358)$

$$= +406$$

$$\beta = \frac{R_{e_2}}{R_{e_2} + R'}$$

$$= \frac{50}{1{,}250} = 0.04.$$

$$(1 + \beta A_i) = 1 + (0.040)(406) = 17.2$$

$$A_i' = \frac{A_I}{1 + \beta A_i} = \frac{406}{17.2} = 23.6$$

$$A_V' = \frac{V_o}{V_s} = \frac{-I_{c_2}.R_{c_2}}{I_s.R_s} = \frac{A_I'.R_{c_2}}{R_s}$$

$$= \frac{(23.6)(0.5)}{1.2} = 9.83$$

$$[A'_V \text{ is also} = \frac{R_{c_2}}{\beta R_s} = \frac{0.5}{(0.040)(1.2)} = 10.4]$$

R_i = Input resistance without feedback

= R in parallel with h_{ie}

$$= \frac{(0.61)(1.1)}{1.71} = 0.394 \text{ k}\Omega$$

$$R'_i = \frac{R_i}{1+\beta A_i} = \frac{394}{17.2} = 23.0\Omega$$

R_{oL} = Output resistance considering R_L,

$= R_o$ in parallel with $R_{c2} = R_{c2}$ $\because$ R_o is large $\frac{1}{h_{ie}} = \alpha$.

R'_L = with feedback considering R_L,

$$= \frac{R_{OL}(1+\beta A_i)}{1+\beta A_i} = R'_0 = R_{c2} = 500$$

When we represent R on the input side and output side and calculate the value of A_I, it is not the current gain with feedback. Because, R' is represented on the input side leaving E_2 terminal open and on the output side shorting B_1 to ground (to make I_o and $I_s = 0$) . So thus A_I will not be the same if R' is actually connected between E_2 and B_1. We are taking into account the effect of R' and not the feedback effect. In practice for shunt or series feedback, the signal generator will act as a current source or voltage source. Therefore it is capable of supplying current or voltage required .In theory we assume ideal voltage and current sources. Therefore for shunt feedback we must have a current source irrespective of input voltage. The current source will supply sufficient current to drive the resistance. (Fig. 5.29).

R_i = (1+9) 20 kΩ = 200 kΩ

R_s is negligible compared to 200 kΩ.

$$R_o = \frac{r_o}{1+\beta A_o} = \frac{20\Omega}{10} = 2\Omega$$

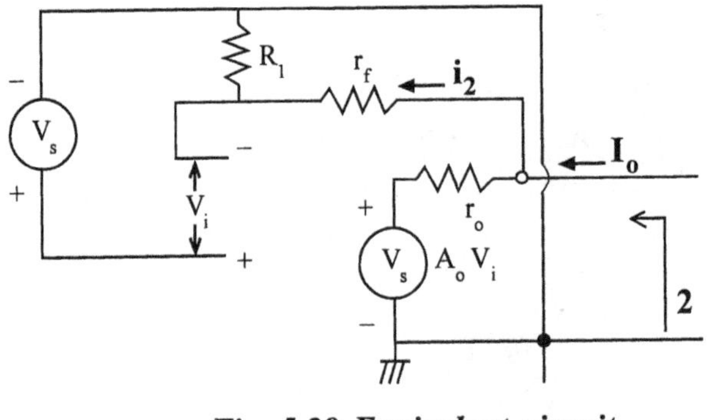

Fig. 5.29 Equivalent circuit.

5.10.3 Current Series Feedback

INPUT RESISTANCE (R_I')

$$V_i = V_i' - V_f$$

$$R_i' = \frac{V_i'}{I_i} = \frac{V_i + V_f}{I_i} = \frac{V_i}{I_i} + \frac{V_f}{I_i} \qquad \therefore \quad V_i' = V_i + V_f.$$

$$\frac{V_i}{I_i} = R_i;$$

$$V_f = I_e R_e \; ; \; I_i = I_e - I_c \qquad [\because \text{ negative current series feedback}] \; (I_b)$$

$$\therefore \qquad R_i' = R_i + \frac{I_e R_e}{I_e - I_c}$$

$$R_i' = R_i + \frac{R_e}{1 - \frac{I_c}{I_e}}$$

Dividing II term by I_e.

$$\frac{I_C}{I_e} = \alpha = \frac{\beta}{1+\beta}$$

$$\frac{I_C}{I_e} = \frac{I_C}{I_C + I_b}$$

$$= \frac{I_c \,|\, I_b}{\frac{I_c}{I_b} + 1} = \frac{h_{fe}}{1 + h_{fe}}.$$

$$\therefore \qquad R_i' = R_i + \frac{R_e}{1 - \frac{h_{fe}}{1 + h_{fe}}}$$

$$R_i' = R_i + (1 + h_{fe}).R_e$$

But $\qquad R_i = h_{ie}$ for the transistor

$$\boxed{\therefore \; R_i' = h_{ie} + (1 + h_{fe})\, R_e}$$

If we consider the bias resistor also, R_1 and R_2 will come in parallel with R_i'

$$= R_i \parallel R_1 \parallel R_2$$

VOLTAGE GAIN (A_V')

$$A_v = \frac{A_i . R_L}{R_i} \qquad \text{[This is the general expression for } A_V \text{ interms of } A_i]$$

$$\therefore \qquad A_V' = \frac{A_i . R_L}{R_i'}$$

[With feedback, A_i will not change, R_L will not change, but R_i will be R'_i]

$$A_i \simeq -h_{fe}$$

$$\therefore \qquad A'_V = \frac{-h_{fe}.R_L}{h_{ie}+(1+h_{fe})R_e}$$

$$\therefore \qquad A'_V < A_V$$

OUTPUT RESISTANCE (R'_0)

$$R \text{ without feedback} \simeq \frac{1+h_{fe}}{h_{oe}}$$

This will be very large.

Taking into effect, the feedback,

$$R'_0 = R_o \parallel R_L$$

$$\frac{1}{R_o} = h_{oe} - \frac{h_{fe}h_{re}}{h_{ie}+R_s}$$

ACTUAL EXPRESSION FOR β THE FEEDBACK FACTOR :

With negative feedback, $R'_i = R_i(1+\beta.A_V)$

$$A_V = \frac{-h_{fe}.R_L}{R_i};$$

$$\therefore \qquad R'_i = R_i\left[1-\frac{h_{fe}.R_L}{R_i}.\beta\right]$$

$$= R_i - h_{fe}.R_L.\beta \qquad .....(5.7)$$

The expression we get for

$$R'_i = R_i + (1+h_{fe}).R_e \qquad .....(5.8)$$

Comparing (5.7) and (5.8), we find that

$$\beta = \frac{(1+h_{fe})}{h_{fe}}\cdot\frac{R_e}{R_L}$$

This is the actual expression for β

$$\frac{1+h_{fe}}{h_{fe}} \simeq 1.$$

$$\therefore \qquad \beta \simeq \frac{R_e}{R_L}$$

5.10.4 CURRENT SERIES FEEDBACK (TRANSCONDUCTANCE AMPLIFIER)

Consider the circuit shown in Fig. 5.30 output is taken between collector and ground. The drop across R_e is the feedback signal V_f. The sampled signal is the load current I_o and not V_0. This is current series feedback. Because V_f is in series with V_i. It is current series feedback.

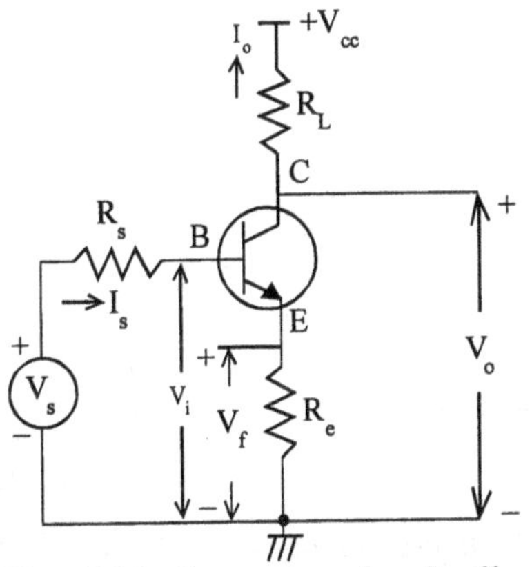

Fig. 5.30 Current series feedback

$$V_f = I_e R_e.$$
$$I_e \simeq I_c = I_o.$$
$$\therefore \quad V_f = I_o R_e.$$

$\because$ R_e is constant,

$$V_f \propto I_o.$$

[β must be independent of R_s or R_L]

$$\beta = \frac{V_f}{V_o} = \frac{-I_o.R_e}{I_o.R_L} = -\frac{R_e}{R_L}$$

The basic amplifier circuit without feedback is shown in Fig.5.31.

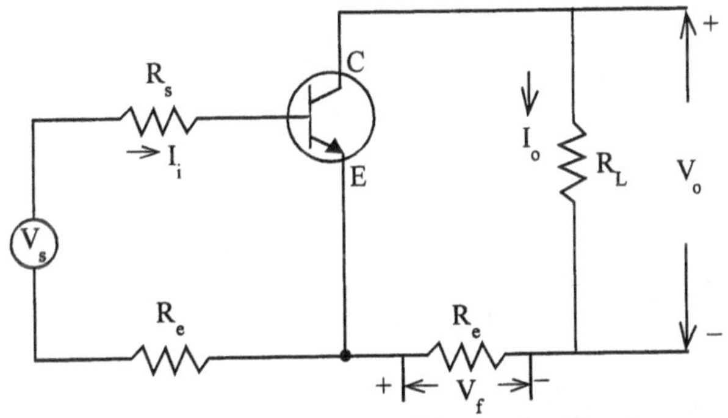

Fig.5.31 Circuit without feedback

The equivalent circuit when the active device is replaced by ***h-parameter*** circuit is shown in Fig.5.32.

This has to be considered as Transconductance Amplifier, Since, β is taken as $\frac{V_f}{V_o}$.

It depends, on R_L, because for this circuit, G_M is considered.

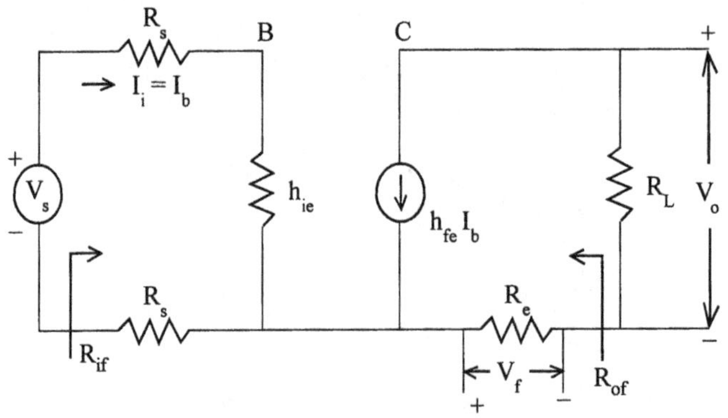

Fig. 5.32 h-parameter equivalent circuit.

$$\beta = \frac{V_f}{I_o} = \frac{-I_o.R_e}{I_o} = -R_e.$$

($\because$ the sampled signal is I_o and not V_o.)

$$G_M = \frac{I_o}{V_i} \quad I_0 = -h_{fe}.I_b \text{ (Input current gain)}$$

Since, $V_i = I_b(R_s + h_{ie} + R_e)$

$\therefore$
$$G_m = \frac{-h_{fe}.I_b}{I_b(R_s + h_{ie} + R_e)}$$

$$= \frac{-h_{fe}}{R_s + h_{ie} + R_e}$$

$$D = 1 + \beta.G_M = 1 + \frac{h_{fe}.R_e}{R_s + h_{ie} + R_e}$$

$$= \frac{R_s + h_{ie} + (1 + h_{fe})R_e}{R_s + h_{ie} + R_e}$$

$$G_{Mf} = \frac{G_M}{D} = \frac{-h_{fe}}{R_s + h_{ie} + (1 + h_{fe})R_e}$$

Voltage gain
$$A_{vf} = \frac{I_o.R_L}{V_s} = \frac{I_o}{V_s} = G_{Mf}$$

$\therefore$
$$A_{vf} = G_{Mf}.R_L$$

$$= \frac{-h_{fe}.R_L}{R_s + h_{ie} + (1 + h_{fe})R_e}$$

$$(1+h_{fe})R_e >> R_s + h_{ie}.$$

Since h_{fe} is a large quantity.

$\therefore$ Denominator $\simeq (1+h_{fe})R_e$.

$$h_{fe} >> 1.$$

$\therefore$ Denominator $\simeq h_{fe} \cdot R_e$.

$$\therefore \quad \boxed{A_{vf} = \frac{-h_{fe}R_L}{h_{fe}.R_e} = -\frac{R_L}{R_e}}$$

$R_i = R_s + h_{ie} + R_e$ (without feedback).

$R_{if} = R_i D = R_s + h_{ie} + (1 + h_{fe}) R_e$.

5.10.5 Voltage Shunt Feedback (Transresistance Amplifier)

This is ***voltage shunt feedback*** because, the feedback current through R_B is proportional to the output voltage or it is in shunt with the input. So it is voltage shunt feedback. [We are not interested whether voltage is fed or current is fed. But the feedback signal is proportional to output voltage. ***So it is voltage feedback***]

The circuit can be written as, shown in Fig. 5.33. This circuit is redrawn in Fig. 5.34.

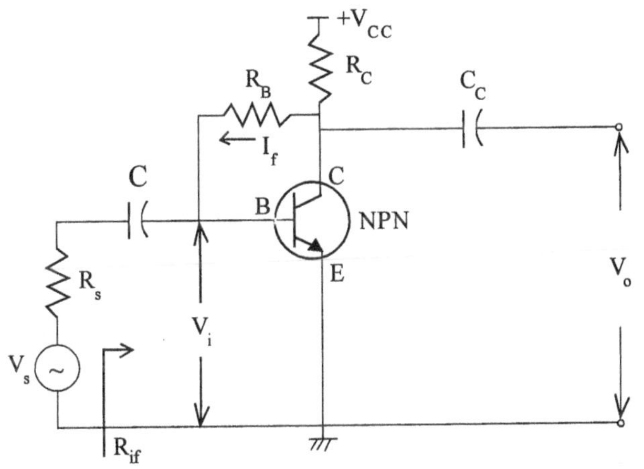

Fig. 5.33 Circuit for voltage shunt feedback.

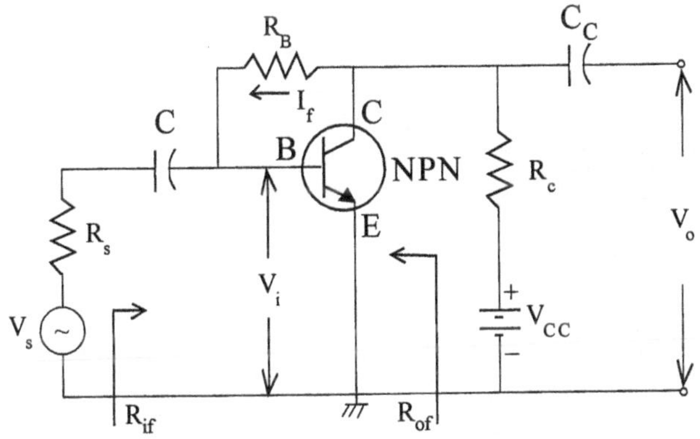

Fig. 5.34 Redrawn circuit for voltage shunt feedback.

The low frequency ***h-parameter*** equivalent circuit is, shown in Fig. 5.35.

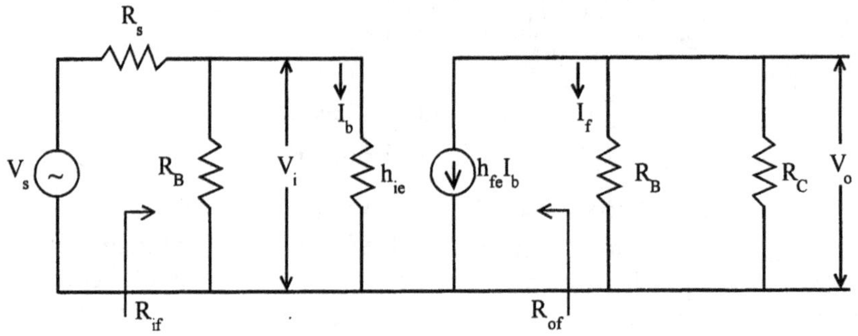

Fig. 5.35 h-parameter equivalent circuit.

$V_0 >> V_i$ $\quad\because$ Amplification is there.

$$I_f \simeq \frac{-V_o}{R_B},$$

$$\therefore \quad I_f \simeq \frac{-V_o}{R_B};$$

$$\therefore \quad \beta = \frac{-I_f}{V_o}$$

$$\beta = \frac{-1}{R_B}$$

$$A_v = \frac{V_o}{V_i} = \frac{-h_{fe} \cdot R_L'}{h_{ie}} \text{ without feedback} \left(\frac{A_I R_L}{R_v}\right)$$

$$R_L' = R_B \parallel R_C.$$

$$A_v{}^1 = \frac{V_o}{V_s} \text{ (with feedback)}$$

$$\boxed{A_v{}^1 = A_v \cdot \frac{R_i'}{R_s + R_i'}}$$

R_i' = Input resistance with feedback

$$A_{vf} = \frac{V_o}{V_s}$$

$$= \frac{V_o}{I_s . R_s} \simeq \frac{1}{\beta R_s} = \frac{-R_B}{R_s}$$

R_i' :

$$I_s = I_f + I_b,\ I_b \text{ is negligible}$$

$$\therefore \quad I_s \simeq I_f \text{ or } \beta = \frac{-1}{R_B}.$$

$$R_i' = h_{ie} \text{ in parallel with } \frac{R_B}{1 - A_v}$$

Transresistance $$R_M = \frac{V_o}{I_s} = \frac{-I_o.R_c}{I_s} = \frac{-h_{fe}.I_b.R_c}{I_s}$$

$$R_C' = R_c \parallel R_B;\ I_s = \frac{(R + h_{ie})}{R} I_b. \qquad \text{Where } R = R_s \parallel R_B.$$

$$\therefore \quad R_M = \frac{-h_{fe}.R_c'.R}{(R + h_{ie})};$$

$$I_b = \frac{-h_{fe}.R_e'.R}{(R + h_{ie})}$$

$$R_M' = \frac{R_M}{1 + \beta R_M}.$$

$$R_i = \frac{R \times h_{ie}}{R + h_{ie}}$$

$$R_i' = \frac{R_i}{1 + \beta R_M}$$

$$R_0' = \frac{R_B}{R_s} \times \frac{R_s + h_{ie}}{h_{fe}}$$

Problem 5.9

For the circuit shown, $R_c = 4\ k\Omega$, $R_B = 40\ k$, $R_s = 10\ k$, $h_{ie} = 1.1\ k\Omega$, $h_{fe} = 50$, $h_{re} = h_{oe} = 0$. Find A_{vf}, R_{if}.

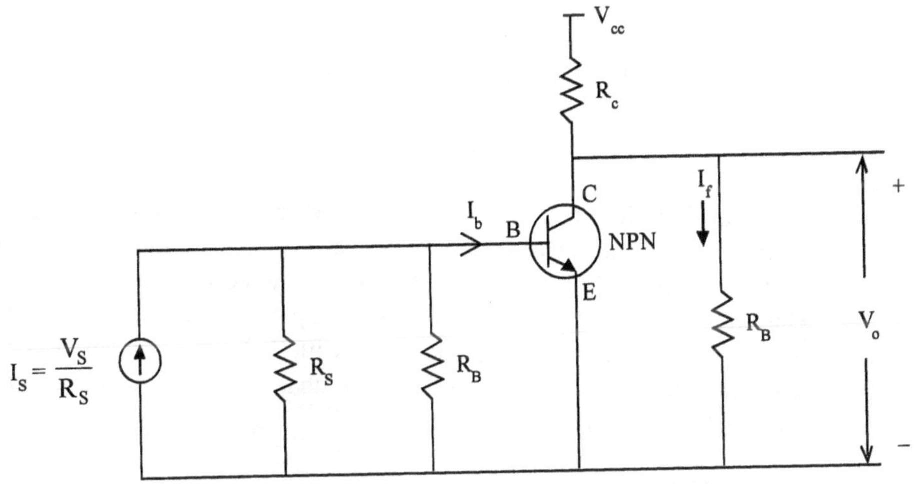

Fig. 5.36 (For Problem 5.9).

Solution

$$R'_C = R_c \,||^{le}\, R_B = \frac{4 \times 40}{4+40} = 3.64 \text{ k}\Omega$$

$$R = R_s \,||^{le}\, R_B = \frac{10 \times 40}{10+40} = 8 \text{ k}\Omega$$

$$\text{Transresistance}\, R_M = \frac{V_o}{I_s} = \frac{-I_C.R_c'}{I_s} = \frac{-h_{fe}.I_b.R_c'}{I_s} \qquad \because\ I_c = h_{fe}\,.\,I_b$$

$$I_s = \frac{(R + h_{ie})}{R}\, I_b; \qquad \therefore\ R_M = \frac{-h_{fe}.I_b\, R'_C\,.\,R}{(R + h_{ie}).I_b}$$

$$R_M = -\,160 \text{ k}$$

$$\therefore \qquad \beta = -\frac{-1}{R_B} = -\,0.025 \text{ mA|V}$$

$$R'_M = \frac{R_M}{1+\beta R_M} = \frac{-160}{5.00} = -\,32.0 \text{ k}\Omega$$

$$A'_V = \frac{V_o}{V_s} = \frac{V_o'}{I_s.R_s}$$

$$= \frac{V_o'}{I_s} = R'_M$$

$$\therefore \qquad A'_V = \frac{R_M'}{R_s} = -\frac{32}{10} = -\,3.2\,.$$

$$R_i = \frac{R \times h_{ie}}{R + h_{ie}} = \frac{8 \times 1.1}{8+1.1} = 0.968 \text{ k}\Omega;$$

$$R'_i = \frac{R_i}{1+\beta R_M} = \frac{968}{5.0} = 193\Omega$$

SUMMARY

- An Amplifier circuit is to provide voltage gain or current gain or both in the form of power gain. But the other desirable characteristics of the amplifier circuits are high Z_i, Low Z_o, Large B.W, low distortion, low noise and high stability. To achieve these characteristics a part of output signal is feedback and coupled to input, to oppose in phase with (V_i). So it is negative feedback. Though gain reduces due to negative feedback, it is employed in amplifier circuits to get the other advantages.

- Feedback factor $\beta = V_f|V_o$. The product βA is called return ratio. $(1 + \beta A)$ is called return difference D or desensitivity factor.
- The different types of feedback are (i) voltage series (ii) current series (iii) voltage shunt and (iv) current shunt.
- With voltage feedback (series or shunt) output resistance R_o decreases.
- With current feedback (series or shunt) R_o increases.
- With series feedback (current or voltage) R_i increases.
- With shunt feedback (current or voltage) R_o decreases.
- With negative feedback, distortion, noise, gain reduce by a factor $(1 + \beta A)$. Bandwidth, f_2, stability improve by $(1 + \beta A)$.
- If the feedback signal is proportional to voltage, it is voltage feedback. If the feedback signal is porportional to current, it is current feedback. If the feedback signal V_f is coming in series with input signal V_i, it is series feedback. If V_f is in parallel with V_i it is shunt feedback.
- If the feedback signal V_f is out of phase with V_i, opposing it is negative feedback. If V_f is in phase with V_i, adding to it or aiding V_i, it is positive feedback.

OBJECTIVE TYPE QUESTIONS

1. The disadvantage of negative feedback is ______________.
2. The expression for sensitivity of an amplifier with negative feedback is ______________.
3. The expression for Desentivity of negative feedback amplifier is D = ______________.
4. The relation between bandwidth of an amplifier without feedback and with negative feedback is ______________.
5. Relation between upper cut-off frequency f_2' with negative feedback and f_2 without negative feedback is f_2' = ______________.
6. Negative feedback is also called as ______________.
7. For Ideal transconductance amplifier R_i = ______________. R_o = ______________.
8. For practical transresistance amplifier, it is desirable that R_i is ______________ and R_o is ______________.
9. Voltage sampling is also known as ______________.
10. Characteristics of ideal voltage amplifier are : ______________.
11. Desirable characteristics of practical current amplifier are = ______________.
12. βA in feedback amplifier circuits is called ______________.
13. With voltage feedback, (series or shunt), output resistance R_o of an amplifier ______________.

14. With series feedback, (voltage or current), input resistance R_i of an amplifier ________________.

15. Expression for reverse transmission factor or feedback factor β = ________________.

16. Identify the type of feedback.

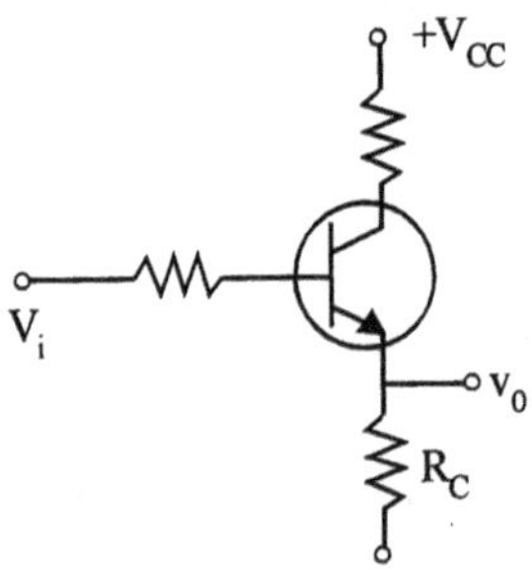

(i)

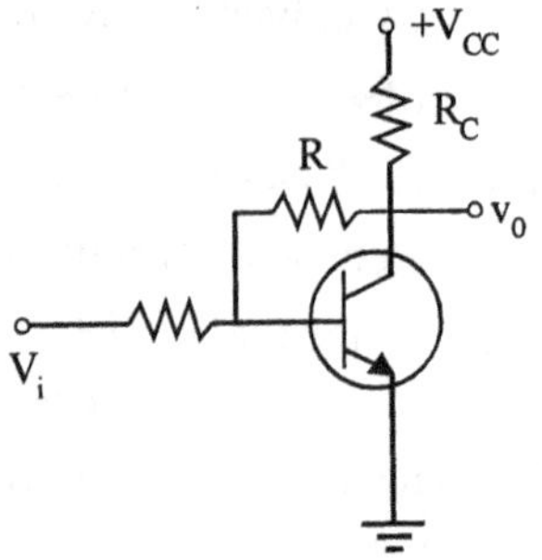

(ii)

17. In the case of Voltage - series feedback, expression for output impedance with feed back z_{of} is ....

18. In the case of Voltage shunt feedback amplifier, expression for input impedance with feedback is Z_{if} = ...

19. In the case of current series feedback amplifier, expression for output impedance with feedback z_{of} is ....

20. Expression for z_{if} with current shunt feedback is ....

ESSAY TYPE QUESTIONS

1. Explain the concept of feedback as applied to electronic amplifier circuits. What are the advantages and disadvantages of positive and negative feedback ?

2. With the help of a general block schematic diagram explain the term feedback.

3. What type of feedback is used in electronic amplifiers ? What are the advantages of this type of feedback ? Prove each one mathematically.

4. Define the terms Return Ratio, Return Difference feedback factor, closed loop voltage gain and open loop voltage gain. Why negative feedback is used in electronic amplifiers eventhough closed loop voltage gain decreases with this type of feedback ?

5. Give the equivalent circuits, and characteristics of ideal and practical amplifiers of the following types (i) Voltage amplifier, (ii) Current amplifiers, (iii) Transresistance amplifier, (iv) Transconductance amplifier.

6. Derive the expression for the input resistance with feedback R_{if} (or R_i') and output resistance with feedback R_{0f} (or R_i') in the case of

   (a) Voltage series feedback amplifier.

   (b) Voltage shunt feedback amplifier.

   (c) Current series feedback amplifier.

   (d) Current shunt feedback amplifier.

7. In which type of amplifier the input impedance increases and the output impedance decreases with negative impedance ? Prove the same drawing equivalent circuit.

8. Draw the circuit for Voltage series amplifier and justify the type of feedback. Derive the expressions for A_V', β, R_i' and R_0' for the circuit.

9. Draw the circuit for Current series amplifier and justify the type of feedback. Derive the expressions for A_V', β, R_i' and R_0' for the circuit.

10. Draw the circuit for Voltage shunt amplifier and justify the type of feedback. Derive the expressions for A_V', β, R_i' and R_0' for the circuit.

11. Draw the circuit for Current shunt amplifier and justify the type of feedback. Derive the expressions for A_V', β, R_i' and R_0' for the circuit.

ANSWERS TO OBJECTIVE TYPE QUESTIONS

1. Gain is reduced

2. $S = \dfrac{1}{1+\beta A}$

3. $D = (1 + \beta A)$

4. $(B.W)_f = (1 + \beta A)\ B.W$

5. $f_2' = f_2\,(1 + \beta A)$

6. Degenerative feedback

7. $R_i = \infty$; $R_0 = \infty$

8. Low, Low

9. Node sampling

10. $A_V = \infty$ $R_i = \infty$ $R_0 = 0$

11. High current gain, Low R_i, High R_0

12. Return ratio

13. Decreases

14. Increases

15. $\beta = V_f | V_0$ = feedback signal/output signal

16. (i) voltage-series (ii) voltage-shunt

17. $Z_0/(1 + \beta A)$

18. $Z_i/(1 + \beta A)$

19. $Z_0(1+\beta A)$

20. $Z_i/(1 + \beta A)$

6 Oscillators

In this Chapter,

- Basic principle of oscillator circuits is explained. Generation of sinusoidal waveforms by the oscillator circuits without external A.C. input is explained.
- Barkhausen criteria to be satisfied for generation of oscillations is given.
- R-C phase shift oscillator circuit, Hartley oscillator, Colpitts oscillator, crystal oscillator, Resonant oscillator circuits are given.

6.1 OSCILLATORS

Oscillator is a source of AC voltage or current. We get A.C.output from the oscillator circuit. In alternators (AC generators) the thermal energy is converted to electric energy at 50Hz.

In the oscillator circuits that we are describing now, the electric energy in the form of DC is converted into electric energy in the form of AC. 'Invertors' in electrical engineering convert DC to AC, but there, only output power is the criterian and not the actual shape of the wave form.

An amplifier is different from oscillator in the sense that an amplifier requires some A.C. input which will be amplified. But an oscillator doesn't need any external AC signal. This is shown in Fig. 6.1.

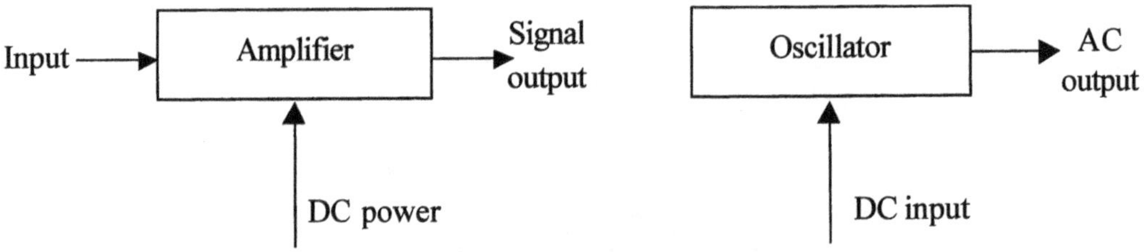

(a) Amplifier circuit *(b) Oscillator circuit*

Fig. 6.1

For an amplifier, the additional power due to amplification is derived from the DC bias supply. So an amplifier effectively converts DC to AC. But it needs AC input. Without AC input, there is no AC output. In the oscillator circuits also DC power is converted to AC. But there is no AC input signal. So the difference between amplifier and oscillator is in amplifiers circuits, the DC power conversion to AC is controlled by the AC input signal. But in oscillators, it is not so.

There are two types of oscillators circuits :

1. Harmonic Oscillators
2. Relaxation Oscillators.

Harmonic Oscillators produce sine waves. Relaxation Oscillators produce sawtooth and square waves etc. Oscillator circuits employ both active and passive devices. Active devices convert the DC power to AC. Passive components determine the frequency of oscillators.

6.1.1 Performance Measures of Oscillator Circuits

1. ***Stability :*** This is determined by the passive components. R,C and L determine frequency of oscillations. If R changes with T, f changes so stability is affected. Capacitors should be of high quantity with low leakage. So silver mica and ceramic capacitors are widely used.
2. ***Amplitude stability :*** To get large output voltage, amplification is to be done.
3. ***Output power :*** Class A, B and C operations can be done. Class C gives largest output power but harmonics are more.

   Class A gives less output power but hormonics are low.
4. ***Harmonics :*** Undesirable frequency components are harmonics. An elementary sinusoidal oscillator circuit is shown in Fig.6.2.

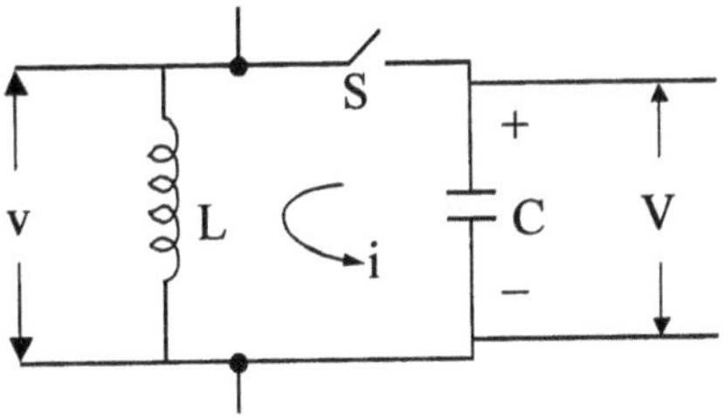

Fig. 6.2 LC tank circuit

L and C are reactive elements. They can store energy. The capacitor stores energy whenever there is voltage across its plates. Inductor stores energy in its magnetic field whenever current flows. Both C and L are lossless, ideal devices. So quality factor is infinity ∞. Energy is introduced into the circuit by charging capacitor to 'V' Volts. If switch is open, C cannot discharges because there is no path for discharge current to flow.

Suppose at $t = t_0$, switch 'S' is closed. Then current flows. Voltage across L will be V. At $t = t_0$, the Voltage across C is V volts. When switch is closed, current flows. So the charge across Capacitor C decreases. Voltage across C decreases, as shown in the waveform. So as the energy stored in Capacitor decreases, the energy stored in inductor L increases, because current is flowing through L. Thus total energy in the circuit remains the same as before. When V across C becomes 0, current through the inductor is maximum. When the energy in C is 0, energy in L is maximum. Then the current in L starts charging C in the opposite directions. So at $t = t_1$ current in L is maximum and for $t > t_1$, the current starts charging C in the opposite direction. So V across C becomes negative as shown in Fig. 6.3. Thus we get sinusoidal oscillations from LC circuit.

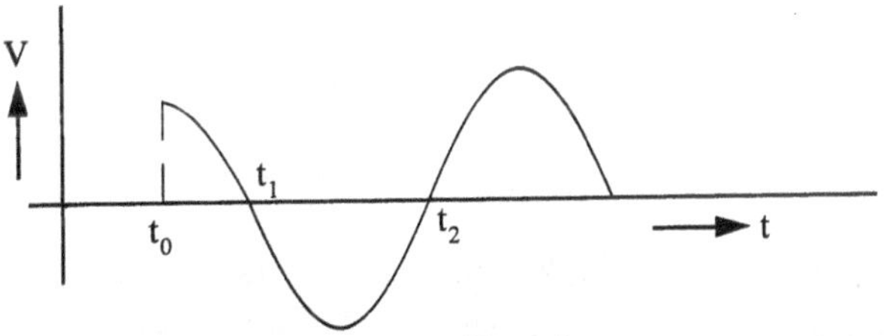

Fig. 6.3 Waveform

Thus we are getting sinusoidal variations without giving any one input (Fig. 6.3). What we have done is depositing some charge on C, so that the circuit operates on its own. But why should we get sinusoidal wave and not triangular or square wave ? Sinusoidal function is the only function that satisfies the conditions governing the exchange of the energy in the circuit.

But the above circuit is not a practical circuit. Because we have to take output from the circuit; i.e. energy has to be extracted, from the circuit. As we draw energy from the circuit, the energy stored in C and L decreases. So output also decreases or the voltage across C and L decreases so we get damped oscillation as shown in Fig. 6.4.

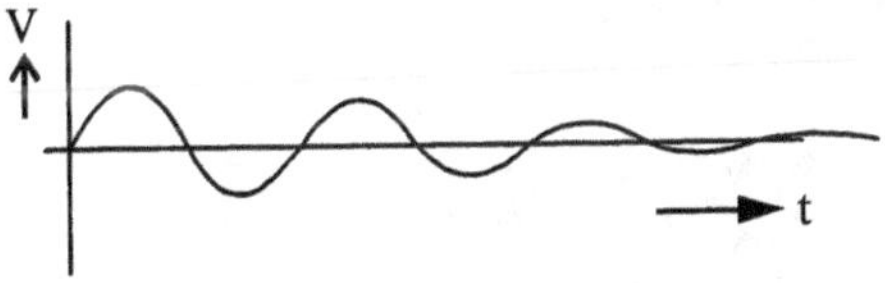

Fig. 6.4 Damped oscillations

The energy lost by the elements must be replenished, so that oscillations are obtained continuously. If a negative resistance R is connected in the circuit, it replenishes, whatever energy that is lost in the circuit. Certain devices like tunnel diode, UJT, etc., exhibit negative resistance. The energy supplied by the negative resistance to the circuit actually comes from the DC bias supply. (Fig. 6.5).

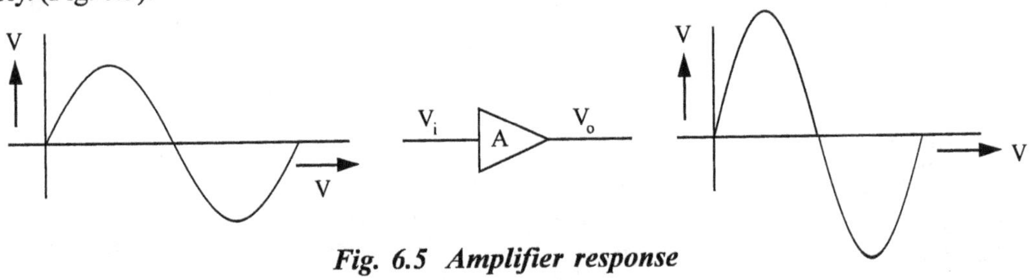

Fig. 6.5 Amplifier response

Another method of producing sinusoidal oscillations is :

Suppose we have an amplifier with gain A_V and phase shift 180°, $A_V >> 1$.

If we connect V_o through a feedback network to V_i as shown in Fig.6.6 so that after feedback, the feedback signal $V_X = V_i$, and also, the feedback network provides 180°, phase shift, after feedback the feedback signal V_X will be removed. Thus, without any input we get sinusoidal output.

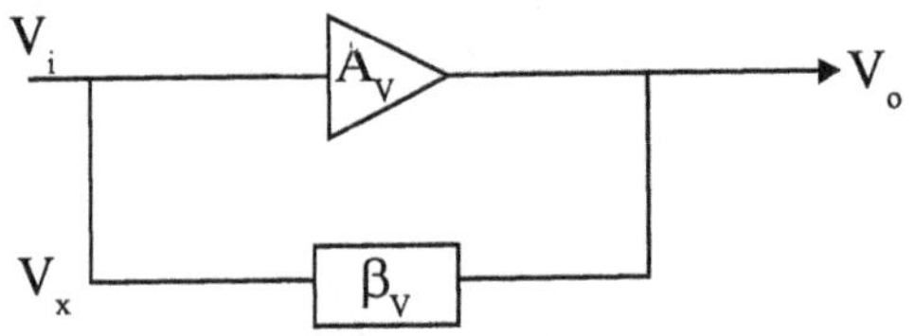

Fig. 6.6 Feedback network

V_i is provided from the feedback signal of V_o itself. To get initially V_o, the noise signal of transistor after switching itself is sufficient. Thus without external AC input we get sinusoidal oscillators.

$$\beta_v = \frac{V_x}{V_o}.$$

$$A_v = \frac{V_o}{V_i}.$$

$$V_X = V_i$$

$$\therefore \quad \beta . A_v = \frac{V_x}{V_o} \times \frac{V_o}{V_i}$$

$$= \frac{V_x}{V_i} \quad \text{or} \quad V_x = V_i$$

$$\therefore \quad \beta A_V = 1$$

Total phase shift = 360° (180 + 180). Therefore, to get sustained oscillations,

1. The loop gain must be unit 1.
2. Total Loop phase shift must be 0° or 360°. (Amplifier circuit produces 180° phase shift and feedback network another 180°.

6.2 SINUSOIDAL OSCILLATORS

Figure 6.7 shows an amplifier, a feedback network and input mixing circuit not yet connected to form a closed loop. The amplifier provides an output signal X_o, as a consequence of the signal X_i applied directly to the amplifier input terminal. Output of feedback network

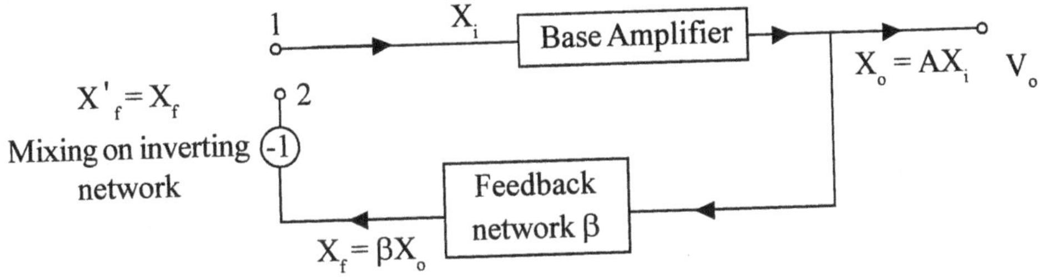

Fig. 6.7 Block schematic.

is $X_f = \beta X_o = A\beta X_i$ and the output of the mixing circuit,

is $X_f' = -X_f = -A\beta X_i$

$$\text{Loop gain,} \quad = \frac{X_f'}{X_i} = \frac{-A\beta X_i}{X_i} = -\beta A.$$

If X_f' were to be identically equal to X_i, input signal, $-\beta A$ should be = 1, so the output will be X_0, even if the input source is removed. The condition that $-\beta A = 1$ means, the loop gain must be 1.

6.3 BARKHAUSEN CRITERION

We make an asssumption that the circuit operates only in the linear region, and the amplifier feedback network contains reactive elements. For a sinusoidal wave form, if $X_i = X_f'$, the amplitude, phase and frequency of X_i and X_f' be identical. ***The frequency of a sinusoidal oscillator is determined by the condition that loop gain, Phase shift is zero at that frequency.***

For oscillator circuits positive feedback must be there i.e., V_f must be in phase with V_i to get added to V_i. When active device BJT or FET gives 180^0 phase shift, the feedback network must produce another 180^0 phase shift so that net phase shift is 0^0 or 360^0 and V_f is in phase with V_i to make it positive feedback.

Oscillations will not be sustained if, at the oscillator frequency the magnitude of the product of the transfer gain of the amplifier and of β are less than unity.

The conditions $-A\beta = 1$ is called ***Barkhausen criterion***

i.e., $|\beta A| = 1$ and phase of $-A\beta = 0$.

But $A_f = \frac{A}{1+\beta A}$. If $\beta A = -1$, than $A_f \to \infty$.

which implies that, there exists an output voltage even in the absence of an externally applied voltage.

A 1-V signal appearing initially at the input terminals will, after a trip around the loop and back to the input terminals, appear there, with an amplitude larger than 1V. This larger voltage will then reappear as a still larger voltage and so on. So if $|\beta A|$ is larger than1, the amplitude of oscillations will continue to increase without limit. But practically , to limit the increase of amplitude of oscillations, nonlinearity ability of the circuit will be set in. Though a circuit has been designed for $|\beta A| = 1$, since circuit components and transistor change characteristics with age, temperature, voltage etc. $|\beta A|$ will become larger or smaller than 1. If $\beta A < 1$, the oscillations will stop. If $\beta A >$ 1, the amplitude practically will increase. So, to achieve a sinusoidal osciallation practically $|\beta A| >$ 1 to 5%, so that with incidental variations in transistor circuit parameters, $|\beta A|$ shall not fall below unity.

The type of noise in electronic circuits and the causes are :

1. ***Johnson noise or Thermal Noise :*** Due to temperature.
2. ***Schottky noise or Shot noise :*** Because of variation is concentration in the Semiconductor Devices.

6.4 R – C PHASE-SHIFT OSCILLATOR (USING JFET)

This is voltage series feedback. FET amplifier is followed by three cascaded arrangements of a capacitor C, resistor R. The output of the last RC combination is returned to the gate. This forms the feedback connection. The FET amplifier shifts the phase of voltage appearing at the gate by 180°. The RC network shifts the phase by additional amount. At some frequency, the phase-shift introduced by this network will be exactly 180°. The total phase-shift at this frequency, from the gate around the circuit and back to the gate is +180 – 180 = 0°. At this particular frequency, the circuit will oscillate (Fig. 6.8).

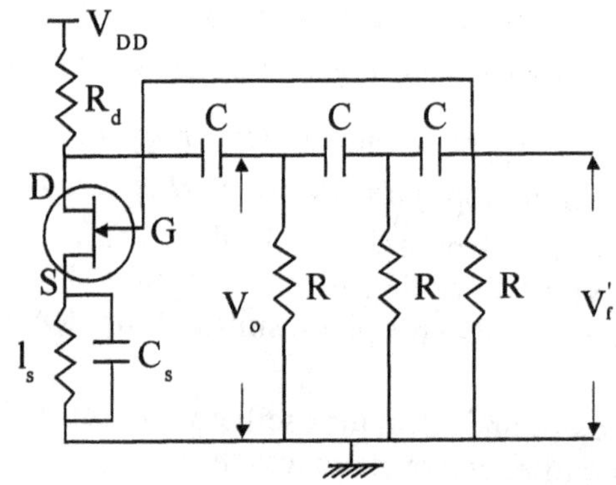

Fig. 6.8 JFET R-C phaseshift oscillator circuit

$\therefore$ At that frequency, $V_f' = -V_f$ $\because$ 180° out of phase.

The equivalent circuit is, show below in Fig. 6.9.

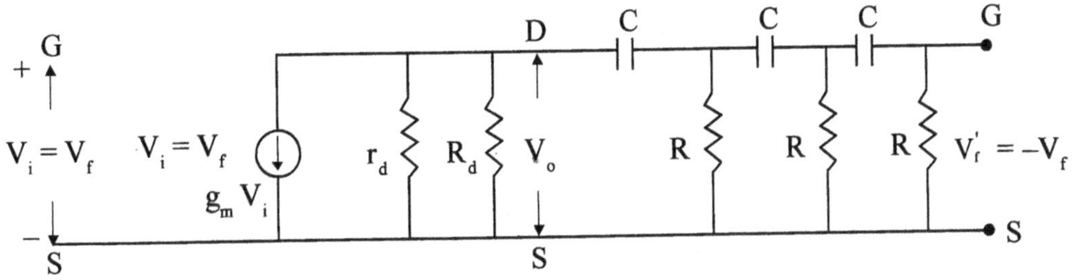

Fig. 6.9 Equivalent circuit

The transformation of RC network is $\frac{V_f}{V_o} = -\beta$; $\therefore\ \beta = \frac{V_f}{V_o}$ and $V_f' = -V_f$

$$\therefore \qquad -\beta = \frac{V_f'}{V_o} = \frac{1}{1 - 5\alpha^2 - j(6\alpha - \alpha^3)}.$$

where $\qquad \alpha = \frac{1}{\omega RC}.$

The phase-shift of $\frac{V_f'}{V_o}$ is 180^0, for $\alpha^2 = 6$ or

β must be real. Therefore $j(6\alpha - \alpha^3) = 0$.

$$\therefore \qquad f = \frac{1}{2\pi RC\sqrt{6}}. \quad \because \quad 6 = \frac{1}{\omega^2 R^2 C^2}.$$

$$\omega^2 = \frac{1}{6R^2C^2}.$$

$$\therefore \qquad \boxed{f = \frac{1}{2\pi RC\sqrt{6}}.}$$

when $\qquad \alpha^2 = 6, -\beta = \frac{1}{1 - 5\times 6 - 5(6\alpha - 6\alpha)} = \frac{-1}{29}$

or $\qquad \boxed{\beta = \frac{1}{29}.}$

In order that $|\beta A|$ is not les than unity, A should be atleast $|29|$. So select F E T whose μ is atleast 29.

6.4.1 To Find the β of the RC Phase-Shift Network (JFET)

Each RC network introduces a phase-shift of 60^0. Therefore, total phase-shift = 180^0 (See Fig.6.10).

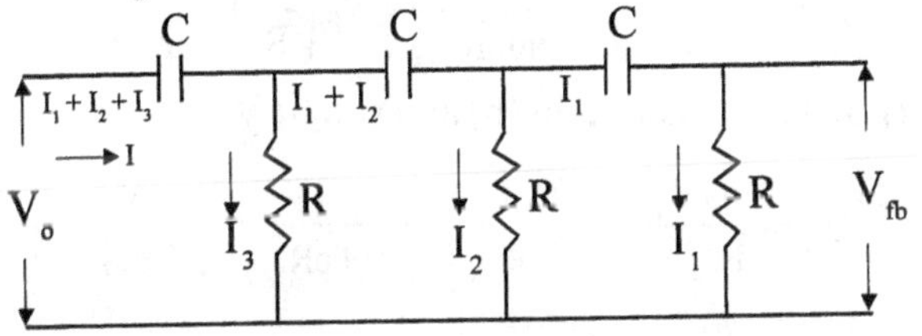

Fig. 6.10 RC phase shift network

$$V_{fb} = I_L . R \quad .....(6.1)$$

$$I_2 .R = \frac{I_1}{j\omega C} + I_1 . R \quad .....(6.2)$$

$$I_3.R = \frac{(I_2 + I_1)}{j\omega C} + I_2.R \quad .....(6.3)$$

$$V_o = \frac{(I_1 + I_2 + I_3)}{j\omega C} + I_3.R \quad .....(a)$$

But,

$$I_3.R = \frac{(I_2 + I_1)}{j\omega C} + I_2.R$$

$$I_2 .R = \frac{I_1}{j\omega C} + I_1.R$$

$$I_1 R = V_{FB}$$

$$\therefore \quad I_3.R = \frac{(I_2 + I_1)}{j\omega c} + \frac{I_1}{j\omega c} + V_{fb}.$$

By substituting (6.1), (6.2), and (6.3) in (a) we get,

$$V_o = \frac{(I_1 + I_2 + I_3)}{j\omega c} + \frac{(I_2 + I_1)}{j\omega c} + \frac{I_1}{j\omega c} + V_{fb}.$$

$$V_o = \frac{3I_1 + 2I_2 + I_3}{j\omega c} + V_{fb}. \quad .....(b)$$

To eliminate I_1, I_2, and I_3,

$$I_1 = \frac{V_{fb}}{R} \quad .....(c)$$

$$I_2 = \frac{I_1}{j\omega cR} + I_1 = \frac{V_{fb}}{j\omega cR^2} + \frac{V_{fb}}{R}$$

$$I_2 = V_{fb}\left(\frac{1}{R} + \frac{1}{J\omega CR^2}\right) \quad .....(d)$$

$$I_3 = \frac{(I_2 + I_1)}{j\omega CR} + I_2$$

$$I_3 = \frac{V_{fb}}{j\omega cR}\left(\frac{2}{R} + \frac{1}{j\omega cR^2}\right) + V_{fb}\left(\frac{1}{R} + \frac{1}{j\omega CR^2}\right) \quad .....(e)$$

Substitute (c), (d), and (e) equation in (b) and simplify

$$V_o = \frac{3V_{fb}}{R(j\omega c)} + \frac{2V_{fb}}{j\omega c}\left(\frac{1}{R} + \frac{1}{j\omega cR^2}\right) + \frac{V_{fb}}{(j\omega c)^2 R}\left(\frac{2}{R} + \frac{1}{j\omega cR^2}\right)$$

$$+ \frac{V_{fb}}{j\omega c}\left(\frac{1}{R} + \frac{1}{j\omega cR^2}\right) + V_{fb}$$

$$\frac{V_o}{V_{fb}} = \frac{1}{\beta} = 1 - \frac{5}{\omega^2 c^2 R^2} + j\left(\frac{1}{\omega^3 C^3 R^3} - \frac{6}{\omega CR}\right)$$

Let, $\alpha = \dfrac{1}{\omega CR}$

$\therefore$ $\dfrac{1}{\beta} = 1 - 5\alpha^2 + j(\alpha^3 - 6\alpha)$

or $\beta = \dfrac{1}{1 - 5\alpha^2 - j(6\alpha - \alpha^3)}$

$\beta A = 1$

But β is a complex number. Therefore equating imaginary part to zero.

$$6\alpha - \alpha^3 = 0.$$

$$6\alpha = \alpha^3. \text{ or } \alpha = \sqrt{6}.$$

But $\alpha = \dfrac{1}{\omega CR}.$

$\therefore$ $\dfrac{1}{\omega CR}. = \sqrt{6}.$

or $\omega = \dfrac{1}{\sqrt{6}CR}$

or $\boxed{f = \dfrac{1}{2\pi CR\sqrt{6}}}$

The gain A, corresponding to this frequency will be 29.

6.5 TRANSISTOR RC PHASE-SHIFT OSCILLATOR

For transistor circuit, voltage shunt feedback is employed, because the input impedance of transistor is small. If voltage series feedback is employed, the resistance of feedback network will be shunted by the low 'R' of the transistor (Fig. 6.11(a) and Fig.6.11(b)).

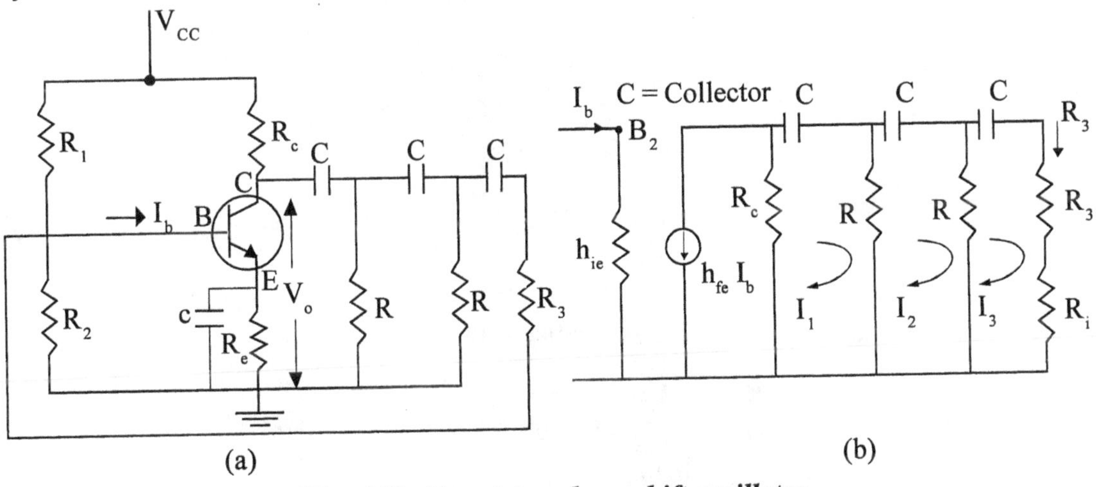

Fig. 6.11 Transistor phase shift oscillator

The value of $R_3 = R - R_i$, where $R_i \simeq h_{ie}$, the input resistance of transistor. ∴ The three RC sections of the phase-shifting network are identical R_1 R_2 and R_e are biasing resistors. (Fig. 6.12).

The feedback current $x_f = -I_3$ Input current $x_i = I_b$ (Negative sign is because it is negative feedback)

$$\therefore \text{ Loop current gain } = \frac{-x_f}{x_i} = \frac{I_3}{I_b}$$

By writing K V L for the three nodes (Fig. 6.12), $\frac{I_3}{I_b}$ can be found out.

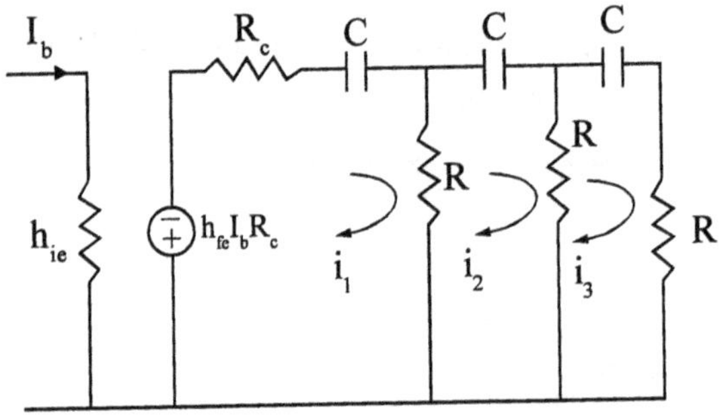

Fig. 6.12 RC equivalent circuit

$$\textit{Loop 1}: (R_C + R - \frac{j}{\omega c}) I_1 - R I_2 = -h_{fe} I_b R_C \qquad .....(6.4)$$

$$\textit{Loop 2}: -RI_1 + (2R - \frac{j}{\omega c}) I_2 - RI_3 = 0 \qquad .....(6.5)$$

$$\textit{Loop 3}: -RI_2 + (2R - \frac{j}{\omega c}) I_3 = 0 \qquad .....(6.6)$$

By taking 'R' as common, the modified given by equations are

$$\left(\frac{R_C}{R} + 1 - \frac{j}{\omega_{RC}}\right) I_1 - I_2 = -h_{fe} I_b \frac{R_C}{R} \qquad .....(6.7)$$

$$-I_1 + \left(2 - \frac{j}{\omega RC}\right) I_2 - I_3 = 0 \qquad .....(6.8)$$

$$-I_2 + \left(2 - \frac{j}{\omega RC}\right) I_3 = 0 \qquad .....(6.9)$$

Let $\frac{R_C}{R} = K$ and $\alpha = \frac{1}{\omega RC}$, the equations are

$$(1 + K - j\alpha) I_1 - I_2 = -h_{fe} I_b = 0 \qquad .....(6.10)$$

$$-I_1 + (2 - j\alpha) I_2 - I_3 = 0 \qquad .....(6.11)$$

$$-I_2 + (2 - j\alpha) I_3 = 0 \qquad .....(6.12)$$

$$\begin{bmatrix} 1+k-j\alpha & -1 & 0 \\ -1 & 2-j\alpha & -1 \\ 0 & -1 & 2-j\alpha \end{bmatrix}\begin{bmatrix} I_1 \\ I_2 \\ I_3 \end{bmatrix}=\begin{bmatrix} -h_{fe}I_bk \\ 0 \\ 0 \end{bmatrix}$$

$$I_3 = \frac{1}{\Delta}\begin{vmatrix} 1-k-j\alpha & -1 & -h_{fe}I_bK \\ -1 & 2-j\alpha & 0 \\ 0 & -1 & 0 \end{vmatrix} = \frac{1}{\Delta}\left[-h_{fe}I_bK\right]$$

$$\Delta = [1 + K - j\alpha]\,[(2 - j\alpha)^2 - 1] + [-(2 - j\alpha)]$$
$$= [1 + K - j\alpha]\,[4 - \alpha^2 - j4\alpha - 1] - 2 + j\alpha$$
$$= - j\alpha^3 - j6\alpha - j4K\alpha - 5\alpha^2 - \alpha^2K + 3K + 1$$
$$\Delta = -(5 + K)\,\alpha^2 + 3K + 1 + j\,[\alpha^3 - (6 + 4K)\,\alpha]$$

$$\therefore\ I_3 / I_b = \frac{-h_{fe}K}{-(5+K)\alpha^2 + 3K + 1 + j\,[\alpha^3 - (6+4K)\alpha]} \qquad .....(6.13)$$

The Barkhausen condition that the loop gain $\frac{I_3}{I_b}$ phase shift must equal zero. The phase shift equals zero provided imaginary part is zero.

Hence $\alpha^3 = (6 + 4K)\,\alpha$

$$\alpha^2 = 6 + 4K$$

$$\alpha = \sqrt{6+4K}$$

Since $\alpha = \frac{1}{\omega RC}$, $\omega = \frac{1}{RC\sqrt{6+4K}}$

$\therefore$ The frequency of oscillation f is given by

$$f = \frac{1}{2\pi RC\sqrt{6+4K}} \qquad .....(6.14)$$

For mantaining the oscillations at the above frequency, $|I_3 / I_b| > 1$

$$\left|\frac{I_3}{I_b}\right| = \left|\frac{-h_{fe}K}{-(5+K)(6+4K)+3K+1}\right| > 1$$

$$|-h_{fe}K| > |-(5+K)(6+4K)+3K+1|$$

$$|-h_{fe}K| > |-4K^2 - 23K - 29|$$

$$h_{fe}\,K > 4K^2 + 23K + 29$$

$$h_{fe} > 4K + 23 + \frac{29}{K} \qquad .....(6.15)$$

To determine the minimum value of h_{fe}, the optimum value of K should be determined by differentiating h_{fe} w.r.t K and equate it to zero.

$$\frac{dh_{fe}}{dk} > \frac{d}{dk}\left(4K + 23 + \frac{29}{K}\right)$$

$$0 > 4 - \frac{29}{K^2}$$

$$\therefore K^2 < \frac{29}{4}, \qquad \therefore \mathbf{K < 2.7} \qquad .....(6.16)$$

Substitute the optimum K = 2.7 in eq. (6.15)

$$h_{fe} > 4 \times 2.7 + 23 + \frac{29}{2.7}$$

$$\therefore h_{fe} > 44.5 \qquad .....(6.17)$$

The BJT with a small-signal common-emitter short-circuit gain h_{fe} lessthan 44.5 cannot be used in this phase shift oscillator.

In R–C phase shift oscillator circuit, each RC network produces 60° phase shift. Thus 3 sections produce the required 180° phase shift. If there are 4-sections, each sections must produce 45° phase shift. But more number of components are to be used. If only 2 sections are these, 90° phase shift must be produced, which is not possible for practical R-C network.

6.6 A GENERAL FORM OF LC OSCILLATOR CIRCUIT

Many Oscillator Circuits fall in to the general form as shown in Fig.6.13 (a) and (b).

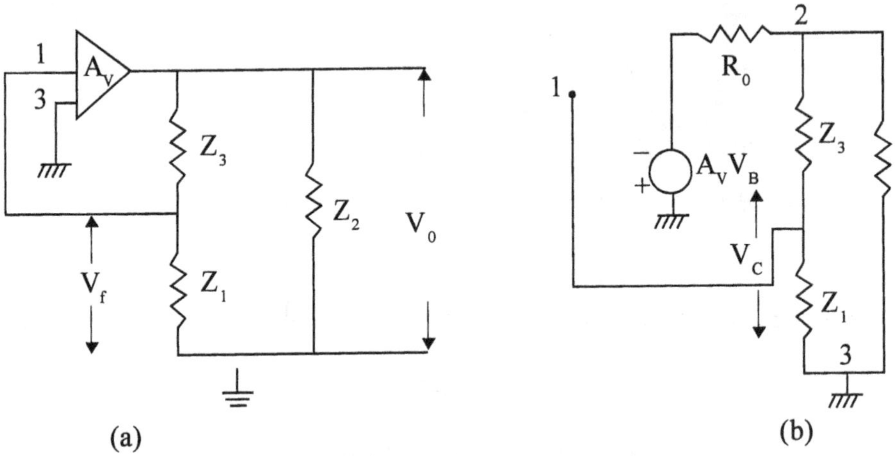

Fig. 6.13 General form of oscillator circuit

The active device can be FET, transistor or operational amplifier, Fig.6.13(b) shows the equivalent circuit using an amplifier with negative gain A_v and output resistance R_0. This is Voltage Series Feedback.

6.7 LOOP GAIN

$$\beta = -\frac{V_f}{V_o}$$

$$\therefore \quad \beta = \frac{-V_f}{V_o} = -\frac{Z_1}{Z_1 + Z_3}.$$

The load impedance, Z_L = (Z_3 in series with Z_1) parallel with Z_L.
R_0 is the output resistance.(Fig. 6.14).

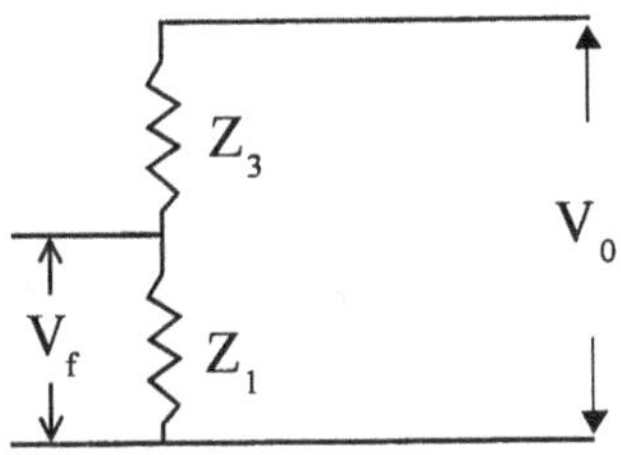

Fig. 6.14 Potential divider network.

$$\text{Gain without feedback } A = \frac{-A_V.Z_L}{Z_L + R_o}$$

$$\therefore \quad -\beta\, A = \frac{-A_V.Z_1Z_L}{(Z_L + R_o)(Z_1 + Z_3)}; \; Z_L = \frac{Z_2(Z_1 + Z_3)}{Z_2 + Z_1 + Z_3}$$

$$= \frac{-A_V.Z_2(Z_1 + Z_3)Z_1}{(Z_1 + Z_2 + Z_3)\dfrac{(Z_2(Z_1 + Z_3)}{(Z_1 + Z_2 + Z_3)} + R_o)(Z_1 + Z_3)}$$

$$-\beta\, A = \frac{-A_V.Z_1\; Z_2}{R_o(Z_1 + Z_2 + Z_3) + Z_2(Z_1 + Z_3)}$$

If the impedances are pure reactances, then $Z_1 = j\, x_1, \quad Z_2 = jx_2; \; Z_3 = jx_3$

$$\therefore \quad -A\,\beta = \frac{A_v X_1\; X_2 (j)^2}{jR_o(X_1 + X_2 + X_3) - X_2(X_1 + X_3)}$$

If $\beta\, A = 1$, or for zero phase-shift, imaginary part must be zero.

$$jR_o\,(X_1 + X_2 + X_3) = 0.$$

$$\text{or} \quad X_1 + X_2 + X_3 = 0$$

$$-A\beta = \frac{A_v X_1\; X_2}{-X_2(X_1 + X_3)} = -\frac{A_v\; X_1}{X_1 + X_3}$$

But $X_1 + X_2 + X_3 = 0 \;\therefore\; X_1 + X_3 = -X_2$

$$\therefore \quad -A\beta = \frac{A_V X_1}{X_2}.$$

$\therefore$ $-A\beta$ must be positive, and at least unity in magnitude. Than X_1 and X_2 must have the same sign.

So if X_1 and X_2 are capacitive, X_3 should be inductive and vice versa.

If X_1 and X_2 are capacitors, the circuit is called ***Colpitts Oscillator (Fig. 6.15(a))***
If X_1 and X_2 are inductors, the circuit is called ***Hartley Oscillators (Fig. 6.15(b))***

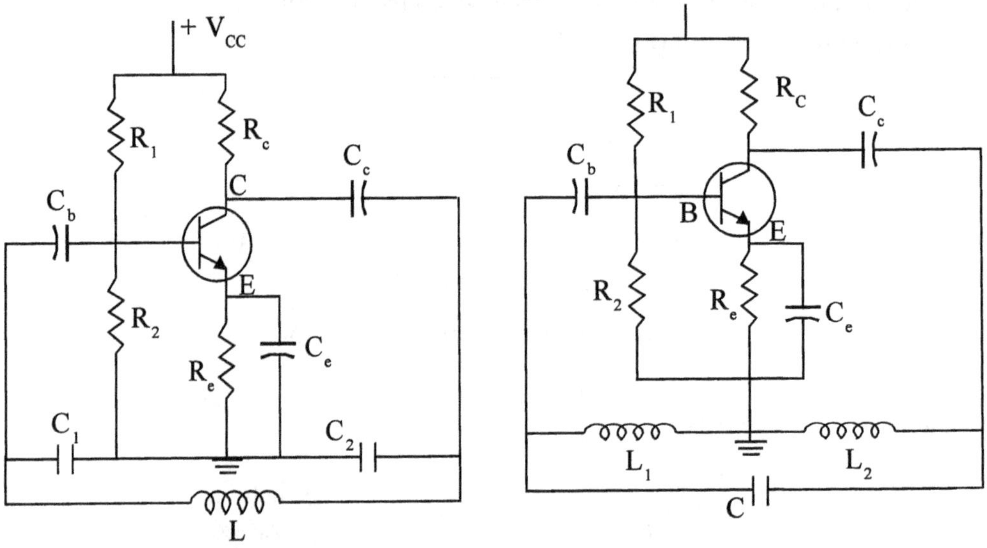

(a) Colpitts oscillator *(b) Hartley oscillator circuit*

Fig. 6.15

6.7.1 For Hartley Oscillator

Condition for oscillations, $X_1 + X_2 + X_3 = 0.$

where $X_1 = j\omega L_1 \;\; X_2 = j\omega L_2;\; X_3 = +\dfrac{1}{j\omega C}$

$\therefore$ $$j\omega L_1 + j\omega L_2 + \frac{1}{j\omega C} = 0 \;\text{ or }\; \omega L_1 + \omega L_2 = \frac{1}{\omega C}$$

$$\omega^2(L_1 + L_2) = \frac{1}{C};\; \frac{1}{\sqrt{(L_1 + L_2)C}} = \omega$$

$$\boxed{f = \frac{1}{2\pi\sqrt{(L_1 + L_2)\,C_3}}}$$

6.7.2 For Colpitts Oscillator

$$X_1 = -\frac{j}{\omega C_1};\; X_2 = \frac{-j}{\omega C_2};\;\; X_3 = j\omega L.$$

$$X_1 + X_2 + X_3 = 0;$$

$$\frac{-j}{\omega C_1} - \frac{-j}{\omega C_2} + j\omega L = 0 \;\text{ or}$$

$$\omega L = \frac{1}{\omega C_1} + \frac{1}{\omega C_2}$$

$$\omega L = \frac{\omega C_2 + \omega C_1}{\omega C_1\ C_2} = \frac{C_2 + C_1}{\omega C_1\ C_2} \quad \text{or} \quad \omega = \left(\frac{1}{\sqrt{L_3}}\sqrt{\frac{C_2 + C_1}{C_1\ C_2}}\right)$$

$$\boxed{f = \frac{1}{2\pi\sqrt{LC_T}}}$$

where $$C_T = \frac{C_1 C_2}{C_1 + C_2}$$

6.8 WIEN BRIDGE OSCILLATOR

In this circuit, a balanced bridge is used as the feedback network. The active element is an operational amplifier. It employs lead-lag Network. Frequency f_0 can be varied in the ratio of 10 : 1 compared to 3 : 1 in other oscillator circuits.

External voltage V_o' is applied betweeen 3 and 4, as shown in Fig.6.16.

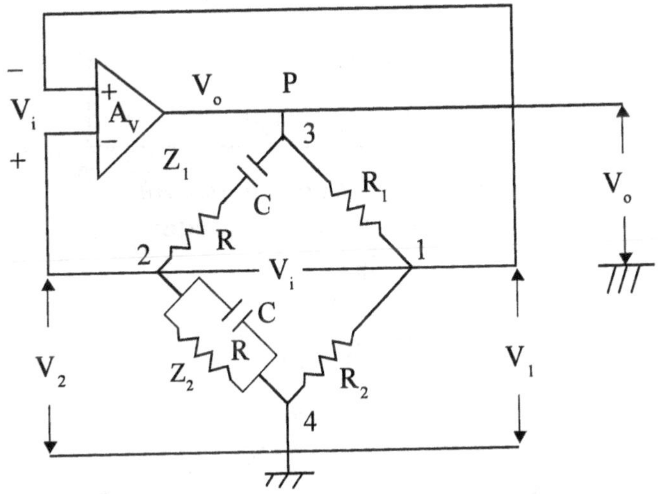

Fig. 6.16 Wien bridge oscillator circuit

To find loop gain, – βA, (–sign because phase-shift feedback).

Lead and lag networks are shown in Figs. 6.17 (a) and (b).

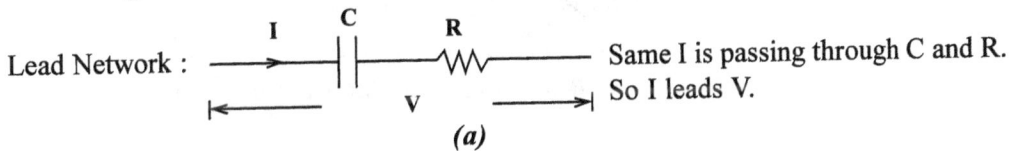

Lead Network :

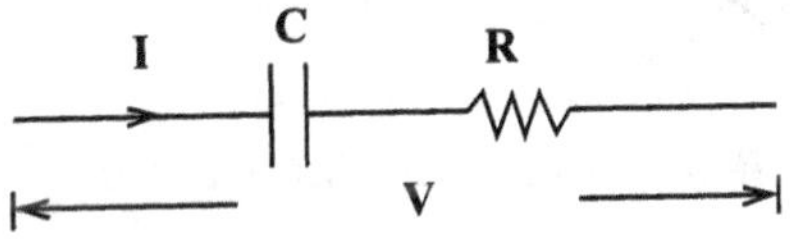

Same I is passing through C and R. So I leads V.

(a)

Lag Network :

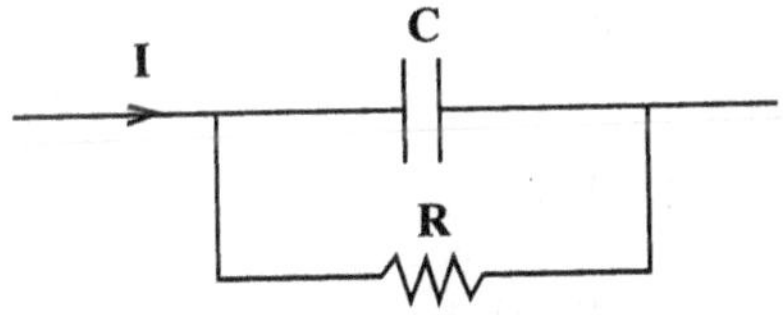

I lags with respect to V.

(b)

Fig. 6.17 Lead and lag networks

Frequency variation of 10 : 1 is possible in Wein Bridge compared to 3 : 1 in other oscillator circuits.

$$V_o = A_v V_i;\ V_i \text{ is } (V_2 - V_1);\ V_i = V_f.$$

Loop gain $$= \frac{V_o}{V_o'} = \frac{A_V . V_i}{V_O'} = -\beta A. \qquad .....(6.18)$$

V_1 and V_2 are auxillary voltages. $V_i = V_2 - V_1$.

$$-\beta = \frac{V_i}{V_o'} \quad \because V_f = V_i \quad \therefore A = A_v.$$

$$-\beta = \frac{V_i}{V_o'} = \frac{V_2 - V_1}{V_o'} = \left[\frac{Z_2}{Z_1 + Z_2} - \frac{R_L}{R_1 + R_2}\right]$$

The lead-lag network is shown in Fig. 6.18.

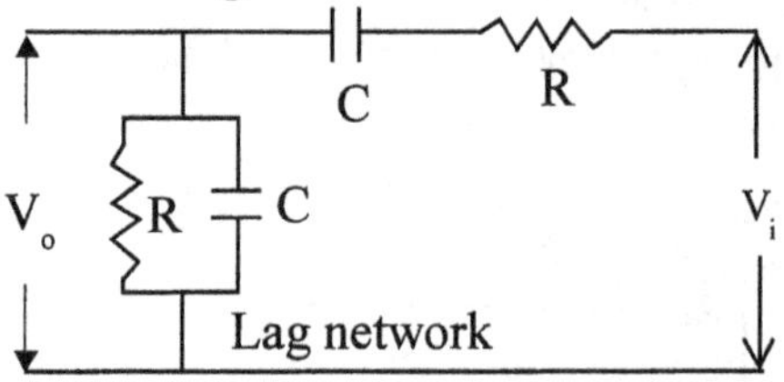

Fig. 6.18 Lead-lag network

6.9 EXPRESSION FOR *f* of Wien Bridge Oscillator

The Wien bridge oscillator circuit is shown in Fig. 6.19.

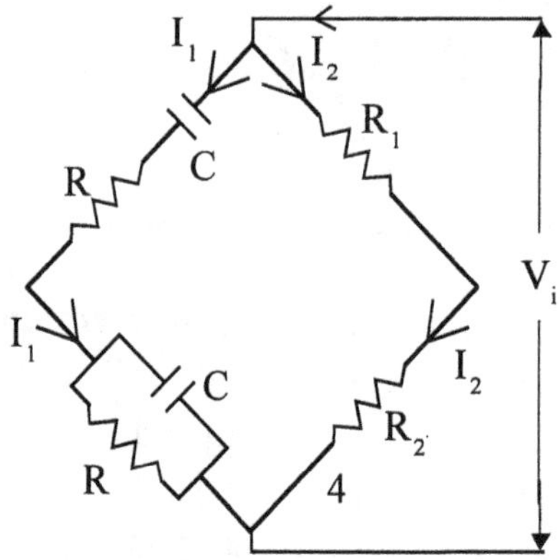

Fig. 6.19 Wien bridge oscillator circuit.

$$I_1 = \frac{\left(R + \frac{1}{j\omega c}\right)}{I_1\left(\frac{R}{1 + j\omega cR}\right)} = \frac{I_2 R_1}{I_2 R_L}$$

$$R_1 = \frac{\left(R_2 R + \frac{R_2}{j\omega C}\right)(1 + j\omega CR)}{R};$$

$$R_1R = \left(R_2\,R + \frac{R_2}{jwC}\right)(1 + jwCR)$$

$$R_2\,.\,R + \frac{R_2}{j\omega C} = \left(\frac{R_1\,R}{1 + j\omega CR}\right)$$

$$[(R\,R_2)\,(j\omega C) + R_2]\,R_2\,j\omega CR - \omega^2\,C^2\,R^2\,R_2 = R_1\,R\,j\omega C$$

$$(R_1\,R_2)\,(j\omega C) + R_2 + R_2\,j\omega CR - \omega^2 c^2 R^2 R_2 = R_1\,R\,j\omega C$$

Equating imaginary parts,

$$R_1\,R_2\,\omega C + R_2\,\omega CR = R_1\,R\omega C$$

Equating real parts,

$$A = 1 + \frac{R_1}{R_2} + \frac{C_1}{C_2},\quad R_2 = \omega^2\,C^2\,R^2\,R_2$$

$$\therefore \quad \omega^2 = \frac{1}{C^2R^2} \quad \text{or} \quad \boxed{f = \frac{1}{2\pi RC}}$$

$$R_1 = \frac{R_2.R}{R} + \frac{R_2}{j\omega CR} + j\omega CRR_2 + R_2$$

$$2R_2 = R_1 \quad \text{or} \quad \boxed{\frac{R_2}{R_1 + R_2} = \frac{1}{3}}.$$

So the minimum gain of the amplifier must be 3.

$$\frac{-R_2}{\omega CR} = \omega CRR_2$$

$$\omega^2 = \frac{1}{(RC)^2}\;;\quad \text{or} \quad \boxed{f = \frac{1}{2\pi RC}}$$

This is the frequency at which the circuit oscillates. Continuous variation of frequency is accomplished by using the capacitors 'C'.

6.10 THERMISTOR

Conductivity of Germanium and Si increases with temperature. A semiconductor when used in this way, taking advantage of this property is called a ***Thermistor***. Ex : NiO, Mn_2O_3 etc. These are used for temperature compensation in oscillator circuits.

6.11 SENSISTOR

A heavily doped semiconductor can exhibit a positive temperature coefficient of resistance because under heavy doping, semiconductor acquires the properties of a metal. So R increases because mobility decreases. Such a device is called ***Sensistor***. These are also used for temperature compensation like thermistors.

6.12 AMPLITUDE STABILIZATION

The amplitude of oscillations can be stabilized by replacing R_2 with a senistor. If β is fixed, as A increases, amplitude of oscillations increases. If a senistor is introduced, as its 'R' changes with temperature, it changes β, so that βA is always constant, (when A changes).

By equating the imaginary part to zero we get $f = \frac{1}{2\pi RC\sqrt{6+4K}}$ where $K = \frac{R_C}{R}$

Forward Current Gain $\boxed{h_{fe} = 4k + 23 + \frac{29}{K}.}$

Practically RC phase-shift oscillators can be used from several hertz to several hundred kilohertzs. In the mega hertz range, tuned LC circuits are more advantageous. Frequency of oscillators can be changed by changing R and C. Amplitude of oscillations will not vary if any C is varied, because X_C varies, but the imaginary part will be zero. Phase-shift oscillator is operated in class A in order to keep distortion minimum.

6.13 APPLICATIONS

Elevation levelling systems, Burglar detection: frequency of oscillators change as a result of local disturbance. The change in frequency causes further electronic action or alarm signal.

6.14 REASONANT CRICUIT OSCILLATORS

Initial transient due to switching will initiate electrical signals. These are feedback to the transistor as input. The input is amplified and obtained as output. Oscillations are sustained. Output is coupled to the base of transistor through L_1. LL_1 is a transformer. R represents the resistance in series with the winding in order to account for the losses in the transformer shown in Fig. 6.20. If 'r' is very small, then at $\omega = \frac{1}{\sqrt{LC}}$ the Z is purely resistive. Then voltage drop across inductor is 180^0 out of phase with the applied input voltage to the FET. If the direction of winding of the secondary connected to the gate is such that it introduces another 180° phase-shift, the total phase-shift is zero.

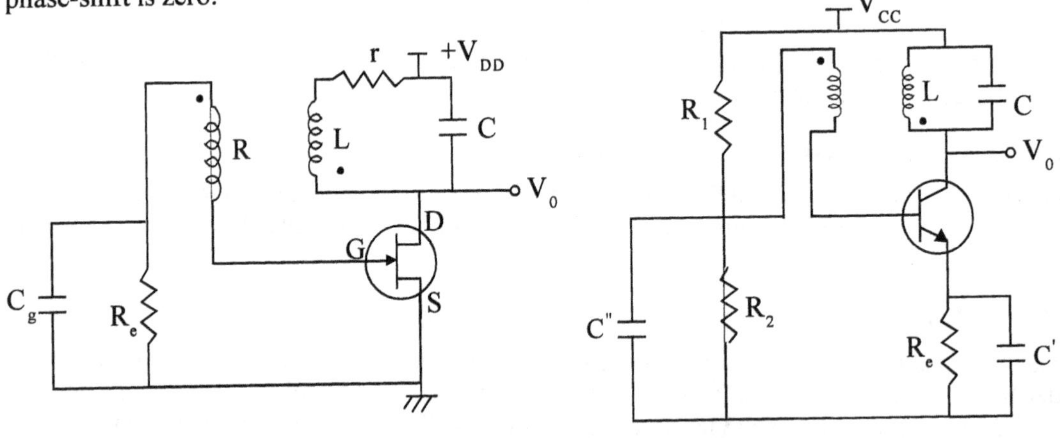

Fig. 6.20 Resonant circuit oscillator.

The ratio of the amplitude of secondary to the primary voltage is M/L where M is the inductance.

$$\text{i.e., } \frac{V_f}{V_o} = \frac{M}{L} = \beta$$

Voltage gain $A_v = -\mu$

$$\beta A_v = -1$$

$$\therefore \quad -\mu\beta = -1;$$

$$\mu = \frac{1}{\beta}$$

$$\therefore \quad \mu = \frac{L}{M}$$

$$\therefore \quad \mu = L/M$$

If we consider resistance 'r' also, $\omega^2 = \frac{1}{LC}(1 + \frac{r}{r_d})$

$$\boxed{g_m = \frac{\mu rC}{\mu M - L}.}$$

6.15 CRYSTAL OSCILLATORS

When certain solid materials are deformed, they generate within them, an electric charge. This effect is reversible in that, if a charge is applied, the material will mechanically deform in response. This is called ***Piezoelectric effect***.

Naturally available materials : 1. Quartz 2. Rochelle salt.

Synthetic materials : 1. Lithium sulphate 2. Ammonium-di-hydrogen phosphate, PZT (Lead Zirconate Titanate), $BaTiO_3$ (Barium Titanate).

If the crystal is properly mounted, deformations take place within the crystal, and an electro mechanical system is formed which will vibrate when properly excited. The resonant frequency and Q depend upon crystal dimensions etc. With these, frequencies from few KHz to MHz and Q in the range from 1000s to 100,000 can be obtained. Since Q is high, and for Quartz, the characteristics are extremely stable, with respect to time, temperature etc., very stable oscillators can be designed. The frequency stability will be $\pm$ 0.001%. It is same as $\pm$ 10 parts per million (10 ppm).

The electrical equivalent circuit of a crystal is as shown in Fig.6.21. L,C,R are analogous to mass, spring constant, and viscous damping factor of the mechanical system.

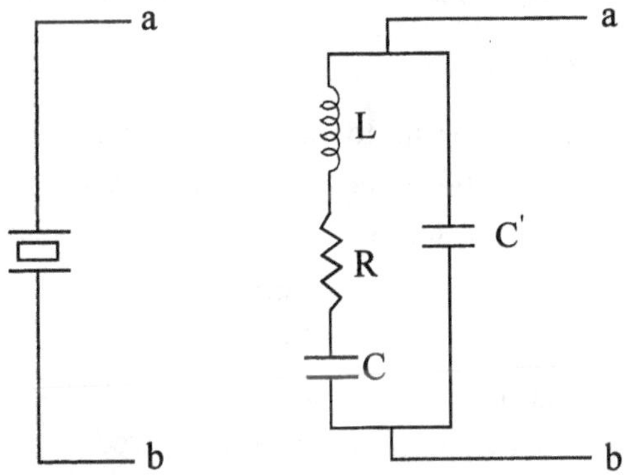

Crystal ***Electrical model***

Fig. 6.21

Values for a 90 kHz crystal are L = 137 H, C = 0.023pF; R = 15kΩ corresponding to Q = 5,500. The dimension of a crystal will be 30 × 4 ×1.5 mm. C' is the electostatic capacitance between electrodes with the crystal as a dielectric. C' = 3.5 pF and is larger than C.

When the crystal slabs are cut in proper directions, with regard to the crystal axis, a potential difference exists between the faces of the crystal slab when pressure is brought to bear on them. And if the slab is placed in an electrostatic field, the slab undergoes deformation. (Fig.6.22). If the electric field is an alternating one, with a frequency which sets the slab into mechanical resonance, the slab will physically vibrate vigorously. Such a crystal can be employed to maintain an oscillation of great frequency stability. When the L.C. circuit in the plate circuit is tuned close to the crystal resonant frequency, steady oscillations will be established. These are maintained by C whose oscillations value is small. By placing crystal between the gate and source, of the FET amplifier, and feeding back a small A.C. voltage from the output, to keep crystal vibrating, the circuit becomes an oscillator with precise stability. Accuracy ≃ 0.01% '*f*' range is 0.1 to 20 MHz. By keeping crystal in an oven, accuracy can be improved to 0.001%.

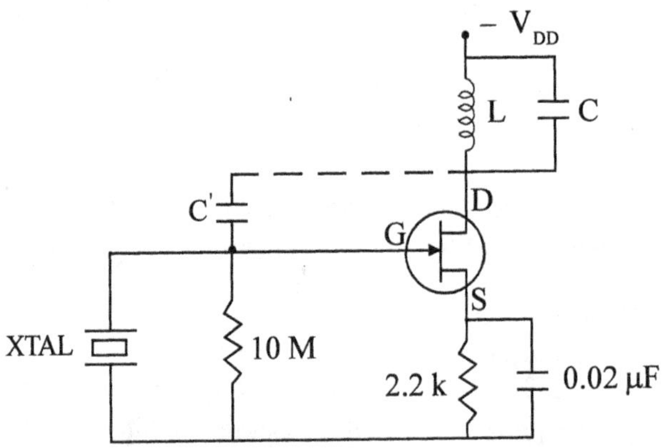

Fig. 6.22 Crystal oscillator circuit.

6.16 FREQUENCY STABILITY

It is a measure of the ability of the circuit to maintain exactly the same frequency for which it is designed over a long time interval. But actually in many circuits the '*f*' will not remain fixed but it drifts from the designed frequency continuously. This is because of variations of number of parameters, circuit components, transistor parameters, supply voltages temperature, stray capacitances etc. In order to increase the stability the factors which effect the '*f*' largely should be taken care of. If '*f*' depends only on R and C high precision R and C should be employed. Also temperature compensating elements are to be employed.

Effect of temperature on inductors and capacitors amounts to more than 10 parts per million per degree change.

Crystal will have mass, elasticity and damping. Crystal will have very high mass to elastic ratio $\left(\frac{L}{C}\right)$ and to a high ratio of mass to damping (high Q). Its value is of the order of 10,000 to 30,000.

The crystal is coupled with external C, and the LC circuit oscillates. In the oscillator circuits, instead of external L and C, crystal is connected.i.e. crystal L and C are being used. So 'f' is fixed by crystal itself. 'f' will not vary with temperature. To change 'f' another crystal is used.

6.17 FREQUENCY OF OSCILLATIONS FOR PARALLEL RESONANCE CIRCUIT

$$Z = \frac{(R - j\omega C)\times \frac{1}{j\omega C}}{(R + j\omega C) + \frac{1}{j\omega C}} = \frac{R + j\omega C}{(1-\omega^2 LC) + j\omega RC}$$

$$Y_1 = \frac{1}{R + j\omega C} = \frac{R - j\omega C}{R^2 + \omega^2 C^2} \quad Y_2 = j\omega C$$

Total admittance $Y = Y_1 + Y_2 \quad = \frac{R}{R^2+\omega^2 C^2} - j\left(\frac{\omega C}{R^2+\omega^2 C^2} - \omega C\right)$

At resonance, imaginary part is zero $\quad \therefore \quad R \rightarrow \infty$.

$$\therefore \quad = \frac{\omega_2}{R^2+\omega^2 C^2} = \omega C \text{ or } \frac{L}{C} = R^2 + \omega^2 C^2$$

$$\omega^2 C^2 = \frac{L}{C} - R^2 \text{ or } \omega^2 = \frac{1}{LC} - \frac{R^2}{L^2}$$

or $$f = \frac{1}{2\pi}\sqrt{\frac{1}{LC} - \frac{R^2}{L^2}}$$

This is the expression for frequency of oscillations.

Amplitude of oscillator will be very small (v is Less) for series resonance circuit. So parallel tuned circuits are employed.

6.18 1-MHz FET CRYSTAL OSCILLATOR CIRCUIT

In the basic configuration of oscillator circuit,

$Z_1 + Z_2 + Z_3 = 0.$

Z_1 is crystal Z_2 is L and C combination.

Z_3 is C_{gd}. The frequency of oscillator essentially depends up on crystal only as shown in Fig. 6.23. Z_2 and Z_3 values are insignificant compared to Z_1 of the crystal. Therefore the stability is high.

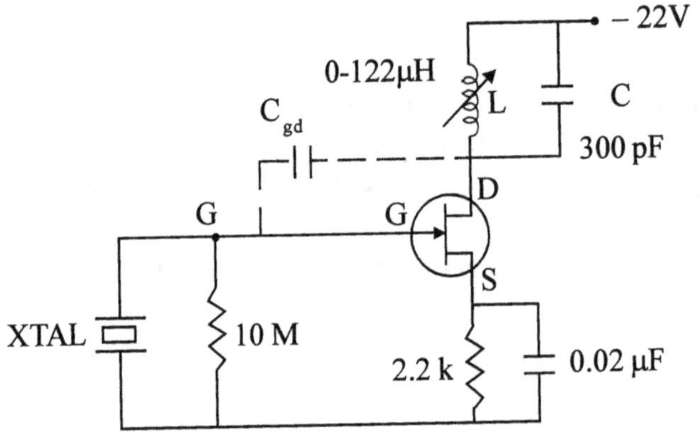

Fig. 6.23 Crystal oscillator circuit.

When electrical input is given, mechanical vibrations are set in the crystal, due to piezoelectric effect. But the crystal is being considered as an inductor capacitor combination only. So it acts like a |Z| R_o L, C combination. Frequency of oscillations depend only on crystal. Other L and Cs are insignificant, as their values are less compared to L, C and R of the crystal.

SUMMARY

- Oscillators generate A.C. output without external A.C input by using Noise A.C signal generated in switching. D.C power from V_{CC} or V_{DD} is converted to A.C. power. The range of frequency signals generated by the circuit can be high and output power is small.
- For sustained oscillations, the conditions known as Barkhausen criterion to be satisfied are (i) $|\beta A| \geq 1$ (ii) Total loop phase shift must be 0^0 or 360^0 (or phase shift of feedback network must be 180^0 when the amplifying device produces another 180^0).
- In the case of JFET, R - C phase shift oscillator circuit, $f_o = \dfrac{1}{2\pi RC\sqrt{6}}$
- In the case BJT, R - C phase shift oscillator circuit, $f_o = \dfrac{1}{2\pi RC\sqrt{6+4k}}$ where

  $K = \dfrac{R_C}{R}$
- For Hartley oscillator circuit which employs two inductors and one capacitor in the feedback network is, $f_o = \dfrac{1}{2\pi\sqrt{(L_1 + L_2)C_3}}$

- For Colpitts oscillator circuit with two capacitors and one inductor in the feedback network, $f_0 = \frac{1}{2\pi}\frac{1}{\sqrt{L_3}}\sqrt{\frac{C_2 + C_1}{C_1 C_2}}$
- For Wein Bridge oscillator circuit, the minimum gain of amplifier must be 3. It can produce variations in f_0 in the ratio of 10 : 1 compared to a variation of 3 : 1 in other types of oscillator circuits.
- Thermistors with (NTCR) and sensistor with positive temperature coefficient of resistance (PTCR) are used for frequency stability in oscillator circuits.
- The specification parameters of oscillator circuits are (i) Amplitude stability (ii) Frequency stability (iii) Frequency range (iv) Distortion in output waveform etc.
- Crystal oscillators produce highly stable output waveform in the high frequency range of MHz also.

OBJECTIVE TYPE QUESTIONS

1. The difference between an amplifier circuit and oscillator circuit is ______________.
2. A.C. output signal is generated by the oscillator circuits without external A.C input, by amplifying ______________ signal.
3. Oscillator circuits employ ______________ type of feedback.
4. One condition for sustained oscillations is total loop phase shift must be ______________ degrees.
5. As per Berkauhsen criteria for sustained oscillations, $|\beta A|$ must be ______________.
6. The range of frequencies over which R - C phase shift oscillator circuit is used is ______________.
7. In the Mega Hertzs frequency range, the type of oscillator circuit used is ______________.
8. The range of frequency over which Wein Bridge oscillator used is ______________.
9. In the feedback network if two inductors and one capacitor elements are used, the oscillator circuit is ______________.
10. The oscillator circuit which employs two capacitors and one inductor in the feedback network, is ______________ oscillator circuit.
11. Naturally occuring materials used for in crystal oscillator circuits, exhibiting piezoelectric effect are 1. ______________ 2. ______________.

12. Synthetic materials which exhibit piezoelectric effect are,

    1. ____________ 2. ____________

    3. ____________ 4. ____________

13. The ratio of frequency variation possible with Wein Bridge Oscillator circuit is ____________ is compared to the ratio of ____________ with others oscillator circuits.

14. Thermistors have temperature coefficient and the materials used are ____________.

15. Typical values of L, C, R and Q of a crystal used in oscillator circuits are L = ____________ C = ____________ R = ____________ Q = ____________.

16. Expression for frequency of Osillations in the case of Phase-shaft oscillator circuit is f_o = ... and the value of β must be at least ....

17. Expression for frequency of oscillations in the case of Wien Bridge Oscillator is f_o = ...

18. In the case of Colpitts oscillator f_0 = ...

19. For Hartley oscillator, f_0 = ...

20. Frequency of oscillations, in the case of uJT oscillator circuit is fo$\simeq$ ....

ESSAY TYPE QUESTIONS

1. Explain the basic principle of generation of oscillations in LC tank circuits. What are the considerations to be made in the case of practical L.C. Oscillator Circuits ?
2. Deduce the Barkausen Criterion for the generation of sustained oscillations. How are the oscillations initiated?
3. Draw the circuit and explain the princple of operaiton of R.C.phase-shift oscillator circuit. What is the frequency range of generation of oscillations ? Derive the expression for the frequency of oscillations.
4. Derive the expression for the frequency of Hartely oscillators.
5. Derive the expression for the frequency of Colpitt Oscillators.
6. Derive the expression for the frequency of Wein Bridge Oscillators.
7. Derive the expression for the frequency of Crystal Oscillators.
8. Explain how better frequency stability is obtained in crystal oscillator ?
9. Draw the equivalent circuit for a crystal and explain how oscillations can be generated in electronic circuits, using crystals.
10. Why three identical R-C sections are used in R-C phase-shift oscillator circuits? Consider the other possible combinations and limitations.

ANSWERS TO OBJECTIVE TYPE QUESTIONS

1. Amplifier needs external A.C. input. Oscillator circuit doesn't need external A.C. input.
2. Internally generated Noise Signal
3. Positive feedback
4. Zero or 360°
5. $|\beta A| \geq 1$
6. Few 100 Hz to KHz range
7. Tuned oscillator circuit
8. 5 Hz to 1 MHz
9. Hartley oscillator
10. Collpitts oscillator circuit
11. Quartz and Rochelle salt
12. 1. Lithium sulphate 2. Rochelle salt
    3. PZT (Lead Zirconate Titanate) 4. Barrium titanate ($Batio_3$)
13. 10 : 1, 3 : 1
14. Positive, Oxides of Nickel and Manganese
15. L = 137 H C = 0.0235 μF R = 15K Q = 5500
16. $f_0 = \dfrac{1}{2\pi RC\sqrt{6}}$; $\beta = \dfrac{1}{29}$
17. $f_o = \dfrac{1}{2\pi\sqrt{R_1C_1R_2C_2}}$
18. $\dfrac{1}{2\pi\sqrt{LCeq}}$

    where $Ceq = \dfrac{C_1C_2}{C_1 + C_2}$
19. $\dfrac{1}{2\pi\sqrt{LeqC}}$ $\text{where} Leq = L_1 + L_2 + 2M$
20. $1/R_TC_T\ l_n\ [1/(1-n)]$

7 Large Signal Amplifiers

In this Unit,

- Power amplifiers - Class A, Class B, Class C, Class AB and other types of amplifiers are analyzed.
- Advantages and Disadvantages of different types are discussed.
- Thermal considerations and use of heat sinks is also explained.

7.1 INTRODUCTION

When the output to be delivered is large, much greater than mW range and is of the order of few watts or more watts, conventional transistor (BJT) amplifiers cannot be used. Such electronic amplifier circuits, delivering significant output power to the load (in watts range) are termed as *Power Amplifiers*. Since the input to this type of amplifier circuits is also *large*, they are termed as *Large Signal Amplifiers*. In order to improve the circuit efficiency, which is the ratio of output power delivered to the load P_0 to input power, the device is operated in varying conduction angles of 360^0, 180^0 less than 180^0 etc. Based on the variation of conduction angle, the amplifier circuits are classified as Class A, Class B, Class C, Class AB, Class D, and Class S.

7.1.1 POWER AMPLIFIER

Large input signals are used to obtain appreciable power output from amplifiers. But if the input signal is large in magnitude, the operating point is driven over a considerable portion of the output characteristic of the transistor (BJT). The transfer characteristic of a transistor which is a plot between the output current I_C and input voltage V_{BE} is not linear. The transfer characteristic indicates the change in i_c when V_b or I_B is changed. For equal increments of V_{BE}, increase in I_C will not be uniform since output characteristics are not linear (for equal increments of V_{BE}, I_C will not increase by the same current). So the transfer characteristic is not linear. Hence because of this, when the magnitude of the input signal is very large, distortion is introduced in the output in large signal power amplifiers. To eliminate distortion in the output, pushpull connection and negative feedback are employed. (Fig. 7.1).

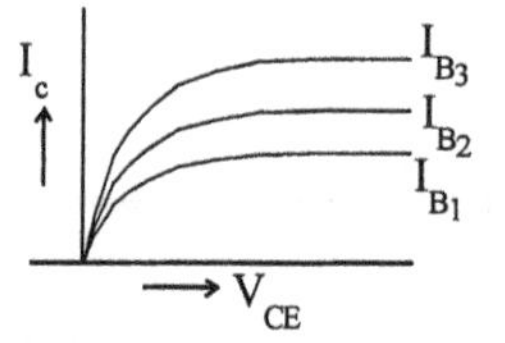

Fig. 7.1 Output characteristics of BJT in CE mode

For simplicity let us assume that the dynamic characteristic of the transistor is linear. (Fig. 7.2).

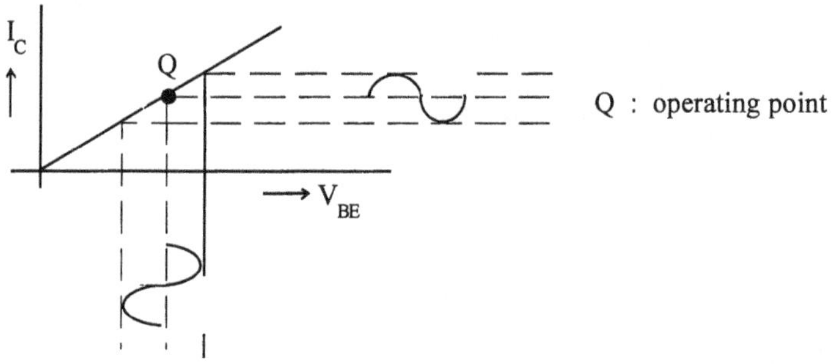

Fig. 7.2 Transfer characteristics of BJT

7.1.2 CLASS A OPERATION

If the *Q point is placed near the centre of the linear region of the* dynamic curve, class A operation results. Because the transistor will conduct for the complete 360°, distortion is low for small signals and conversion efficiency (η) is low.

7.1.3 Class B Operation

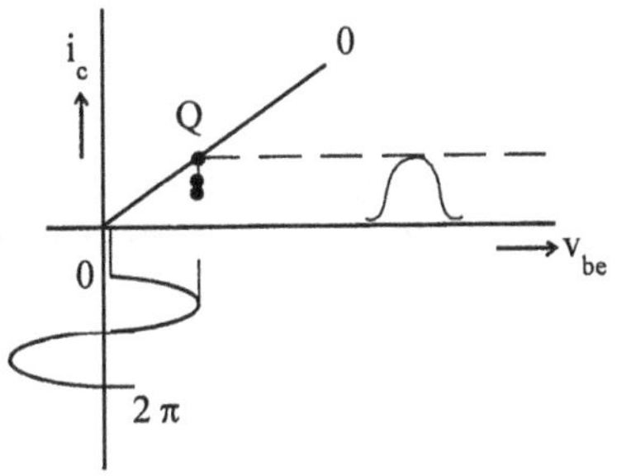

Fig. 7.3 Transfer curve

For class B operation the Q point is set near cutoff (Fig. 7.3). So output power will be more and conversion efficiency (η) is more. Conduction is only for 180^0, from π - 2π. Since the transistor Q point is beyond cutoff, the output is zero or the transistor will not conduct. Output power is more because the complete linear region is available for an operating signal excursion, resulting from one half of the input wave. The other half of input wave gives no output, because it drives the transistor below cutoff.

7.1.4 Class C Operation

Here Q point is set well beyond cutoff and the device conducts for less than *180^0*. The conversion efficiency (η) can theoretically reach 100%. Distortion is very high. These are used in radio frequency circuits where resonant circuit may be used to filter the output waveform. Class A and class B amplifiers are used in the audio frequency range. Class B and class C are used in Radio Frequency range where conversion efficiency is important (Fig. 7.4).

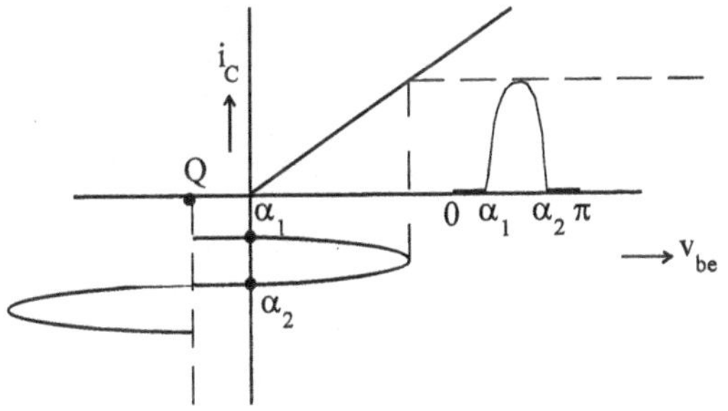

Fig. 7.4 Transfer curve

7.1.5 Signal Amplifiers

With respect to the input signal, the amplifier circuits are classified as

(i) Small signal amplifiers (ii) Large signal amplifiers

7.1.6 Small Signal Amplifiers

Here the magnitude of the input signal is very small, slightly deviating from the operating point. But always the operation is in the active region only. The characteristics of the device can be assumed to be linear. We can draw the equivalent circuit and analyse the performance. The magnitude of the signal may be few mV, in single digits. The operating point or Quiescent point Q swings with the input signal. Because the input signal magnitude is small, the operating point is in the active region only.

7.1.7 Large Signal Amplifiers

Here the magnitude of the input signal is very large and deviation from the operating point on both sides is very wide. So because of this, the device performance cannot be assumed to be linear. Because of the large swing of the input signal, the non linear portion of the transistor characteristics are also to be considered. Hence the linear equivalent circuit analysis is not valid. So for large signal amplifiers only graphical analysis is employed.

Power amplifiers, class A, class B, class C amplifiers, push-pull amplifier are of this type. Large signal amplifiers are used where the output *power requirement is large.* If we use small signal amplifiers, the number of stages to be cascaded will be large, complicating the circuit.

Factors to be considered in large signal amplifiers :

1. Output power
2. Distortion
3. Operating region
4. Thermal considerations
5. Efficiency (η)

Amplifier circuits may be classified in terms of the portion of the cycle for which the active device conducts.

Class A : It is one, in which the active device conducts for the full 360^0. The device is biased in that way.

Class B : Conduction for 180°

Class C : Conduction for $< 180^\circ$

Class AB : Conduction angle is between 180° and 360°

7.2 CLASS A POWER AMPLIFIER

The circuit for class A amplifier considering only load resistance R_L is as shown, in Fig. 7.5.

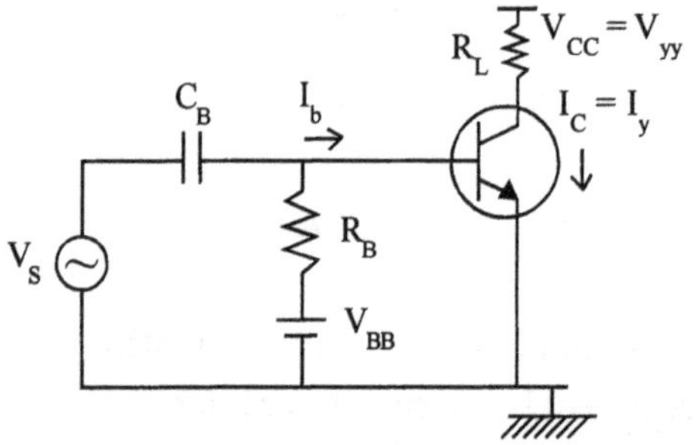

Fig. 7.5 Class A power amplifier

There are two types of operations :

1. Series fed 2. Transformer coupled

7.2.1 Series Fed

There is no transformer in the circuit. R_L is in series with V_{cc}. There is DC power drop across R_L. Therefore efficiency (η) = 25% (maximum).

7.2.2 Transformer Coupled

The load is coupled through a transformer. DC drop across the primary of the transformer is negligible. There is no DC drop across R_L. Therefore η = 50% maximum (Fig. 7.6).

V_y and I_y are the root mean square (rms) values of voltage and current.

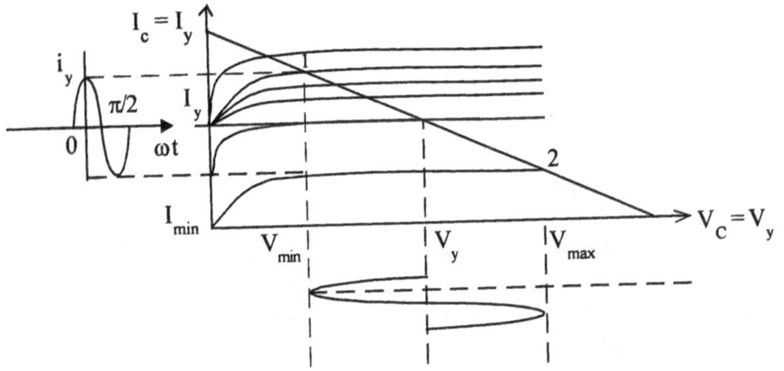

Fig. 7.6 Output characteristics

In class A amplifier, the conduction is for full 360⁰. Therefore the operating point lies in the active region only. Let us assume that the static output characteristics of the transistor are ideal and linear. So if the input is a sinusoidal signal, then the output will also be sinusoidal.

Let us use the subscripts y for output and x for input. Therefore $i_c = i_y$. Output is on y-axis. So subscript y is used. Input is on x-axis. So subscript 'x' is used.

The output power P_y can be found graphically.

V_y = (rms) output voltage

I_y = (rms) output current

Subscript 'y' for output

Subscript 'x' for input

$$P = V_y I_y = I_y^2 R_L.$$

$$I_m = \text{Peak value fo the current} = \left(\frac{I_{max} - I_{min}}{2}\right)$$

$$I_{rms} = I_m/\sqrt{2} = I_y$$

$$I_y = \frac{I_m}{\sqrt{2}} = \frac{I_{max} - I_{min}}{2\sqrt{2}}$$

$\because$ $(I_{max} - I_{min})$ = Peak to peak value

$$\therefore \quad I_m = I_y = \frac{I_{p-p}}{2\sqrt{2}}$$

$$V_y = \frac{V_m}{\sqrt{2}} = \frac{V_{max} - V_{min}}{2\sqrt{2}}$$

Therefore output power $P = \frac{V_m . I_m}{2} = \frac{(V_{max} - V_{min})(I_{max} - I_{min})}{8}$

$$= \left(\frac{V_m}{\sqrt{2}} . \frac{I_m}{\sqrt{2}}\right)$$

$$= \left(\frac{V_m . I_m}{2}\right)$$

7.2.3 Efficiency of Amplifier Circuits

Let V_{yy} is the DC voltage being supplied to the circuit and I_y is the DC current drawn by the circuit. Therefore the DC power input to the circuit is $V_{yy}. I_y$. Let R_L be the load resistance. Therefore *DC power* absorbed by the load is $(I_y^2 . R_L + I_y V_y)$ where I_y and V_y (with small subscripts y) are the rms current and voltages absorbed by the load and I_Y (capital Y) is the DC current absorbed by the load. In addition to the DC drop across the load and AC drop across the load there is thermal power dissipation P_D across the device, since it gets heated. According to the law of conservation of energy, the input power should be equal to AC power, + DC power loss across the load and thermal dissipation.

$\therefore$ $V_{yy} I_y = I_y^2, R_L + I_y V_y + P_D$

$V_{yy} . I_y$ = Total input power

$I_y^2 . R_L$ = DC power drop in the load

$I_y . V_y$ = AC power in the load

But $V_{yy} = V_y + I_y R_L$

V_Y = DC voltage, I_Y = DC current

$\therefore$ $P_D = (V_y + I_y R_L) I_y = I_y^2 R_L + I_y V_y + P_D$

P_D = Thermal power dissipation.

$\therefore$ $P_D = V_y I_y - v_y i_y$

If the load is not a pure resistance, $V_y I_y$ should be replaced by $V_y I_y \cos\theta$. The total AC input power + Input DC power = DC drop across the load + AC output voltage + thermal dissipated power. Now if there is no AC output power i.e, the device is not conducting, then the rest of the power should be dissipated as heat. Therefore if AC power output is zero, ie., AC input signal is zero, then P_D is maximum. and has its maximum value $V_y I_y$. Therefore the device is cooler when delivering power to a load than when there is no such AC power transfer. When there is power drop across the device itself, it gets heated.

7.3 MAXIMUM VALUE OF EFFICIENCY OF CLASS A : AMPLIFIER

Certain assumptions are made in the derivation, which will simplify the estimation of the efficiency (η). Because of this some errors will be there and the expression is approximate.

The assumption is that the static output characteristic of the transistors are equally spaced, in the region of the load line for equal increments in the base current. If i_b is increased by 1 μA, i_c will increase by 1mA, and if i_b is increased by 3 μA, i_c will increase by 3mA. Thus for the load line shown in the Fig. 7.7 the distance from 1 to Q is the same as that from Q to 2.

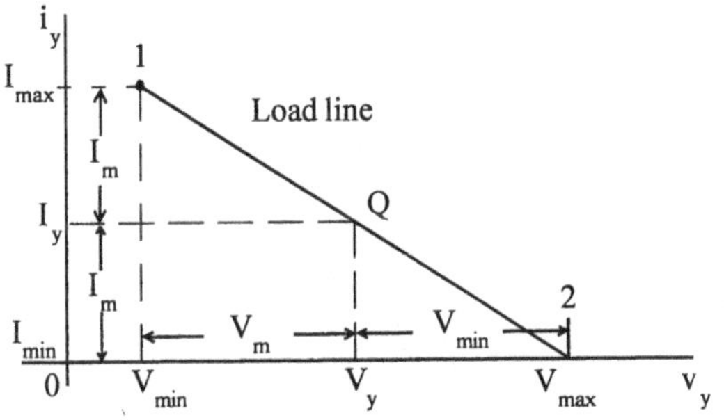

Fig. 7.7 Transfer curve

In the case of transformer coupled amplifier, supply voltage is only V_y and not V_{max}, since the DC drop across transformer can be neglected. In the case of series fed amplifiers, supply voltage V_{yy} is V_{max}.

$$I_y = I_m$$

I_m is the current corresponding to the operating point.
V_m is the voltage corresponding to the operating point.

$$V_m = I_m Z_m = \frac{V_{max} - V_{min}}{2}$$

The general expression for conversion efficiency is

$$\eta = \frac{\text{Signal power delivered to load}}{\text{DC power supplied to output circuit}} \times 100$$

$$P_{ac} = V_m \cdot I_m/2 \qquad P_{DC} = V_{yy} I_y$$

V_m is the peak value or maximum value. Since rms value $= \frac{V_m}{\sqrt{2}}$ $\qquad I_m = \frac{I_m}{\sqrt{2}}$

$$V_{rms} = \frac{V_m}{\sqrt{2}}$$

$$I_{rms} = \frac{I_m}{\sqrt{2}}$$

$$\therefore \quad P_{ac} = \frac{(V_m I_m)}{2}$$

$$\eta = \frac{\frac{1}{2} V_m . I_m}{V_{yy} . I_y} \times 100\% \quad = 50 \frac{V_m\ I_m}{V_{yy}\ I_y} \%$$

$$V_m = \text{Peak value} = \frac{1}{2} \text{ (Peak to peak value)}$$

$$V_m = \left(\frac{V_{max} - V_{min}}{2}\right), \quad V_{rms} = \frac{V_m}{\sqrt{2}}$$

$$V_m \cdot I_m = \left(\frac{V_{max} - V_{min}}{2}\right) I_y \quad \because I_m = I_y, \; I_m = \left(\frac{I_{max} - I_{min}}{2}\right); \; I_{min} = 0$$

$$\eta = \frac{50\left(V_{max} - V_{min}\right) I_y}{V_{yy}\, I_y \times 2}$$

$$= \frac{25\left(V_{max} - V_{min}\right)}{V_{yy}} \%$$

$$\eta = \frac{V_m I_m / 2}{V_{yy}\, I_y} \times 100\%$$

$I_m = I_y$, since transistor will not conduct, if $I_{min} = 0$.

$$P_0 = \frac{\left(V_{CC} - V_{CE}\right) I_C}{2}$$

$$I_C = \frac{V_{CC} - V_{CE}}{R'_L}$$

R_L= Load resistance referred to primary.

$$P_0 = (V_{CC} - V_{CE})^2 / 2 R'_L$$

$$P_{DC} = V_{CC} \times I_C$$

$$= V_{CC} \left(\frac{V_{CC} - V_{CE}}{R'_L}\right)$$

$$\eta = \frac{P_0}{P_{DC}} \times 100$$

$$= 50 \left\{1 - \frac{V_{CE}}{V_{CC}}\right\}$$

If it is a transformer coupled amplifier, V_Y is the DC voltage. Since Q point is chosen in the middle of the load line, graphically,

$$V_y = \frac{V_{max} - V_{min}}{2}$$

for a transformer coupled amplifier, there is no DC drop across the transformer.

$\therefore \quad V_{CC} \simeq V_{CE} = V_y$

$\therefore \quad$ DC input power $= V_y \cdot I_y$

$$V_{yy} = \frac{V_{max} + V_{min}}{2}$$

$$\eta = \frac{25(V_{max} - V_{min}) \times 2}{(V_{max} + V_{min})}$$

$$= \frac{50(V_{max} - V_{min})}{(V_{max} + V_{min})}$$

V_{yy} will be the quiescent voltage itself for transformer coupled amplifier. Since there is no DC voltage drop across the transformer.

If $V_{min} = 0$, maximum efficiency = *50% for class A transformer coupled amplifier.*

For series fed amplifier, $V_{yy} = V_{max} = 2\,V_m$

$$\therefore \quad \eta = \frac{25\,(V_{max} - V_{min})}{V_{max}} \%. \quad \text{If } V_{max} = 0, \text{ maximum } \eta = 25\%$$

Therefore for a transformer coupled amplifier conversion η is twice. (50% compared to 25% for series fed amplifier)

7.4 TRANSFORMER COUPLED AMPLIFIER

In AC amplifier circuits, the input AC signal should be coupled to the amplifier and output of the amplifier should be coupled to the load resistance. The coupling device should be such that, it allows only the AC signal to the amplifier circuits and blocks the DC components present in the signal generator, because we are interested in amplifying only AC signals. For this purpose a capacitor can be used for coupling (Fig. 7.8).

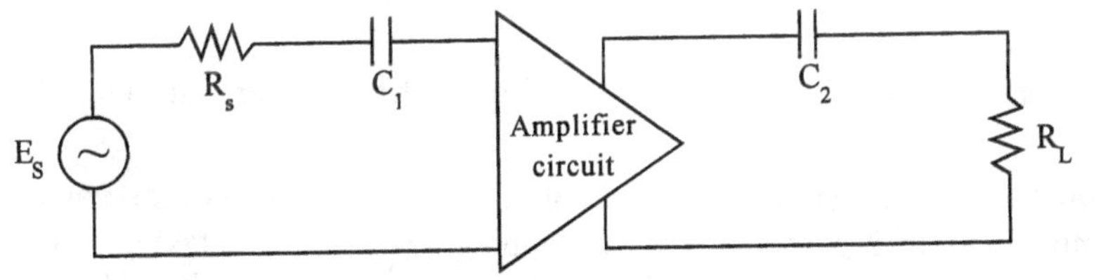

Fig. 7.8 Coupling in amplifier circuits

Types of coupling :

1. Capacitor coupled amplifier 2. Transformer coupled amplifier 3. RC coupled amplifier
4. Direct coupled amplifier 5. Inductor or Tuned amplifier

E_S is the AC signal generator and R_S its source resistance. C_1 and C_2 are the coupling capacitors. They are chosen such that, for the lowest frequency signal to be amplified, X_{C_1} and X_{C_2} are short circuits. But because of these reactive coupling elements, as signal frequency decreases X_C increases. Hence there will be large voltage drop across the capacitor and so the actual input to the amplifier reduces and hence gain decreases. Similarly at high frequencies because of the shunting capacitance C_S, gain falls. Therefore there is a particular frequency range in which gain is of desirable value only.

Instead of capacitors, transformer can also be used for coupling (Fig. 7.9).

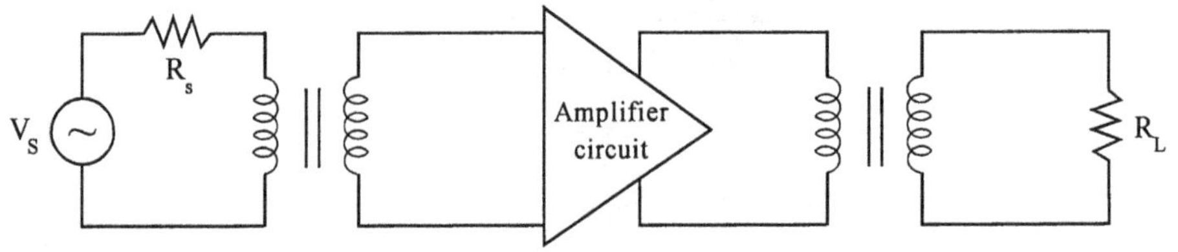

Fig. 7.9 Transformer coupling

Transformer does not respond to DC. Therefore only AC signals from source to the amplifier circuit and from the amplifier to the load will be coupled. But what is the advantage of the transformer coupling ? Suppose, the load resistance R_L is very small $\simeq 4\Omega$, 8Ω or 15Ω as in the case of a loud speaker. The output impedance R_0 of the transistor amplifier is much larger (for common emitter and common base configuration A_V, $A_i >> 1$). Therefore impedance matching will not be there and so maximum power will not be transferred. Even if capacitive coupling is used, impedance matching cannot be achieved. But this can be done using a transformer.

$$R_0 = \left(\frac{N_1}{N_2}\right)^2 .R_L$$

$$\left(\frac{N_1}{N_2}\right) = \text{Turns ratio of transformer}$$

N_1 = Number of turns on primary side.

N_2 = Number of turns on secondary side.

R_0 is much larger than R_L. Therefore $\left(\frac{N_1}{N_2}\right) > 1$ or the transformer that should be used should be a *Stepdown transformer*. Therefore the output voltage at the secondary of the transformer will be much smaller compared to the input voltage since stepping down action is taking place. [This is the case with class A and class B power amplifiers in the case of lab experiment]. But the current amplification will be there and because of Z matching, maximum power will be transferred to the load. Similar to resistive capacitor coupled amplifier, we have the frequency response which depends upon the inductance of the primary and secondary. The transformer on the primary side is chosen such that the source resistance of the generator matches with the input Z of the amplifier circuit. Transformer coupled amplifiers are used in low audio frequency range only because at higher frequencies the X_L of the transformer will be large and so the gain falls.

Another advantage with the transformer coupled amplifiers is the AC current passing through the load resistance R_L results in only wastage of power, since we are interested in only AC output power. Moreover passing DC current through the loudspeaker coil is not desirable since it produces hum or noise. Therefore if transformer coupling is done, DC component of current passing through the transformer can be avoided (Fig. 7.10).

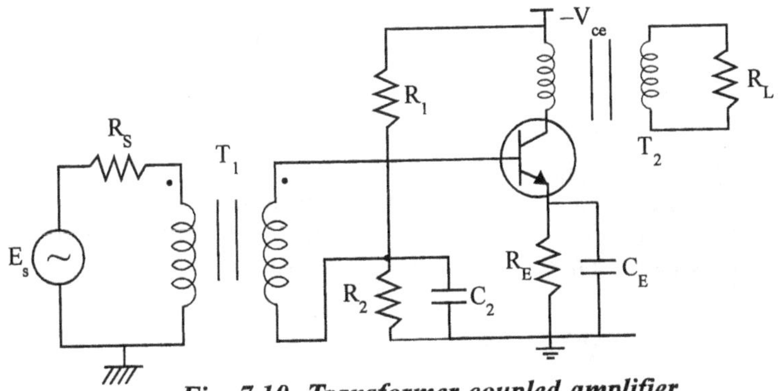

Fig. 7.10 Transformer coupled amplifier

$$\frac{V_1}{V_2} = \frac{I_2}{I_1}; \quad n = \frac{N_2}{N_1} \qquad \text{(For a transformer with usual notation.)}$$

$$V_1 = \left(\frac{N_1}{N_2}\right) V_2; \quad I_1 = \left(\frac{N_2}{N_1}\right) I_2.$$

$$V_1 = \frac{1}{n}.V_2; \; I_1 = n\,.\,I_2 \qquad \text{Therefore } \frac{V_1}{I_1} = \frac{1}{n^2}.\frac{V_2}{I_2}$$

R_1, R_2, R_E are chosen depending upon the biasing point. C_E is emitter bypass resistor. Transformer T_1 is chosen to match R_i of the circuit with R_S and transformer T_2 is chosen for R_0 of the circuit to match with R_L.

$$\therefore \qquad R_L^{'} = \frac{1}{n^2}.R_L$$

C_2 is also a bypass capacitor. For AC it is short circuit.

The equivalent circuit, in terms of h-parameters neglecting the biasing resistors and capacitors, also neglecting the input transformer and considering base and emitter as the input ports.(Fig. 7.11).

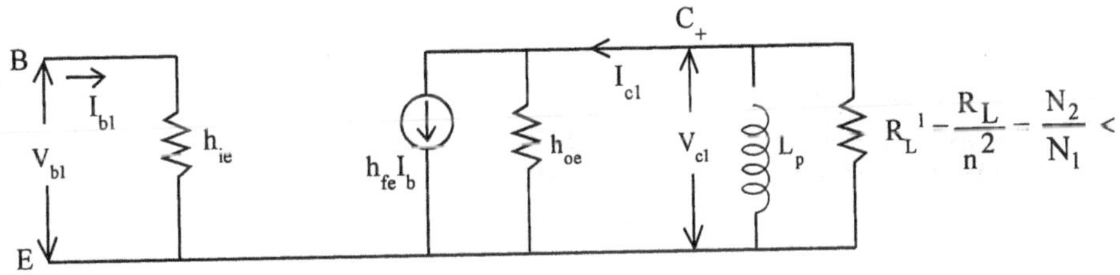

Fig. 7.11 Equivalent circuit

The transistor is replaced by its h-parameter equivalent circuit. The load resistance R_L is referred to primary and so $R_L' = \frac{R_L}{n^2}$ where $n = \frac{N_2}{N_1}$. Since it is a stepdown transformer, $n < 1$; L_p is the inductance of the primary winding (Since we are considering load referred to primary).

7.4.1 Mid Frequency Range

In the mid frequency range, the inductive reactance X_{LP} is high. f is large and so it can be regarded as an open circuit (or very large compared to R_L'). Therefore it will not affect the response. $h_{fe} I_b$ is the current source. When impedance matching is done, the output Z of the circuit and the load resistance R_L will be equal. Therefore the current will get divided between R_L and the circuit equally. The total current is $h_{fe} I_b$. Therefore the current through the primary of the transformer (or in other words the current through the collector circuit) is $\frac{h_{fe}.I_b}{2} = I_{C1}$.

But since T_2 is a step down transformer the current through R_L will be stepped up by $\left(\frac{N_1}{N_2}\right)$ due to transformer action $\frac{N_1}{N_2} > 1$.

$$\therefore \quad I_L = \left(\frac{N_1}{N_2}\right) I_{C1} = \frac{h_{fe}.I_b}{2} \cdot \left(\frac{N_1}{N_2}\right)$$

$$\therefore \text{ Current gain} \quad \frac{I_L}{I_b} = A_I = \frac{N_1}{N_2} \times \frac{h_{fe}}{2}$$

7.4.2 Voltage Gain

$$A_V = \frac{V_L}{V_b}; \; V_b = \text{base voltage input voltage}$$

$$V_{C_1} = \frac{h_{fe}\, I_b}{2} R_L \left(\frac{N_1}{N_2}\right)^2$$

$$\therefore \quad \frac{V_{C1}}{V_{b1}} = -\frac{h_{fe}}{2} \times \frac{R_L}{h_{ie}} \left(\frac{N_1}{N_2}\right)^2$$

But $\quad \frac{V_C}{V_L} = \frac{N_1}{N_2}$

V_C = voltage on primary side of transformer
V_L = voltage on secondary side of transformer

$$\therefore \quad A_V = \frac{V_L}{V_b} = -\frac{h_{fe}}{2} \times \frac{R_L}{h_{ie}} \times \frac{N_1}{N_2}$$

In the low frequency range shunting effect of L_p will reduce the effective load resistance. Lower 3db frequency is reached when,

$$2\pi f_1 L_p = R$$

or

$$f_1 = \frac{R}{2\pi L_p}$$

where R is the parallel combination of $\frac{1}{h_{oe}}$ and $\left(\frac{N_1}{N_2}\right)^2 . R_L$

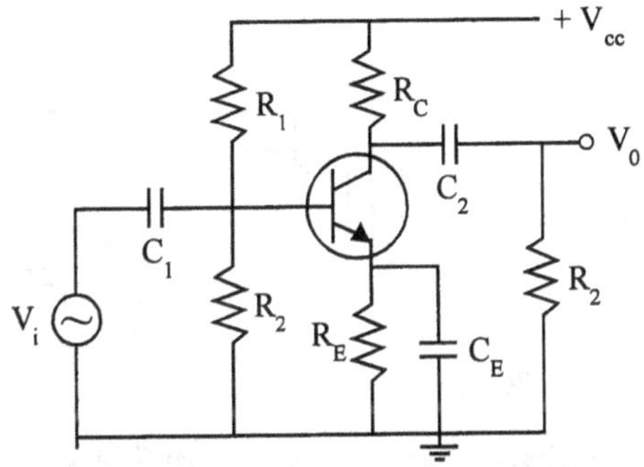

Fig. 7.12 Class A power amplifier circuit

The transformer coupled amplifier circuit is similar to a circuit, like this (Fig. 7.12). Instead of having coupling capacitors C_1 and C_2, we have transformer, coupling. The primary of transformer T_2 acts as R_C. C_2 across R_2 helps in making Emitter of transistor at the ground point for AC. Because of C_E, emitter is at ground potential for AC. Therefore secondary voltage of transformer is applied between base and emitter of transistor.

7.5 TRANSFORMER COUPLED AUDIO AMPLIFIER

Audio amplifier	:	40Hz to 20 KHz
Video	:	5-8 MHz
R.F	:	20 KHz

Classification of Radio Waves.

Very low frequency(VLF)	10-30 K Hz
Low frequency (LF)	30-300 K Hz
Medium frequency (MF)	300-3,000 K Hz
High frequency (HF)	3-30 MHz
VHF	30-300 MHz
Ultra high frequency (UHF)	300-3000 MHz
Super high frequency (SHF)	3000-30,000 MHz

An amplifying system usually consists of several stages in cascade. The input and intermediate stages operate in a small signal class-A mode. Their function is to amplify the small excitation to a large value to drive the final device. This output stage feeds a transducer such as CRT, loud speaker, servo motor etc. So the output stage must be capable of delivering a large voltage or current or large power. Bias stabilization techniques and thermal runaways are very important with power amplifiers.

If the load resistance is connected directly in the output circuit as shown in Fig.7.13 (a), the quiescent current passes through R_L. This results in waste of power since it won't contribute to the AC power signal. In the case of loud speakers it is not desirable to pass DC current through the voice coil. So an arrangement is to be made using an output transformer (Fig. 7.13 (b)).

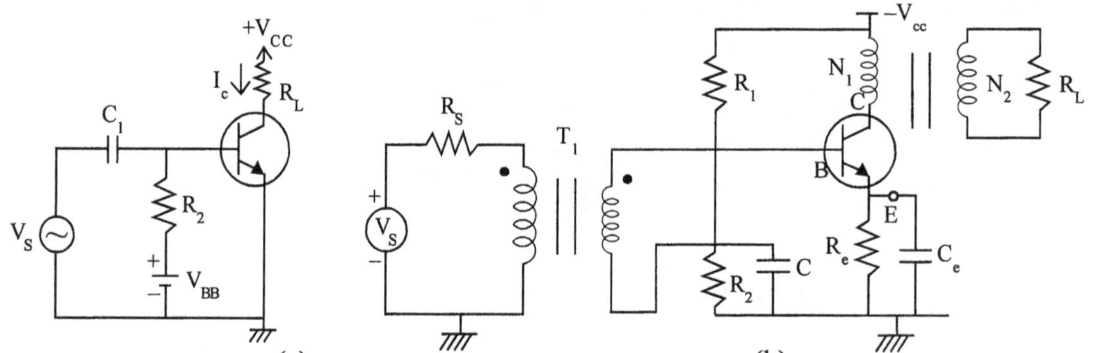

(a) (b)

Fig. 7.13 Transformer coupled amplifiers

7.5.1 Impedance Matching

To transfer significant power to a load such as loud speaker with a voice-coil resistance of 5-15Ω it is necessary to use an output matching transformer. The impedance matching properties of an ideal transformer are :

$$\frac{V_1}{V_2} = \frac{N_1}{N_2}; \quad \frac{I_2}{I_1} = \frac{N_1}{N_2}$$

Let $$\frac{N_2}{N_1} = n \text{ turns ratio.}$$

If $N_2 < N_1$ the transformer reduces the output voltage, and steps up the current by the same ratio.

Impedance matching is required because the internal impedance of the device will be much higher than 5-15Ω of voice coil of a speaker. So power will be lost. Hence output matching transformer is required.

$$V_1 = \frac{N_1}{N_2} \cdot V_2 \qquad I_1 = \frac{N_2}{N_1} \cdot I_2$$

$$V_1 = \frac{1}{n} . V_2 \qquad I_2 = n . I_2$$

Now $$\frac{V_1}{I_1} = \frac{1}{n^2} \cdot \frac{V_2}{I_2}$$

$$\frac{V_1}{I_1} = \text{Effective input resistance } R_L'$$

$$\frac{V_2}{I_2} = \text{Effective output resistance } R_L'$$

$$\therefore \quad R_L' = \frac{1}{n^2} \cdot R_L$$

7.5.2 Maximum Power Output

Consider the Fig. 7.14 (a) and (b).

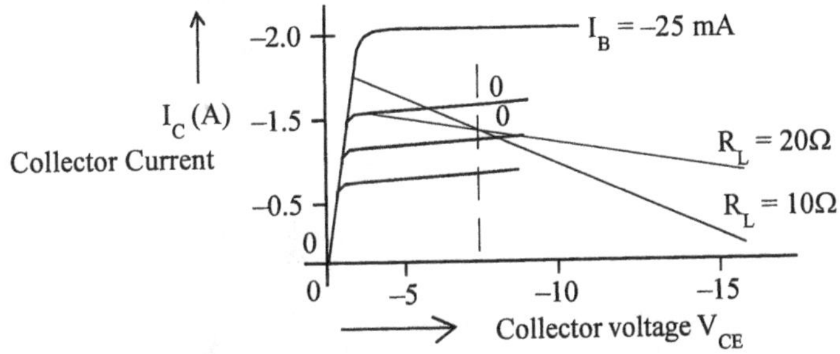

Fig. 7.14 (a) Output characteristics

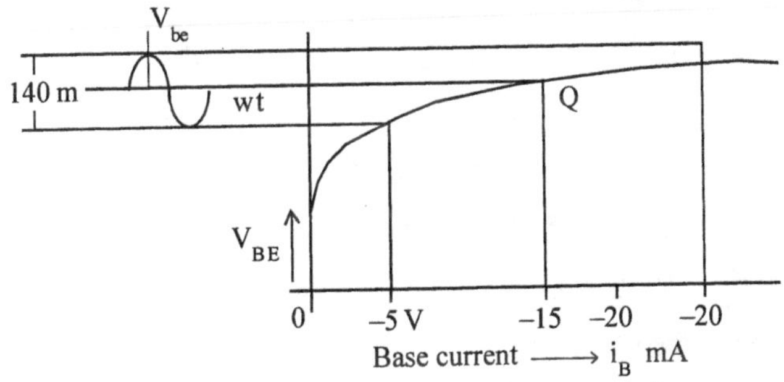

Fig. 7.14 (b) Input characteristics

To find n, for a given R_L, so that power output is maximum is solved graphically. First operating point Q is located $I_C = \frac{V_{CC}}{R_C}$;

P_c = collector dissipation

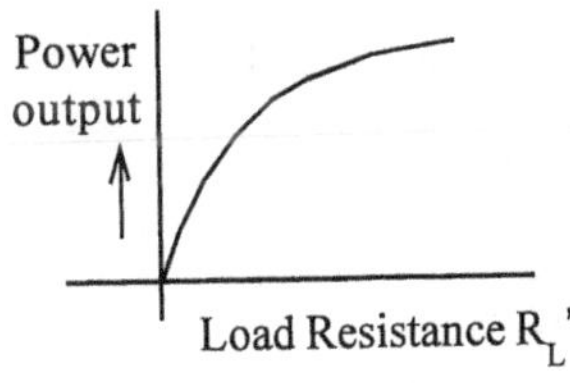

Fig. 7.15 Variation of P_0 with R'_L

V_C = quiescent collector voltage. Peak to peak voltage must be limited to a suitable value such that there is no distortion. From the input characteristic I_{Bmax} to I_{Bmin} can be noted. A series of load lines are drawn through 'Q' point for different values of R_L'. From these two graphs, power output versus load resistance R_L is drawn from the graph, R_L' is chosen that power is maximum and distortion is minimum (Fig. 7.15).

7.5.3 EFFICIENCY

Suppose the amplifier is supplying power to pure resistive load (Fig. 7.16).

Power input from DC supply = $V_{cc}\, I_c$

Power absorbed by output circuit = $I_C^2\, R_1 + i_C^2\, V_C$

R_1 is static load i_C and V_C are rms output current and voltage. If P_D is average power dissipated by the active device.

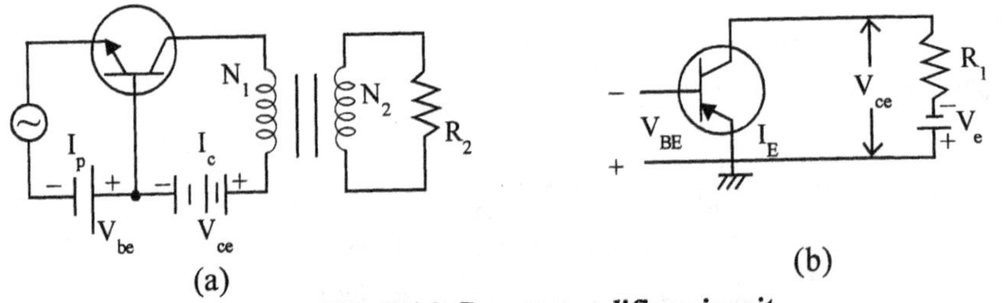

Fig. 7.16 Power amplifier circuit

$$V_{CC}\, I_C = I_C^2\, R_1 + i_C\, v_C + P_D$$

But $$V_{CC} = V_C + I_C\, R_1;$$

$$P_D = V_{CC} \cdot I_C - I_C^2\, R_1 - I_C^2\, V_C$$

∴ $$P_D = V_C\, I_C + R_1\, I_C^2 - i_C\, v_C$$

$$P_D = V_C\, I_C - i_C\, v_C$$

If the load is not pure resistance, $i_C\, v_C$ must be replaced by $i_C\, v_C \cos\theta$, where $\cos\theta$ is power factor of load.

7.5.4 CONVERSION EFFICIENCY, η

An amplifier is essentially a frequency converter, changing DC power to AC power, (Fig. 7.17).

A measure of the ability of an active device to convert DC power of the supply into AC power delivered to the load is called *conversion (η)* or *theoretical efficiency (η)*. It is also called *collector circuit (η)* for transistor amplifier.

$$\eta \equiv \frac{\text{Signal power delivered to load}}{\text{DC power supplied to input circuit}} \times 100\%$$

In general, $$\eta = \frac{\frac{1}{2} \cdot B_1^2 \cdot R_L'}{V_{CC}(I_C + B_0)} \times 100\%$$

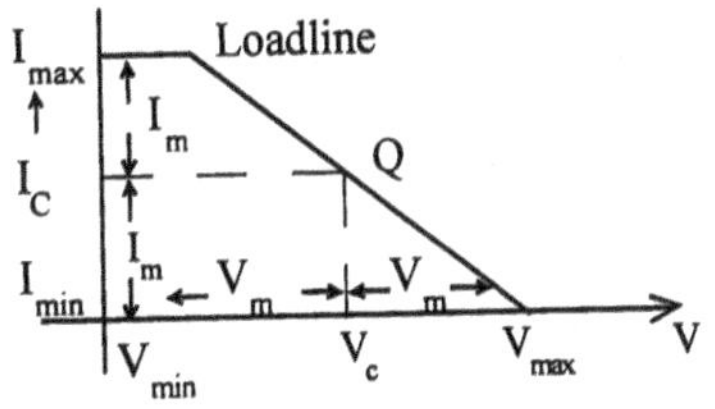

Fig. 7.17 Transfer curve

where B_0 and B_1 are constants in the expression

$$i_C = I_C + B_0 + B_1 \text{ Cos } \omega t + B_2 \text{ Cos } 2\omega t + ...$$

Expression for instantaneous total current.

If distortion components are negligible,

$$\eta = \frac{\frac{1}{2} V_m I_m}{V_{Ce} I_C} \times 100\%$$

$$= \frac{50 V_m I_m}{V_{Ce} I_C} \times 100\%$$

7.5.5 Maximum Value of η

In the case of *series fed amplifiers, the supply voltage* V_{CC} is equal to V_{max}. In the transformer coupled amplifier V_{CC} is equal to the quiescent voltage V_C.

Under ideal conditions,

$$I_C = I_m$$

and

$$V_m = \frac{V_{max} - V_{min}}{2}$$

$\therefore$

$$\eta = 50 \frac{V_m \cdot I_m}{V_{CC} I_C}$$

$$= \frac{50 \times \left(\frac{V_{max} - V_{min}}{2} \right) \times I_C}{V_{CC} I_C}$$

$$\boxed{\eta = \frac{25 (V_{max} - V_{min})}{V_{CC}} \%}$$

For series fed amplifier, $V_{max} = V_{cc}$

$\therefore$ η for series fed amplifiers

$$= \frac{25(V_{max} - V_{min})}{V_{max}} \%$$

$\therefore$ *Maximum possible value = 25%*

In the case of transformer coupled amplifier.

$$V_{ce} = V_c = \frac{V_{max} + V_{min}}{2}$$

$$\therefore \quad \boxed{\eta = 50\left(\frac{V_{max} - V_{min}}{V_{max} + V_{min}}\right)\%}$$

So the Maximum Possible Value of η is 50% for transformer coupled amplifier

Thus transformer coupled amplifier have twice the maximum η compared to series fed amplifiers. For *transformer circuits* occurs near saturation, therefore $V_{min} << V_{max}$ and η can be 50%.

PNP Transistor Amplifier with AC Signal

Emitter is forward biased. Collector is *reverse biased.*

AC is superimposed at the input, we get AC output across R_L. (Fig. 7.18)

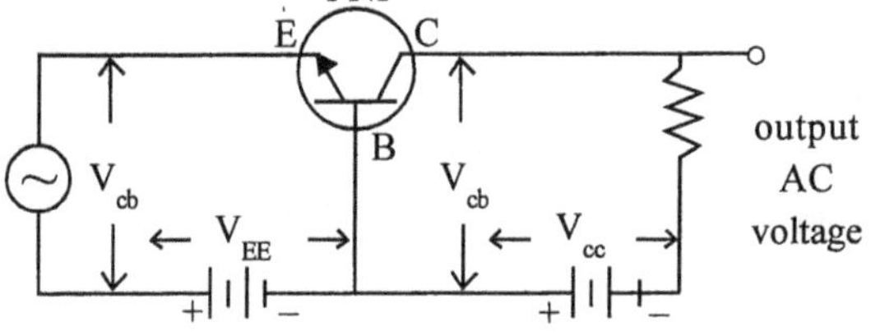

Fig. 7.18 Circuit with PNP transistor

7.6 PUSH PULL AMPLIFIERS

R_1 and R_2 are provided to prevent cross over distortion (Fig. 7.19(a)). Because of R_1 and R_2 the B-E junctions of the two transistors are forward biased so that cut in voltage V_r will not come into the picture. But because of R_1 and R_2, the operation will be slightly class AB operation and not pure class B operation. For a given transistor, the dynamic characteristics are not exactly linear, that is, for some changes in the values of i_b, i_c will not change by the same amount that is. If i_b is increasing by 5 μA, i_c increases by 1mA. For 10 μA increase in i_b, i_c will not increase by 2 mA but something different. So the

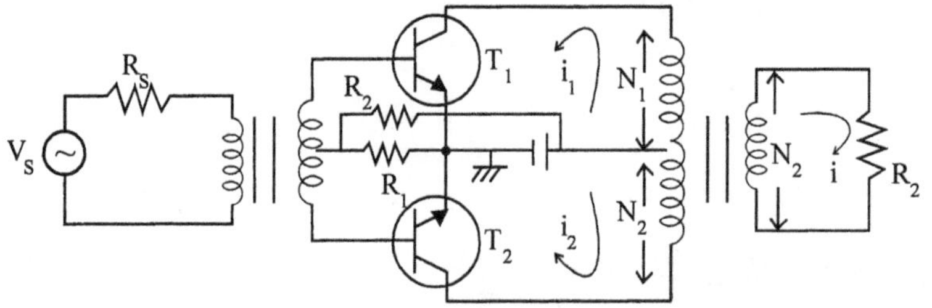

Fig. 7.19 (a) Push pull amplifier circuit

graph of i_b V_S i_c is nonlinear. Therefore for uniform changes in the input, the output will not change uniformly. Hence distortion will be introduced in the output waveform. This can be eliminated by pushpull connection.

The efficiency (η) for class A amplifier is 25% and for transformer coupled class A amplifier is 50%. Therefore class A amplifier because of poor η is used for low output power requirement and where conductance should be for complete 360°. (eg. for the driver stage of the last power stage).

Suppose for the transistor Q_1, the input (base current) is a Cosine wave $x_1 = X_m \cos \omega t$. The output current at the collector $i_1 = I_c + B_0 + B_1 \text{Cos } \omega t + B_2 \text{Cos } 2\omega t + B_3 \text{Cos } 3\omega t$ where I_c is the DC current due to biasing, B_0 is the DC component in the Fourier series of the AC input, B_1 is fundamental component C_m (Transformer T_1 provides phase shift to the inputs. T_2 joins the two outputs. For the second transistor, the input is given from the centre tapped transformer which introduces a phase shift of 180°.

$$\therefore \qquad x_2 = -x_1 = X_m \{\text{Cos } (\omega t + \pi)\}$$

The output current of this transistor i_2 is obtained by replacing ωt by $(\omega t + \pi)$ in the expression for i_1.

i.e.,
$$i_2 (\omega t) = i_1 (\omega t + \pi)$$
$$i_2 = I_C + B_0 + B_1 \text{Cos } (\omega t + \pi) + B_2 \text{Cos } 2 (\omega t + \pi) \text{ .....}$$
$$= I_C + B_0 - B_1 \text{Cos } \omega t + B_2 \text{Cos } 2\omega t - B_3 \text{Cos } 3 \omega t \text{ .....}$$

Therefore i_1 and i_2 are out of phase by 180°. So they flow in the opposite direction through the output transformer primary windings. Therefore the total output current i is proportional to $(i_1 - i_2)$. Since the net output current depends on the turns ratio of the transformer i is the current flowing through R_L.

$$\therefore \qquad i = K (i_1 - i_2) = 2 K (B_1 \text{Cos } \omega t + B_3 \text{Cos } 3 \omega t) \text{ .....}$$

This expression shows that all the even harmonic terms $B_2 \text{Cos } 2 \omega t$, $B_4 \text{Cos } 4 \omega t$ are eliminated. The only harmonic component predominant is $B_3 \text{Cos } 3 \omega t$, the III (third) harmonic terms. Higher harmonics can be neglected. $B_1 \text{Cos } \omega t$ is the original signal. Therefore *Harmonic distortion will be less for pushpull amplifiers*. This is under the assumption that both the transistors have identical characteristics. If not, some even harmonics may also be present.

Pushpull amplifier is said to possess mirror symmetry.

Mirror symmetry means, mathematically,

$$i (\omega t) = - i (\omega t + \pi)$$

The output current i for pushpull amplifier is,

$$i = 2 K (B_1 \text{Cos } \omega t + B_3 \text{Cos } 3 \omega t + \text{.....})$$

If ωt is replaced by $(\omega t + \pi)$, the above equation holds good.

This is also called as halfwave symmetry. It means that the bottom loop of the wave when shifted by 180^0 along the axis will be the mirror image of the top. This is so because only odd harmonic terms are there in the output.

The maximum instantaneous reverse voltage across each transistor occurs when it is not conducting and is equal to 2 V_i. Because when Q_1 is conducting, maximum $V_{CE} = 0$ and so voltage across the upper half winding of output transformer primary is V_{CC}. Due to induction some voltage will appear across the lower half also. Q_2 is not conducting. Therefore the transistor voltage across Q_2 collector and emitter is $V_{CC} + V_{Ce} = 2\ V_{CE}$.

It is called as *push pull amplifier* since, the input to the two transistors are out of phase by 180°. (Since centre tapped transformer is used). Therefore when one transistor is conducting the other is not or when the output current of one transistor is increasing, for the other it is decreasing. This is known as *push pull action.* When one transformer current is being pushed up, the other is being pulled down.

7.6.1 Class B Amplifiers

A transistor circuit is in class B operation, if the emitter is shorted to base (for DC). The transistor will be at cut off. Therefore in the circuit for class B pushpull amplifier, R_2 should be zero. The conduction angle is 180°.

7.6.2 Advantages of Class B Push Pull Circuit AMPLIFIER

1. More output power ; η = 78.5%. Max.
2. η is higher. Since the transistor conducts only for 180°, when it is not conducting, it will not draw DC current.
3. Negligible power loss at no signal.

7.6.3 Disadvantages of Class B Push Pull Circuit AMPLIFIER

1. Supply voltage V_{CC} should have good regulation. Since if V_{CC} changes, the operating point changes (Since I_C changes). Therefore transistor may not be at cut off.
2. Harmonic distortion is higher. (This can be minimized by pushpull connection).

Therefore Class B amplifiers are used in a *system* where the power supply is limited, and is to be conserved such as circuits operating from *Solar cells or battery, Battery Cells, air borne, space and telemetry applications*.

7.6.4 Conversion η

$$P_0 = \frac{I_m}{\sqrt{2}} \times \frac{V_m}{\sqrt{2}} = \frac{I_m\ V_m}{2} = \frac{I_m}{2}\left(V_{CC} - V_{min}\right)$$

$V_m = V_{CC} - V_{min}$ (Since operating point is chosen to be at cut off $V_{CC} = V_{max}$)

Because in pushpull circuit, there are two transistors conducting, each for 180°.

Therefore total conduction is for 360°. $\left(\therefore P_0 = \frac{I_m V_m}{2}\right)$ Corresponding to Q point,

the voltage is V_{CE}. Neglecting the *dissipation across* emitter, $V_{CC} \simeq V_{CE}$.

$P_{DC} = I_{DC} \cdot V_{CC}$. DC current is drawn by the transistor only when it is conducting. V_{CC} is always present. One transistor conducts for 0-π only. I_{DC} drawn by each transistor is the average value of half wave rectified DC (equal to I_m/π). But there are two transistors.

$$\therefore \text{Total} \qquad I_{DC} = \frac{2I_m}{\pi}$$

$$P_{DC} = V_{CC} \cdot \frac{2I_m}{\pi}$$

$$\therefore \quad \eta = \frac{P_o}{P_{DC}} \times 100$$

$$= \frac{I_m}{2} \frac{(V_{CC} - V_{min}).\pi}{V_{CC}.2I_m} \times 100$$

$$= 100 \times \frac{\pi}{4} \frac{(V_{CC} - V_{min})}{V_{CC}}$$

$$\eta = 25 \times \pi \left(1 - \frac{V_{min}}{V_{CC}}\right) \%$$

If $\frac{V_{max}}{V_{min}} = 1$, Max $\eta = 78.5\%$

7.6.5 Dissipation of Transistors in Class B Operation

The DC input power to the transistors in class B configuration is

$$P_{DC} = \frac{2I_m.V_{CC}}{\pi}$$

[Since the transistor is conducting for 180^0 only. So it draws DC current only during that period. Therefore average value of i_C is $\frac{I_m}{\pi}$. There are two transistors each conducting for 180^0, from 0 - π and π - 2π respectively). (Fig. 7.19(b)).

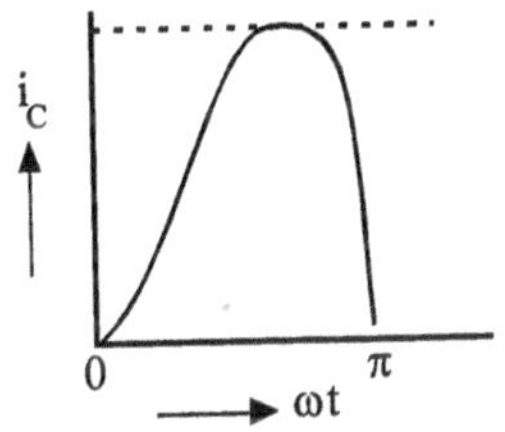

Fig. 7.19 (b) Current cycle

$\therefore$ Total DC current = $2\ I_m/\pi$

$$P_{DC} = \frac{2I_m}{\pi}.V_{CC}$$

But $I_m = \frac{V_m}{R_L'}$ where R_L' is the effective load resistance of the circuit, without considering the secondary of the transformer.

$$\therefore \quad P_i = \frac{2\ V_m.\ V_{CC}}{\pi\ R_L'}$$

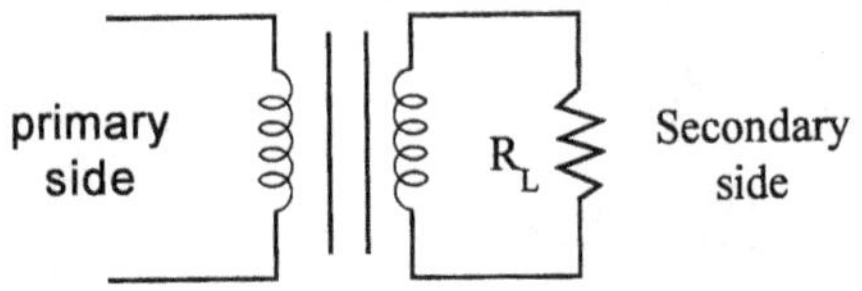

Fig. 7.20 Transformer on load side

The collector dissipation P_C (in both transistors) is the difference between the power input (P_i or P_{DC}) to the collector circuit, and the power delivered to the load. (Both are in watts). Though DC and AC powers (Fig. 7.20).

Output power delivered to load P_0,

$$P_0 = \frac{V^2}{R} = \left(\frac{V_m}{\sqrt{2}}\right)^2 \Big/ R_L'$$

$$= \frac{Vm^2}{2R_L'}$$

$$\therefore \qquad P_C = P_i - P_0 = \left[\frac{2}{\pi} \cdot \frac{V_{CC} V_m}{R_L'}\right] - \left(\frac{V_m^2}{2R_L'}\right)$$

(V_m is the peak value of the AC input).

The above equations shows that at no AC signal, (i.e., $V_m = 0$) the collector dissipation is zero, and as the signal magnitude increases, P_C increases. As V_m increases P_C also increases, P_C is maximum when $V_m = \frac{2V_{CC}}{\pi}$. (Fig. 7.21).

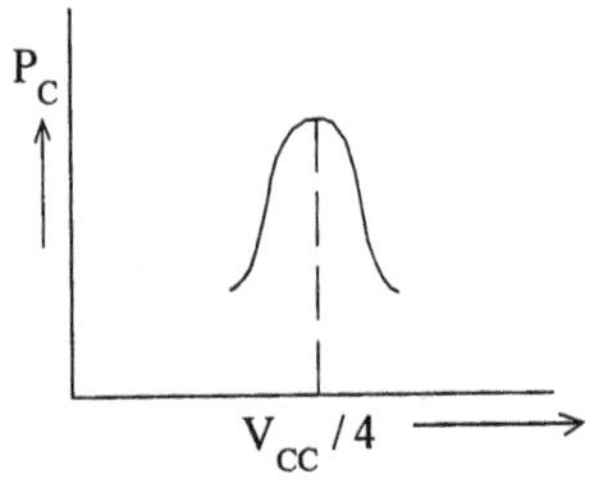

Fig. 7.21 Power output

A graph can be plotted between P_C and V_m

The maximum dissipation

$$P_{C\,max} = \frac{2V_{CC}^2}{\pi^2 R_L'}$$

P_C is maximum, when $V_m = \frac{2V_{CC}}{\pi}$

P_0 is maximum, when $V_m = V_{CC}$

$$\therefore \quad P_{0\,(max)} = \frac{V_{CC}^2}{2R_L'} \qquad \text{Since } P_0 = \frac{V_m^2}{2R_L}$$

P_0 is maximum $V_m = V_{CC}$

$$\therefore \quad P_{C\,(max)} = \frac{4}{\pi^2}(P_0 \max)$$

$$\boxed{P_{C\,(max)} = 0.4\, P_{0\,(max)}}$$

∴ If we want to deliver 10-W of output by a class B pushpull amplifier, the collector of the transistor or the collector dissipation should be $0.4 \times 10 = 4$W. This is for the entire circuit. Therefore each transistor (Since there are two transistors in class B pushpull) should be capable of dissipating 2W of power as heat.

7.6.6 Graphical Construction for Class B Amplifier

for class B, Q point is chosen as shown in Fig. 7.22.

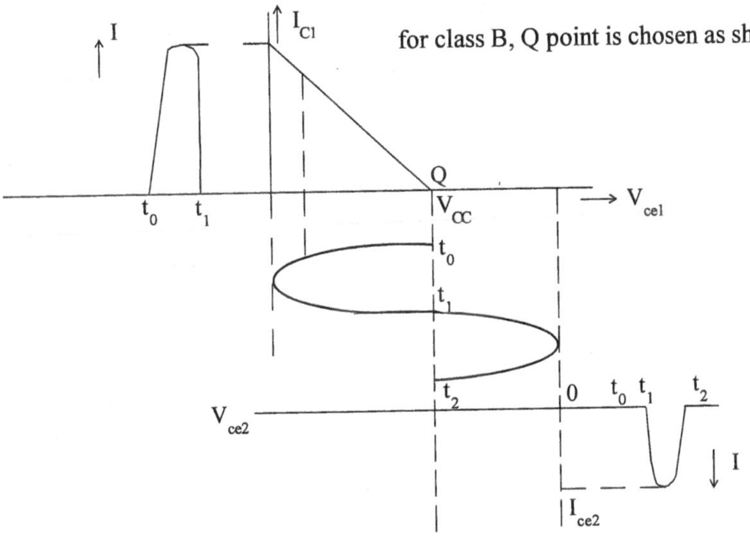

Fig. 7.22 Class B operation

7.6.7 Distortion

Let i_{b_1}, V_C, V_{b_1} be the input characteristic of the first transistor and i_{b_2}, V_S, V_{b_2} is the input characteristic of the second transistor. V_γ is the cut in voltage. These are the two transistors of the class B pushpull amplifier. Now the base input voltage being given to the transistor is sinusoidal, i.e., base drive is sinusoidal. So because of the *cut in voltage*, eventhough input voltage is present, output will not be transmitted or there is distortion in the output current of the transistor. This is known as *crossover distortion.* But this will not occur if the base current drive is sinusoidal. Since in the graphical analysis the input current is taken in the I quadrant. No distortion if the operating point is in the active region. Cross-over distortion can also be eliminated in class AB operation. A small stand by current flows at zero excitation. The input signal is shifted by constant DC bias so that the input signal is shifted by an amount V_γ.

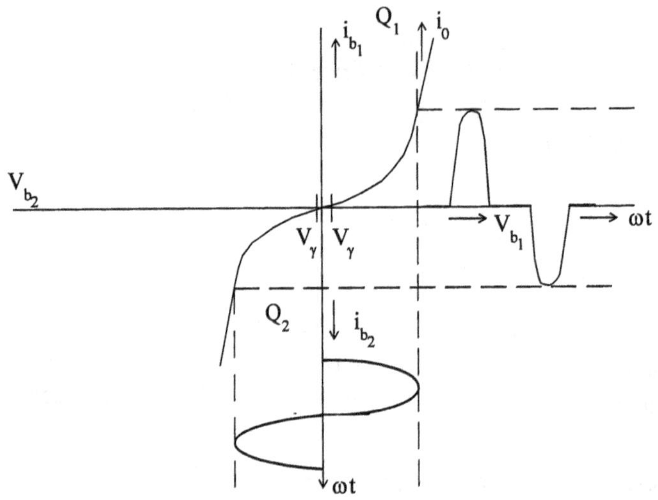

Fig. 7.23 Cross over distortion

Table 7.1 Comparison of amplifiers based on the type of Coupling.

	Direct coupled	R.C. coupled	Transformer coupled
Frequency range :	D.C to medium range	High	A.F. range
f_1 (Lower cutoff frequency	0 Hz (D.C)	50–100 Hz and above	100 Hz and above
f_2 (Upper cutoff frequency)	Limited	Can be more	Limited to A.F range
Cost	Less No R and C No transformer	Medium (Due to R and C)	High (Due to transformer)
Size	Less	Medium	High
Frequency response	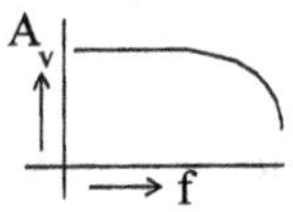	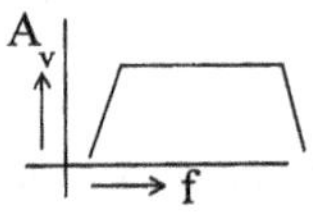	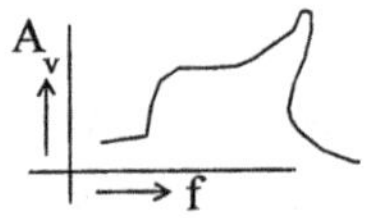 (due to L of tansformer spike occurs)
|Z| matching	Not good	Not good	Excellent

7.7 COMPLIMENTARY SYMMETRY CIRCUITS (TRANSFORMER LESS CLASS B POWER AMPLIFIER)

The standard class B push-pull amplifier requires a centre tapped transformer, since only one transistor conducts for 180^0, so that if two transistors were to conduct for complete 360°, there should be a centre tapped transformer. Otherwise there should be a phase inverter. Complementary symmetry circuits need only one phase. They don't require a centre tapped transformer. But their requirement is a pair of closely matched. Oppositely doped (pnp and npn) transistors. Till recently it was different to get such transistors. But now the technology has improved and pnp and npn transistors with identical circuits, can be manufactured. (Fig. 7.24).

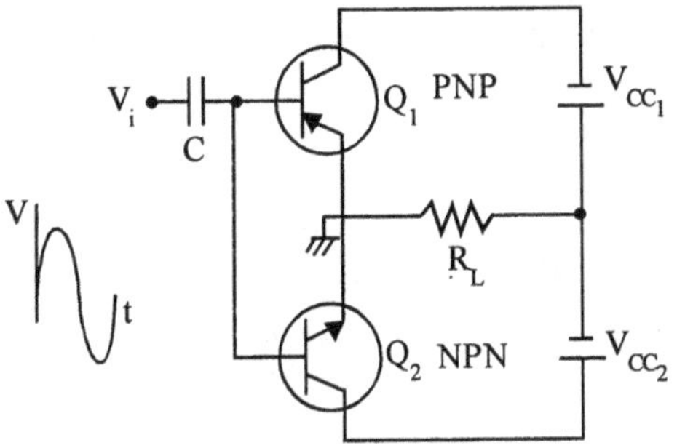

Fig. 7.24 Complimentary symmetry

The circuit shows a basic complimentary circuits in class B. It is class B operation since the operating point is at cutoff. Emitter and base are shorted or $V_{BE} = 0$. The input is capacitance coupled. The output is direct coupled since output is taken directly across R_L. One end of R_L is grounded with no input signal present. Both transistors won't conduct. Therefore current through $R_L = 0$. When the signal (input) is positive going, the transistor Q_1 is cutoff (since it is pnp), base is n type. Therefore. Emitter-Base junction is reverse biased). Q_2 conducts, since it is NPN transistor, base input is positive. So it conducts. (Fig. 7.25).

The resulting current flows through R_L and develops a negative going voltage at point relative to ground. When the signal is negative going Q_2 goes off and Q_1 turns on. Current flows through R_L in such a direction as to make point positive with respect to ground. *There is no DC current through R_L. Hence an electromagnetic load such a loud speaker can be connected* directly without introducing saturation problems.

The difficulty with the above circuit is, the transistor will not conduct till the input signal magnitude exceeds the cut in voltage V_v. So cross over distortion will be present output of the amplifier.

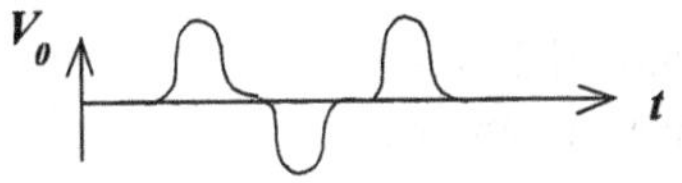

Fig. 7.25 Output with crossover distortion

So DC bias should be provided to overcome the threshold voltage for each base-emitter junction. Therefore the circuit is as shown in Fig. 7.26(a). The voltage developed across R_2 forward biases both the transistors, E-B junctions. R_2 is normally so small as not to produce any significant loss in drive to Q_2.

This circuit needs only one V_{CC} Q_1 is NPN transistor. Therefore its collector is reverse biased, since V_{CE} is positive.

Q_2 is PNP, its collector is negative with respect to V_{CE} since grounded. Therefore its collector is also reverse biased. The drop across R_3 reverse biases the common base junction of Q_2 and the drop across R_1 reverse biases the Common Base junction of Q_1. This circuit requires only one DC supply and is commonly referred to as the "*Totem pole*" configuration. When the input is positive, Q_1 is turned on and Q_2 is turned off. When the input goes negative, Q_1 turns off while Q_2 conducts.

7.8 PHASE INVERTERS

These circuits are used to drive push pull amplifiers since a pushpull amplifier requires two equal inputs with 180^0 phase difference. *The centre tapped* transformers are *bulky and costly.* Therefore Phase inverter circuits with transistors are used (Fig. 7.26(a)).

Phase inverter circuits are also known as *Paraphase amplifiers*. The criterion is, from a single input, we must get two equal outputs with a phase shift of 180^0.

$$V_{01} = -V_{02}$$

The circuit is as shown in Fig. 7.26(b), emitter followers with a collector load R_1 is used.

R_3 and R_4 are bias resistors. R_1 is the collector load. R_2 is the emitter load. Output is taken after the capacitor 'C' to block DC. V_i is AC input. The outputs at points 2 and 1 will be out of phase

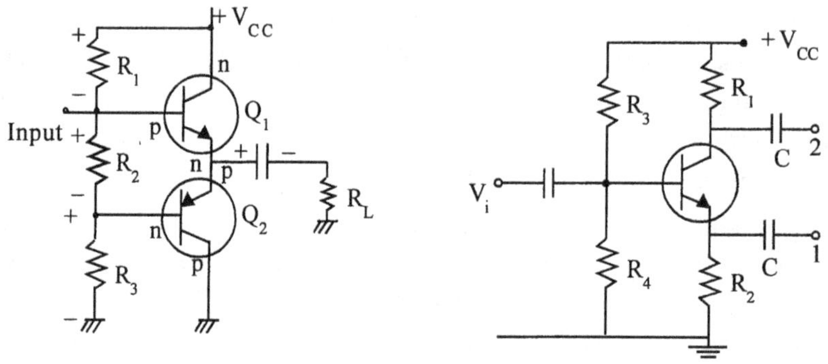

Fig. 7.26 Phase inverter circuits

by 180^0. With $R_1 = R_2$ output voltages will be the same. The output impedance of the circuit from point 1 is that of a common collector configuration (Since output is taken across R_2. So it is common collector configuration).

The output impedance at terminal 2 is that of common emitter configuration. So both will not be the same since $R_{OE} \neq R_{OC}$, since output voltages at points 1 and 2 will not be the same. So for that, another transistor is used to match the gains and source impedances for common emitter collector. There is phase shift of 180^0. Therefore V_{02} will be with phase shift for common capacitor configuration there is no phase shift. Therefore V_{01} is in phase. Therefore V_{01} and V_{02} are out of phase by 180^0.

EXAMPLE : 7.1

The amplifier shown is made up of an NPN and PNP transistors. The h-parameters of the two transistors are identical and are given as $h_{ie} = 1\ k\Omega$, $h_{fe} = 100$, $h_{oe} = 0$ $h_{re} = 0$.

Find overall voltage gain $A_V = V_0/V_i$

Both the transistors are in common emitter configuration. For Q_2 the output is taken across 5 kΩ the collector resistor R_{C_2} which is actually the load resistor. (Fig. 7.27).

$$R_{C_1} = R_{L_1} = 2\ k\Omega$$

$$R_{C_1} = 1\ k\Omega$$

$$A_I = \frac{I_C}{I_b} = -\frac{h_{fe}.I_C}{I_C} = -h_{fe}$$

$$\therefore \quad R_0 = h_{ie} + (1 + h_{fe}) R_e \qquad (1 + h_{fe}) R_e >> h_{ie}$$

$$R_i = \frac{V_i}{I_b} = h_{ie} + (1 + h_{fe}) R_e$$

$$[(1 + h_{fe}) R_e >> h_{ie}]$$

$$\therefore \quad R_i = (1 + h_{fe}) R_e$$

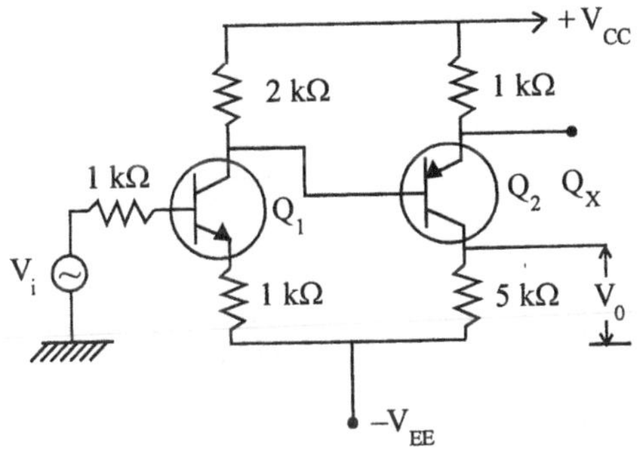

Fig. 7.27 Circuit for Ex : 7.1

$$A_V = \frac{A_I.R_L}{R_i}; \quad A_I = h_{fe}$$

$$\therefore \quad A_V = \frac{-h_{fe}.R_L}{(1+h_{fe})R_e} \quad [1 + h_{fe} \simeq h_{fe}]$$

$$A_V = \frac{R_L}{R_e} = \frac{R_L}{R_e}$$

$$A_{V_2} = \frac{R_{L_2}}{R_{C_2}} = \frac{-5\ k\Omega}{1\ k\Omega} = 5$$

$$A_{V_1} = \frac{2\ k\Omega}{1\ k\Omega} = \frac{R_{L_1}}{R_{C_1}} = 2$$

$$\therefore \quad A_V = A_{V_1} \times A_{V_2} = 5 \times 2 = 10$$

EXAMPLE : 7.2

Design a class B power amplifier to deliver 30W to a load resistor $R_L = 4\Omega$ using a transformer coupling. $V_m = 30V = V_{CC}$. Assume reasonable data wherever necessary.

SOLUTION :

The power to be delivered is 30W. Assume 10% losses in the transformer windings, and design the circuit for 20% over load i.e., even by mistake, if excess current is being drawn or even voltage applied, the transistor must with stand this (Fig. 7.28).

∴ P_0 is taken as 40W. 30 + 7W (overload) + 3W (transformer losses)

∴ The collector dissipation of the transistor

$$P_C\,(max) = 0.4\ P_0\,(max)$$

∴ The transistor to be chosen must be capable of dissipating $P_C\,(max) = 0.4 \times 40 = 16$ W

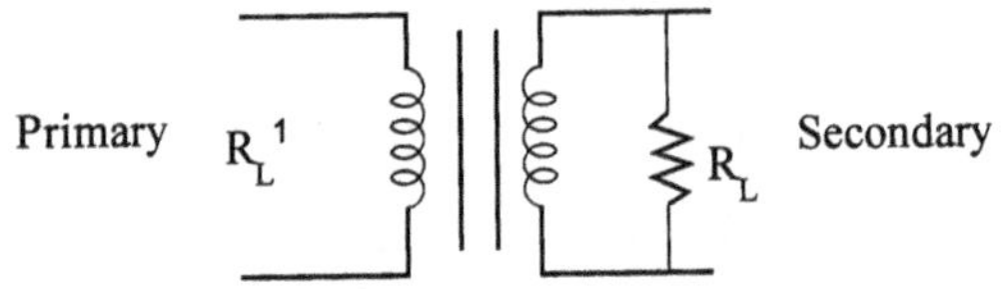

Fig. 7.28 Circuit for Ex : 7.2

$$P_0 = \frac{V_m^2}{2R_L'}$$

$$40 = \frac{(30)^2}{2 \times R_L'}$$

∴ $$R_L' = \frac{(30)^2}{2 \times 40}$$

∴ $$R_L = 11.25\ \Omega$$

∴ R_L is the resistance of transformer secondary referred to primary.

$$\frac{R_L'}{R_L} = \left(\frac{N_1}{N_2}\right)^2$$

∵ $$V_P I_P = V_S I_S$$

$$\frac{V_P}{V_S} = \frac{I_S}{I_P} = \frac{N_1}{N_2}$$

∴ The turns ratio of the output transformer is,

$$\frac{N_1}{N_2} = n = \left(\frac{R_L^1}{R_L}\right)^{1/2} = \left(\frac{11.25}{4}\right)^{1/2} = 1.7$$

Peak collector current swing

$$I_m = \frac{V_m}{R_L} = \frac{V_{CC}}{R_L} = \frac{30}{11.25} = 2.666 \text{ Amperes}$$

EXAMPLE : 7.3

Design a class A transformer coupled amplifier, using the transistor, to deliver 75 mW of audio power into a 4Ω load. At the operating point, $I_B = 250\ \mu A$, $V_{CC} = 16V$. The collector dissipation should not exceed 250 mW. $R_L' = 900\ \Omega$. Make reasonable approximations wherever necessary.

$V_{CE} = \frac{V_{CC}}{2}$ for biasing in normal amplifier. If it is transformer coupled, $V_{CE} \simeq V_{Ce}$

Collection dissipation = $I_C \cdot V_{CE}$ (I_C is the collector current at operating point) (Fig. 7.29).

$$V_{CE} \cdot I_C -= P_{D\ max}$$

or $$V_{CC} \cdot I_C \simeq P_{D\ max}$$

$$I_C \simeq \frac{P_{D\ min}}{(V_{CC} - IV)};$$

Because, the DC. drop across the transformer $V_{CC} \simeq V_{CE}$ can be neglected. The drop across R_E is small $\simeq$ IV. Therefore $V_{CE} \simeq V_{CC} - V = 15$ V.

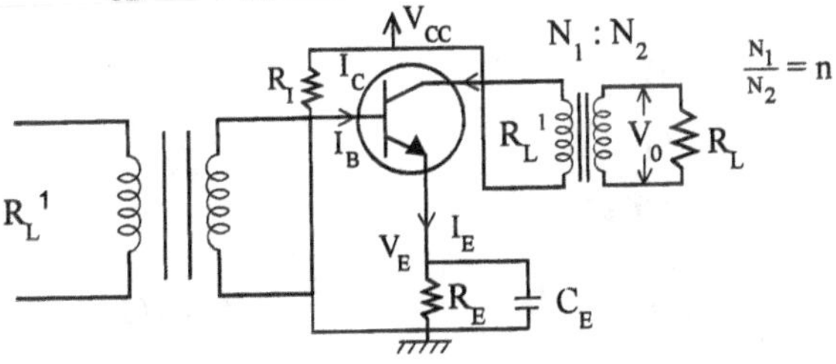

Fig. 7.29 Circuit for Ex : 7.3

$V_{CE} = 16V$. There will be some voltage drop across R_E and the primary winding of the transformer. Therefore V_{CE} can be approximately taken as 15V.

$$\therefore \quad I_C = \frac{P_{D\ min}}{V_{CE}} = \frac{250}{15}$$

$$= 16.66 \text{ mA}$$

Assuming that transformer primary resistance negligible,

$$V_E = V_{CC} - V_{CE}$$

$$= 16 - 15 = IV \text{ (DC)}$$

$$I_E \simeq I_C = 16.66 \text{ mA (DC)}$$

$$\therefore \quad R_B = \frac{V_E}{I_E} = \frac{IV(DC)}{16.66 mA} = 60\Omega$$

At $f = 50$ Hz, $X_E = \frac{R_E}{10} = \frac{60}{10} = 6\Omega$

$\therefore$ $C_E = \frac{1}{2\pi f X_E} = \frac{1}{2\times3.14\times50\times6} \simeq 53\ \mu f$

$R_L' = 900\ \Omega$ $R_L = 4\ \Omega$

$\therefore$ Transformer turns ratio

$$= \sqrt{\left(\frac{R_L'}{R_L}\right)}$$

$$n = \frac{N_1}{N_2} = \sqrt{\left(\frac{900}{4}\right)} = 15$$

$\because$ It is Germanium transistor, $V_{BE} = 0.25V$

$\therefore$ $V_B = V_E + V_{BE}$

$= 1 + 0.25 = 1.25V$

Assuming that the current through R_1 is 10 I_B,

$I_{R_1} = 250\ \mu A \times 10$

$= 2.5$ mA

Neglecting the loading effect due to base of the transistor and assuming that I_B, flows through R_L also

$$R_2 \simeq \frac{V_B}{I_{R_1}} = \frac{1.25}{2.5\,mA}$$

$$= 0.5\ K\Omega$$

$$R_1 = \frac{V_{CC} - V_B}{I_{R_1}} = \frac{16 - 1.25}{2.5\ mA} = 5.9\ k\Omega$$

7.9 CLASS D : OPERATION

These are used in *transmitters* because their efficiency (η) is high $\simeq$ 100%. (Fig. 7.30).

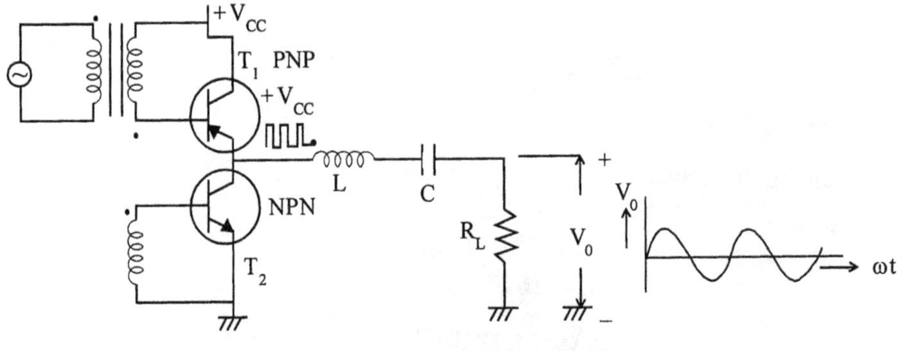

Fig. 7.30 Class D amplifier circuit

A pushpull connection of two transistors in common emitter configuration of complementary transistors (one pnp the other npn) is employed. When the input is positive, T_1 is cut off and T_2 saturates. During the negative half cycle of the input, T_1 saturates and T_2 is cutoff. Therefore the output voltage is a square wave with voltage changing between 0 and V_{CC}.

The dot convention for transformer is, when input is positive, the dotted end of the primary is positive. At the same time, the dotted end of upper secondary winding is positive and dotted end of lower secondary winding is positive. So when the input is positive, T_1 base which is n type (pnp) gets positive voltage. So T_1 is cutoff. Therefore $V = V_{CC}$. When input is negative, T_2 base which is p type (npn) will get negative voltage. So T_2 is cutoff T_1 saturates.

In this circuit, each transistor is saturated for almost 180^0 of the cycle. So each transistor acts like a switch rather than like a current source. When the transistor saturates, the power dissipation.

$$P_D = V_{CE\,(sat)}\ I_{C\,(sat)}$$

It is very small, since $V_{CE\,(sat)}$ is near zero. When the transistor is cutoff, $P_D \simeq 0$. Therefore average power dissipation over the cycle is very small. Therefore $\eta \simeq 100\%$.

The output of the collectors of transistors, is a square wave with 0 - V_{CC} voltages. This is given to a *series resonant* circuit. So the output will be a sine wave (like oscillator circuits).

7.10 CLASS S : OPERATION

Switching regulators are based on class 'S' operation.

In class S operation, a string of pulses are used as the input signal. The pulses have a width 'W', and a period 'T'. Therefore duty cycle $= \dfrac{W}{T} = D.$

7.10.1 CIRCUIT

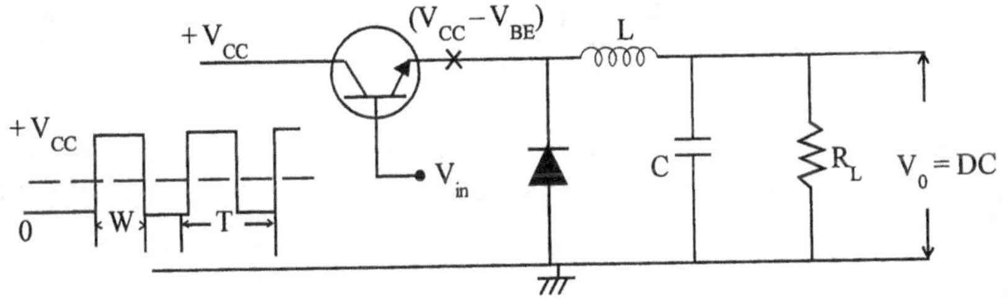

Fig. 7.31 Class S amplifier

The transistor is an emitter follower driven by a train of pulses (Fig. 7.31). Because of the V_{BE} drop, the voltage driving the LC filter is a train of pulses with an amplitude of

$$V_{CC} - V_{BE}. \qquad \text{If } X_L > X_C$$

$$V_{DC} = D\,(V_{CC} - V_{BE})$$

where $$D = \frac{W}{T} = \text{Duty cycle}$$

The higher, the duty cycle, the larger, the DC output. By varying the duty cycle, we can control the a.c. output. So this is class 'S' operation. Because the transistor is cutoff or in saturation its power dissipation is much lower than that in a series regulator. So heat sinks can be small (Fig. 7.32).

Diode rectifies and L, C combination filters the output. So the output is rectified and filtered.

Class A : Conduction of plate current is for complete 360^0, it depends upon operating point.

Class B : Conduction of plate for only 180^0 because the grid is more negative during negative cycle of the signal.

Class C : Conduction is for less than 180^0.

Class AB : Conduction is between 360 and 180^0

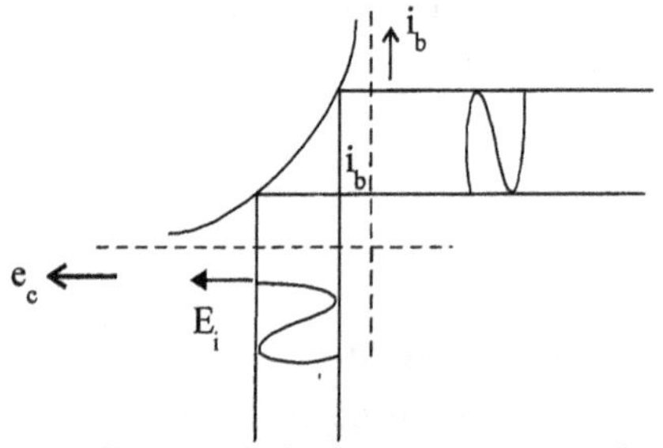

Fig. 7.32 Input output waveforms

Class A	Class B
Less power	More power
Lesser η	More η upto 78.5%
Less Harmonic distortion	Harmonic distortion is more

Example : 7.4

Design a class A power amplifier to deliver 5V rms to a load of 8 Ohms using a transformer coupling. Assume that a supply of 12V is available. The resistance of the primary winding of the transformer also should be considered.

Solution :

1. **First select a suitable transistor**

The power output required is

$$= \frac{(V_{orms})^2}{R_L} = \frac{5 \times 5}{8} = \frac{25}{8} = 3.125\text{W}$$

Assuming a transformer efficiency, η, of 90%, we have, the power required of the amplifier

$$= P_0 / \eta$$
$$= 3.125/0.9 = 3.47\text{W}$$

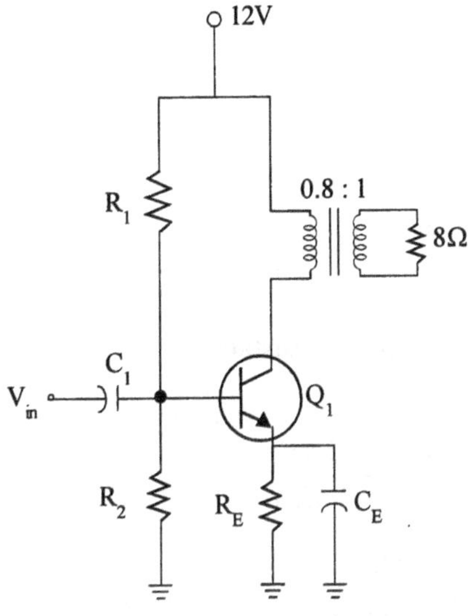

Fig. 7.33 Class A power amplifier

Therefore, we shall have to design the amplifier for 3.47 W. Since the maximum efficiency of the transformer-coupled power amplifier is 50%, the power dissipation capability of the transistor should be at least 3 to 4 times the power required to be developed (Fig. 7.33).

For the transistor, therefore, the $P_{d(max)}$ should be

$$= 3.47 \text{ W} \times 3$$

$$= \text{about } 10.41 \text{ W}$$

Let us select a transistor, EC3054, for the purpose.

This transistor has

$$P_{d(max)} = 30 \text{ W at } 25\ ^{\circ}\text{C}$$

$$I_{c(max)} = 4\text{A}$$

$$V_{CE(sat)} = 1\text{V}.$$

2. Choosing Q-point

For transformer coupled amplifier, ideally, $V_{CEQ} = V_{CC}$. We shall assume the voltage across the resistance R_E as about 20% of the supply voltage, i.e.,

$$V_E = 0.2 \times 12 = 2.4\text{V}$$

Since $V_{CE\,(sat)} = 1\text{V}$, and also to avoid the distortion near the saturation region, we shall take the quiescent point voltage.

$$= V_{CEQ} = \text{about 2/3rd } V_{CC}, \text{ giving us}$$

$$V_{CEQ} = 8\text{V}$$

The maximum swing available will be about 1V less ($V_{CE\,(sat)} = 1\text{V}$) than the supply voltage of 12 volts.

Hence for a power of 3.47 W, we have,

$$3.47 = \frac{V_P}{\sqrt{2}} \times \frac{I_P}{\sqrt{2}}$$

giving I_P = 1.21 Amps.

Therefore, the Q-point is at 8V, 1.21A.

3. Choosing R_E

We have assumed voltage across the resistance R_E as equal to 2.4V, being about 20% of the supply voltage V_{CC}.

Therefore, $$R_E = \frac{V_{RE}}{I_{CQ}} = \frac{2.4V}{1.12A}$$

$$= 2.2\ \Omega.$$

which is the nearest available standard value of the resistance. Let us recalculate the voltage across the resistance R_E.

The voltage $V_{RE} = 1.21\ A \times 2.2\ \Omega = 2.662$ Volts

The power dissipation of the resistance

$$R_E = (1.21\ A)^2 \times 2.2\Omega$$

$$= 3.22 \text{ Watts.}$$

Hence, we select the resistance

$$R_E = 2.2 \text{ Ohms, } 10 \text{ Watts.}$$

4. Turns Ratio of Transformer

Secondary voltage = 6V (rms).

Let us calculate the primary voltage and, hence, the turns ratio. At Q-point, the DC voltage across the primary is

$$= V_{CC} - V_{CEQ} - (I_{CQ} \times R_E)$$

$$= 12 - 8 - (1.21 \times 2.2)$$

$$= 1.4V.$$

Giving DC resistance of the transformer

$$R_{primary} = 1.4V/1.21A$$

$$= 1.16 \text{ ohms.}$$

The equivalent resistance on the primary of the transformer is equal to $R_{ac} - R_{primary}$

$$= \frac{V_P}{I_P} - R_{primary}$$

$$= (11 / 1.21) - 1.16$$

$$= 7.93 \text{ Ohms}$$

The turns ratio $= \dfrac{6}{7.93}$

$= 0.76 \simeq 0.8.$

5. Choosing Resistance R_1 and R_2

Assuming R_B = 10 times R_E, for good stability, we have,

$$\frac{R_1 \times R_2}{R_1 + R_2} = 10\ R_E \qquad .....(7.1)$$

Also, since $V_E = 2.4V,\ V_B = 3V$

we have $3 = \dfrac{R_2}{R_1 + R_2} \times 12$

or $R_1 = = 3\ R_2$ .....(7.2)

From equations (7.1) and (7.2) above, we have

$R_1 = 88\ \Omega.$

$R_2 = 28.6\ \Omega.$

We select the nearest available values, as

$R_1 = 100\ \Omega.$

$R_2 = 33\ \Omega.$

The power rating of these resistances are as under,

For $R_1 = \dfrac{(V_{R1})^2}{R_1} = \dfrac{(18-3)^2}{100} = 2.25W$

For $R_2 = \dfrac{(V_{B2})^2}{R_2} = \dfrac{3^2}{33} = 0.27\ W$

Hence, we select, R_1 = 100 Ohms, 5 Watts.

R_2 = 33 Ohms, 1 Watts.

Let us calculate the maximum undistorted power available, which is equal to $\left(V_p/\sqrt{2}\right) \times \left(I_p/\sqrt{2}\right)$

$= \left(11/\sqrt{2}\right) \times \left(1.21/\sqrt{2}\right)$

= 6.66 Watts, which is more than the required values.

The circuit efficiency :

Useful power output = 6.66 Watts.

$$\text{Power input} = V_{CC} \times I_{CQ} + \frac{(V_{CC})^2}{R_1 + R_2}$$

$$= 14.52 + 1.082 = 15.60 \text{ Watts.}$$

The circuit efficiency, is, therefore,

$$= (6.66 / 15.6) \times 100$$

$$= 42.7\%$$

Likewise, let us calculate the transistor power dissipation when no signal is applied, which is

$$= V_{CEQ} \times I_{CQ}$$

$$= 8V \times 1.21A$$

$$= 10W \simeq \text{Watts.}$$

The power dissipation when the rated power is delivered

$$= 10W - 6.66W$$

$$= 4W$$

7.11 POWER TRANSISTOR

Power transistors are higher power version of normal conventional small signal junction transistors. Like conventional small signal junction transistors, the power transistor is available in both n-p-n and p-n-p forms. However n-p-n power transistors are available with relatively higher current and voltage ratings than that of p-n-p. In present days the power transistors are available upto 1200 V, 800 A with maximum switching frequency of 10 kHz. It is already a known fact that electrons have higher mobility than that of holes, this feature enables n-p-n transistors to occupy less area than that of p-n-p transistors to provide same equivalent performance. Hence the usage of n-p-n transistors is more common than that of p-n-p transistors.

7.11.1 Structure of Power Transistor

The structure of power transistor is given in Fig. 7.34. As shown in Fig. 7.34 the power transistor have chosen vertically oriented structure to maximize the cross sectional area through which the current in the device flows. This vertical orientation of power transistor also aids in reducing the on-state resistance of power transistor, this reduction in on-state resistance of power transistor further helps in reducing the on-state power dissipation of the transistor. The maximization of cross sectional area of transistor incorporate feature of minimizing the thermal resistance of transistor and this results in improved cooling of the transistor.

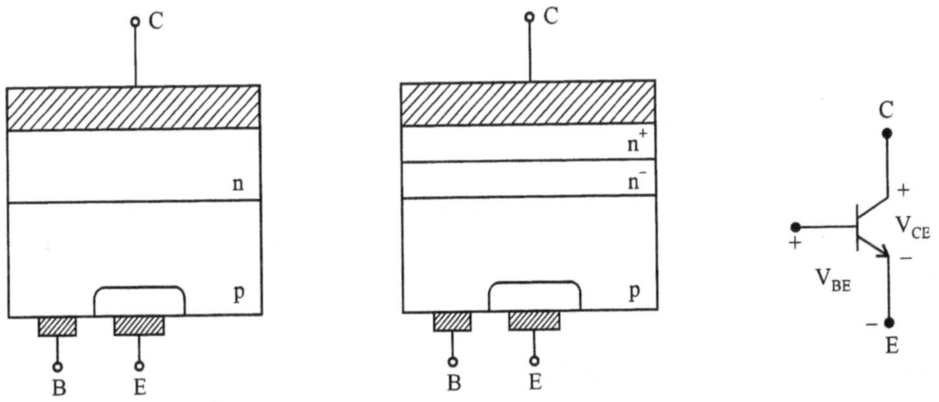

Fig. 7.34 (a) Double diffused structure (b) Triple diffused structure (c) Circuit symbol.

The Fig. 7.34(a) shows double diffused structure of power transistor which is generally employed for power transistor with low voltage rating. The Fig. 7.34(b) shows triple diffused structure of power transistor that can be employed for manufacturing a power transistor with high voltage rating. In triple diffusion structure, a low resistivity n-region labelled n^+ is formed on top collector layer of double diffusion structure of power transistor. Just like in conventional small signal transistor, the collector of power transistor has larger area than that of base and emitter. The base of power transistor is smallest in area. The n-region in triple diffused structure is called as collector drift region, that forms a part of collector that has light doping level. The n-region of collector has same doping level as that of emitter. The thickness of drift region determines break down voltage of the transistor.

7.11.2 Input Characteristics of Power Transistor

In general the power transistors are operated in CE configuration as this offers light input impedance and high current gain. The Fig. 7.35 shows the input characteristics of C.E. configuration.

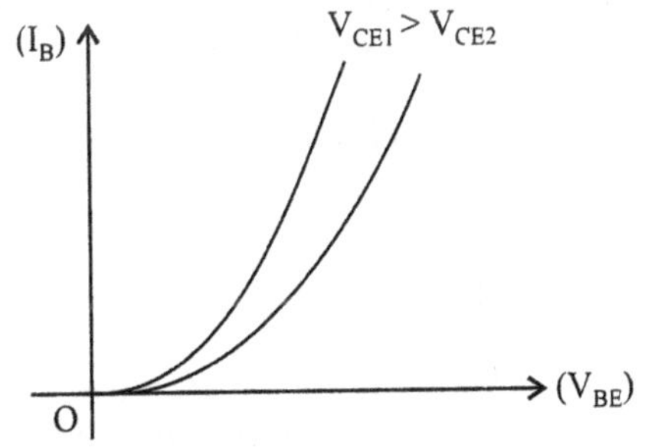

Fig. 7.35 Input characteristics of power transistor connected in CE configuration.

The input characteristics of CE configuration power transistor are similar to that of small signal transistor connected in CE configuration. In other words the input characteristics of power transistor is a plot drawn between base current (I_B) and base emitter voltage (V_{BE}). Keeping V_{CE} constant (for different values of V_{CE}).

The base emitter junction works like a normal p-n junction diode and the current I_b starts flowing when $V_{BE} > V_T$, where V_T is threshold voltage of the base-emitter junction (p-n diode). It is generally 0.6 for silicon. The value of base current is given by following equation.

$$I_B = \frac{V_{BB} - V_{BE}}{R_B}$$

If base supply is reversed i.e., $V_{BE} < 0$, the base emitter junction will be reverse biased, resulting in negligibly small I_b. The reverse voltage blocking of capability of BE junction less and hence it is essential to ensure that this is not exceeded.

7.11.3 Output Characteristics

The Fig. 7.36(a) shows power transistor connected in CE configuration and Fig. 7.36(b) shows output characteristics of power transistor connected in CE configuration. The output characteristics of power transistor connected in CE configuration is plot draw between I_C (on Y-axis) and V_{CE} (on X-axis) for different values of I_B (I_B = constant).

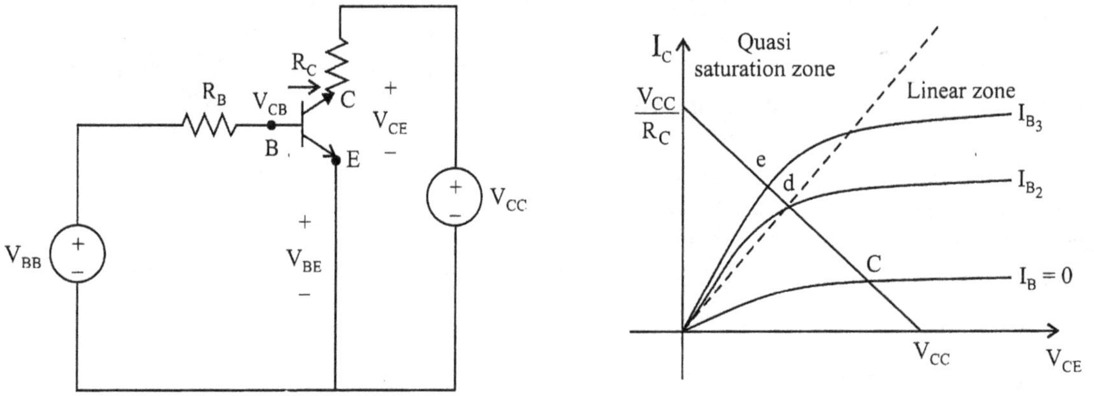

(a) *(b)*

Fig. 7.36 (a) CE configuration of power transistor (b) Output characteristics of power transistor in CE configuration

By applying KVL to output circuit of Fig. 7.36(a) we get

$$I_C = \frac{V_{CC} - V_{CE}}{R_C} = \frac{V_{CC}}{R_C} = \left(\frac{1}{R_C}\right) V_{CE}$$

The above equation is represented by load line in Fig. 7.36(b). For any given value of I_B the operating point of the transistor is the point of intersection of load line with $I_C - V_{CE}$ curve corresponding to the value of I_B. If $I_B = 0$ the transistor is in cut off region i.e., both junction of transistor B-E junction and C-B junction are reverse biased.

If I_B is increased to I_{B_3}, then the operating point is 'e'. The voltage across transistor is very small i.e., the transistor is in saturation (ON switch). Now both junctions (E-B & C-B) are forward biased. Hence the equation for collector current (I_C) is given by following equation

$$I_C = \frac{V_{CC} - V_{CE(sat)}}{R_C}$$

If the value of I_{B_3} is reduced to I_{B_2} the transistor enters into quasi saturation zone. Here operating point is 'd'. In the region between I_{B_1} and I_{B_3} the B-E junction of transistor is forward biased and C-B junction of transistor is reverse biased. In this region I_C is linear function of I_B, hence it is termed as linear region (or) active region of power transistor. The voltage across the transistor in active region is higher than that of voltage across it in saturation region.

7.12 HEAT SINKS

The purpose of heat sinks is to keep the operating temperature of the transistor low, to prevent thermal breakdown. Due to increase in temperature, I_{CO} increases. Due to increase in I_{CO}, I_C increases and hence power dissipation increases. Due to this, temperature increases and thus it is a cummulative process. Due to this, the transistor will fail or breakdown occurs. To prevent this, heat sinks are used to dissipate power to the surroundings and keep the temperature low.

The heat is transferred from the die to the surface of the package or casing of the ambient by convention, from the surface to the ambient by convention and radiation. If heat sink is used, the heat is transferred from the package to heat sink and from heat sink to the ambient. Heat sink expedites the power dissipation and prevents breakdown of the device.

The rise in temperature due to power dissipation is expressed as *Thermal Resistance* expressed in ^{0}C/w, and is symbolically represented as θ. It is the rise in temperature in ^{0}C due to 1W of power dissipation. The equations governing this are,

$$\theta_{ja} = \theta_{jc} + \theta_{cn} + \theta_{na}$$
$$\theta_{jc} = (T_j - T_c) / P$$
$$\theta_{cs} = (T_c - T_s) / P$$
$$\theta_{sa} = (T_s - T_a) / P$$

θ_{ja} = Junction to ambient thermal resistance
θ_{jc} = Junction to casing thermal resistance
θ_{cs} = Casing to heat sink thermal resistance
θ_{sa} = Heat sink to ambient thermal resistance
T_j = Average junction temperature
T_c = Average case temperature
T_{sa} = Average heat sink temperature
T_a = Ambient temperature
P = Power dissipated in Watts.

EXAMPLE : 7.5

What is the junction to ambient thermal resistance for a device dissipating 600 mW into an ambient of 60^{0}C and operating at a junction temperature of 120^{0}C.

Solution :

Here heat sink is not considered.

$$\therefore \quad \theta_{ja} = \theta_{jc} + \theta_{ca}$$

$$\theta_{ja} = \frac{T_j - T_C}{P} + \frac{T_C - T_a}{P}$$

or

$$\theta_{ja} = \frac{T_j - T_C + T_c - T_a}{P} = \frac{T_j - T_a}{P}$$

$$\theta_{ja} = \frac{120 - 60}{0.6} = \frac{60}{0.6}$$

$$= \mathbf{100\ ^oC/W}$$

For Transistor devices, the heat sinks are broadly classified as :

1. Low Power Transistor Type.
2. High Power Transistor Type.

Low Power Transistors can be mounted directly on the metal chassis to increase the heat dissipation capability. The casing of the transistor must be insulated from the metal chassis to prevent shorting.

Beryllium oxide insulating washers are used for insulating casing from the chassis. They have good thermal conductivity.

Zinc oxide film silicon compound between washer and chassis, improves the heat transfer from the semiconductor device to case to the chassis.

High Power Transistor heat sinks.

re TO-3 and TO-66 types. These are diamond shaped. For power transistors, usually, the ease itself in the collector convention and radiation is shown in Fig. 7.37. The thermal resistance of the heat sinks will be typically 3^0C/W (Fig. 7.38).

Fig. 7.37 Fin-type heat sink

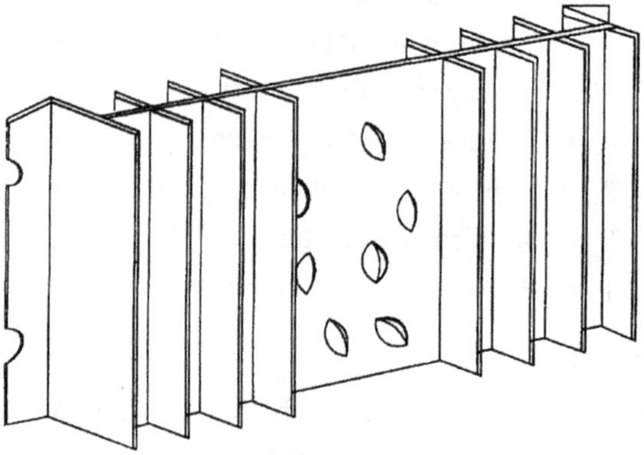

Fig. 7.38 Power transistor heat sink

SUMMARY

- If the signal swing is large, such amplifiers are called as large signal amplifiers.
- Class A, B, C, AB are the commonly used large signal amplifiers.
- Depending on the type of connection to the load, these are also called as series fed and transformer coupled amplifiers.
- Maximum efficiency of series fed Class A amplifier is 25%, for transformer coupled amplifier, it is 50%. These expressions are derived.
- Class B, push pull amplifier has conversion efficiency of 78.5% (theoretical). It has negligible power loss at no signal.
- Comparison of direct coupled, R-C coupled and transformer coupled amplifiers is given.
- Heat sinks are used to keep the temperature of the device low. θ is called thermal resistance of a heat sink and is expressed is °C/W.

OBJECTIVE TYPE QUESTIONS

1. Amplifiers are classified based on
   (a) (b) (c) (d)
2. If the magnitude of signal is small and operating point swing is within the active region, that amplifier is classified as ______________.
3. Different types of coupling employed in amplifier circuits are
   (a) (b) (c) (d)
4. Due to the input signal swing, if the operating point shifts into cut off and saturation regions, that amplifier is classified as ______________.
5. Conduction angles of large signal amplifiers are ____________________.
   (a) Class A ______________ (b) Class B ______________
   (c) Class AB ______________ (d) Class C ______________
6. In class A power amplifiers the operating point Q is in ______________ of dynamic transfer curve of the active device.
7. In class B amplifiers, the Q point is set ______________.
8. In class C amplifiers the operating point is set ______________.
9. Maximum theoretical efficiency of series fed amplifiers is ______________.
10. Maximum efficiency of transformer coupled amplifiers is ______________.
11. The frequency range in which transformer coupled amplifiers are used is ______________.
12. The maximum theoretical efficiency of class B push pull amplifier is ______________.

13. For mirror symmetry or half wave symmetry, the mathematical equation is ______________.
14. In push pull configuration, type of harmonics eliminated are ______________.
15. In class B amplifiers, relation between maximum collector power dissipation P_C (Max) and maximum output power dissipation P_O (Max) is ______________.
16. Cross over distortion occurs because of ______________ characteristic of E - B junction of the transistors.
17. Transformerless class B power amplifier circuit is ______________.
18. Complimentary symmetry circuit is so named because ______________.
19. The complimentary symmetry circuit with single d.c. bias supply circuit is also called ______________.
20. Phase Inverter circuits are also called ______________.
21. What is the mode of operation of a last stage in a cascade ?
22. Derive an expression for second harmonic distortion interms of I_{max}, I_{min}, I_e ?
23. What is the expression for total harmonic distortion interms of second, third harmomic ______________.
24. What is impedance matching ?
25. Why do we go for transformer coupled power amplifier ?
26. What is the equivalent load resistance of a transformer coupled amplifier interms of turms ratio ?
27. Is the equivalent load resistance increasing or decreasing if n greaterthan 1 ?
28. What is conversion efficiency ?
29. What is the maximum value of efficiency for the series fed load ?
30. What is the maximum value of efficiency for a transformer coupled load ?
31. How will the input signals be in a push pull amplifier ?
32. What are the advantages of a push pull configuration ?
33. Draw the waveforms to explain the class B operation.
34. What is the maximum efficiency of a class B amplifier ?

ESSAY TYPE QUESTIONS

1. What are the different methods of clarifying electronic amplifiers ? How are they classified, based on the type of coupling ? Explain.

2. Compare the characteristic features of Direct coupled, resistive capacitor coupled, and Transformer coupled amplifiers.

3. Distinguish between small signal and large signal amplifiers. How are the power amplifiers classified ? Describe their characteristics.

4. Derive the general expression for the ouput power in the case of a class A power amplifier. Draw the circuit and explain the movement of operating point on the load line for a given input signal.

5. Derive the expressions for maximum. Theoretical efficiency for maximum.

   (i) Transformer coupled

   (ii) Serves fed amplifier what are thier advantages and disadvantages.

6. Show that in the case of a class A transforms coupled amplifier, with inpedence matching, the expression for voltage gain AV is given as

$$A_V = -\left(\frac{h_{fe}}{2}\right) \cdot \frac{R_L}{h_{ie}} \cdot \frac{N_1}{N_2} \text{ with usual notation}$$

7. What are the advantages and disadvantages of transformer coupling ?

8. Show that class B push pull amplifiers exhibit halfwave symmetry.

9. Derive the expression for Max. Theoretical efficiency in the case of class B push pull amplifier. Why is it named so ? What are its advantages and disadvantages ?

10. Draw the circuit for composite tune amplifiers and explain its operation.

11. What are phase inverter circuits ? Draw a typical circuit and explain its working.

12. Draw the pentode pushpull amplifier and explain its operation.

13. Explain about Class D and Class S power amplifiers. Mention their salient features and applications.

14. How are the tuned amplifiers classified ? Explain the salient features of each one of them.

15. Draw the circuit for single tuned capacitance coupled amplifier explain its operation.

ANSWERS TO OBJECTIVE TYPE QUESTIONS

1. (a) Frequency range (b) Type of coupling
   (c) Output - power/conduction angle (d) Magnitude of signal.
2. Small signal amplifier
3. (a) Direct coupling (b) R - C coupling
   (c) Transformer coupling (d) L - C tuned coupling (e) series fed
4. Large signal amplifier
5. (a) Class A 360° (b) Class B 180°
   (c) Class AB 180 to 360° (d) Class C $< 180^{o}$.
6. The centre of linear region of the
7. Near cut off of the active device.
8. Beyond cut off
9. 25 %
10. 50 %
11. Audio frequency range 20 Hzs to 20 KHzs.
12. 78.5 %
13. $i\,(wt) = -\,i\,(wt + \pi)$
14. Even Harmonics
15. $P_C\,(Max) = 0.4\,P_O\,(Max)$
16. Cut in voltage or threshold voltage.
17. Complimentary symmetry circuit.
18. Both PNP and NPN transistors are used.
19. Totempole circuit
20. Paraphase amplifiers.
21. Since for the last stage, the input signal has a high amplitude, the mode of operation will be other than class A.
22. Refer to the derivation.

23. $T_D = \sqrt{D_2^2 + D_3^2 + D_4^2 + ...}$

24. For maximum power transfer to the load, the load impedance should be conjugate of the effective impedance.

25. For impedance matching and hence higher efficiency.

26. $R_L' = \frac{1}{n^2} R_L$ where n is the turns ratio.

27. R_L' decreases

28. $\eta = \frac{\text{Signal power delivered to the load}}{\text{dc power absorbed.}}$

29. $\eta = 25\%$

30. $\eta = 50\%$

31. The input signals are both 180° out of phase.

32. Less harmonic distortion, More efficiency, Ripples in power supply are reduced, Magnetic effects are reduced.

33.

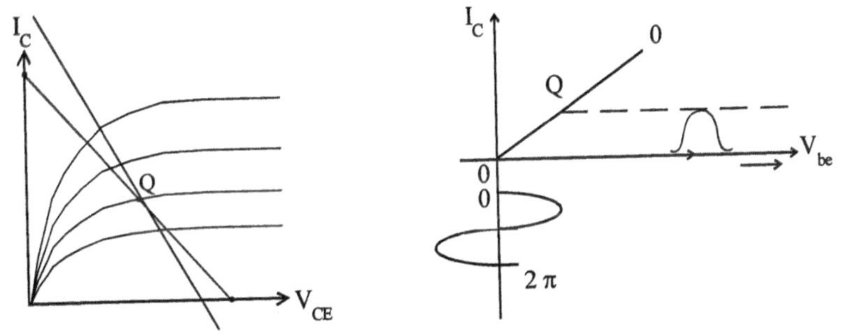

34. 78.5%

8 Tuned Amplifiers

In this Unit,

- Different types of Tuned amplifier circuits are analyzed.
- Equivalent circuits of the output stages are given.
- FET Tuned R.F. amplifier circuits, wideband amplifier circuits, shunt compensation aspects are also explained.
- Single stage amplifiers in the three configurations of C.E, C.B, C.C, with design aspects are given.
- Using the design formulae for A_V, A_I, R_i, R_o etc, the design of single stage amplifier circuits is to be studied.
- Single stage JFET amplifiers in C.D, C.S and C.G configurations are also given.
- The Hybrid - π equivalent circuit of BJT, expressions for Transistor conductances and capacitances are derived.
- Miller's theorem, definitions for f_β and f_T are also given.
- Numerical examples, with design emphasis are given.

8.1 INTRODUCTION

A tuned amplifier is one, which uses a parallel tuned circuit, as its load impedance. A parallel tuned *circuit, is also known as anti resonant circuit.* The characteristics of such an anti resonant circuit is that its |Z| is high, at the resonant frequency, and falls off sharply as the frequency departs from the resonant frequency. So the gain versus frequency characteristics of a tuned amplifier will also be similar to the |Z| characteristics of the resonant circuit. When |Z| is maximum, V_0 will also be maximum. This is for AC. |Z| is considered. (Fig. 8.1).

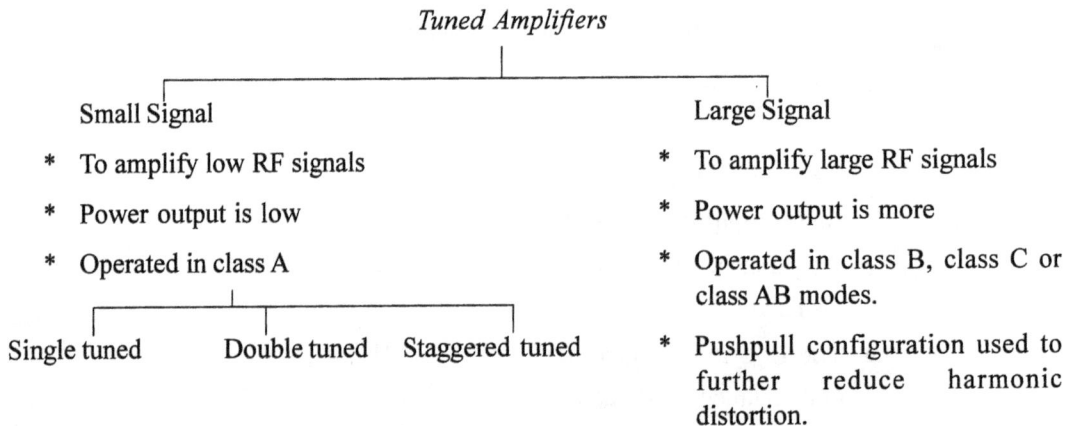

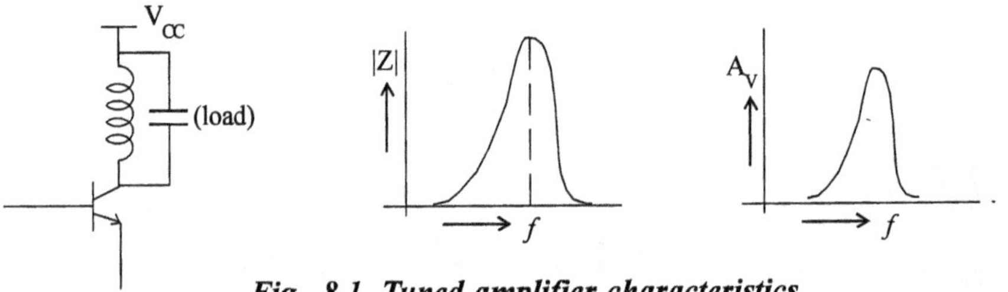

Fig. 8.1 Tuned amplifier characteristics

8.1.1 APPLICATIONS

Tuned amplifiers are used to amplify a single radio frequency or narrow band of frequencies.

So basically they are used in 1. RF amplifiers 2. Communication receivers

Tuned amplifiers use variable Inductance (L) or variable Capacitance (C) to vary the resonant frequency.

In tuned amplifiers, *harmonic distortion is very small,* because the gain of the amplifiers is negligibly small for frequencies other than f_0 (the resonant frequency). So *Harmonics which are of higher frequencies will have very low gain* and hence harmonic distortion *will be less for tuned amplifiers.*

8.1.2 CLASSIFICATION

Tuned Amplifiers

Small Signal

* To amplify low RF signals
* Power output is low
* Operated in class A

Single tuned — Double tuned — Staggered tuned

Large Signal

* To amplify large RF signals
* Power output is more
* Operated in class B, class C or class AB modes.
* Pushpull configuration used to further reduce harmonic distortion.

This classification is similar to the classification of power amplifiers.

Small Signal Tuned Amplifiers

- Single Tuned Amplifiers
- Double Tuned Amplifiers
- Staggered Tuned Amplifiers

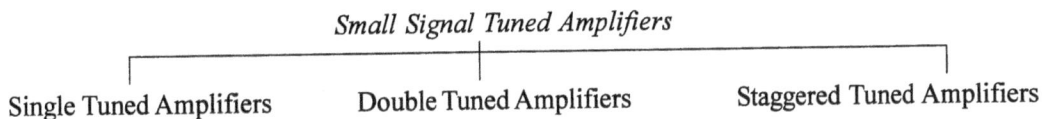

8.1.3 Single Tuned Amplifier

Uses one parallel tuned circuit as the load |Z| in each stage and all these tuned circuits in different stages are tuned to the same frequency. To get large A_V or A_P, multistage amplifiers are used. But each stage is tuned to the same frequency, one tuned circuit in one stage.

8.1.4 Double Tuned Amplifier

It uses two inductively coupled tuned circuits, for each stage of the amplifier. Both the tuned circuits are tuned to the same frequency, two tuned circuits in one stage, to get sharp response.

8.1.5 Stagger Tuned Amplifier

This circuit uses number of single tuned stages in cascade. The successive tuned circuits are tuned to slightly different frequencies.

Single tuned amplifiers are further classified as :

- Capacitive coupled
- Transformer coupled or inductive coupled

8.2 SINGLE TUNED CAPACITIVE COUPLED AMPLIFIER

L, C tuned circuit is not connected between collector and ground because, the transistor will be short circuited at some frequency other than resonant frequency. (Fig. 8.2).

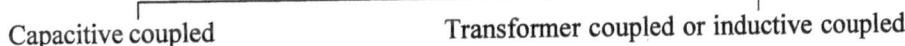

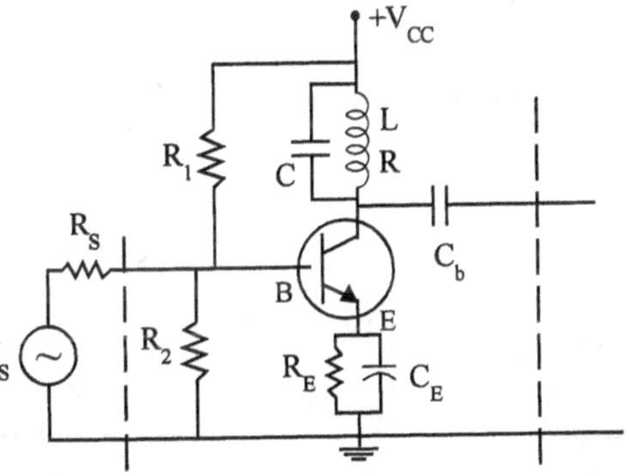

Fig. 8.2 Single tuned capacitive coupled amplifier

The output of the tuned circuit is coupled to the next stage or output device, through capacitor C_b. So this circuit is called single tuned capacitive coupled amplifier.

R_1, R_2, R_E, C_E are biasing resistors and capacitors. The tuned circuit formed by Inductance (L) and capacitor (C) resonates at the frequency of operation.

Transistor hybrid π equivalent circuit must be used since the transistor is operated at high frequencies. Tuned circuits are high frequency circuits. (Fig. 8.3).

R_i = input resistance of the next stage.

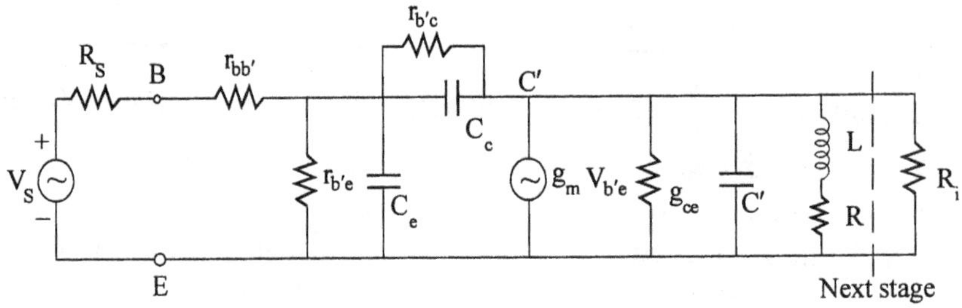

Fig. 8.3 Equivalent circuit

Modified equivalent circuit using Miller's Theorem (Fig. 8.4).

According to Miller's theorem, the feedback capacitance C_C is C_C (1 - A) on the input side and $C_C\left(\frac{A-1}{A}\right)$ on the output side. But where as resistance is $\frac{r_{b'c}}{(1-A)}$ on the input side $\frac{r_{b'c}}{\left(\frac{A-1}{A}\right)}$ on the output side.

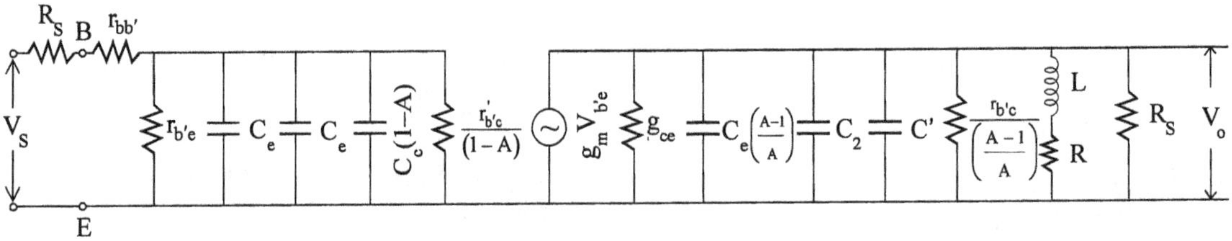

Fig. 8.4 Equivalent circuit (applying Miller's theorem)

The equivalent circuit after simplification, neglecting $\frac{r_{b'c}}{\left(\frac{A-1}{A}\right)}$ is shown in Fig. 8.5.

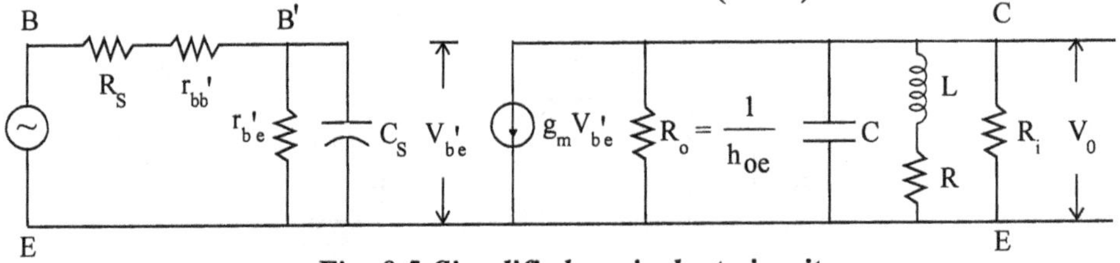

Fig. 8.5 Simplified equivalent circuit

$$Y_i = \frac{1}{R + j\omega L} = \frac{R - j\omega L}{R^2 + \omega^2 L^2} = \frac{R}{R^2 + \omega^2 L^2} - j\frac{\omega L}{R^2 + \omega^2 L^2}$$

Input admittance as seen by II stage.

Instead of L and R being in series, they are being represented as equivalent shunt elements, R_P and L_P for parallel

$$= \frac{1}{R_P} + \frac{1}{j\omega L_P}$$

where $$R_P = \frac{R^2 + \omega^2 L^2}{R}$$

$$L_P = \frac{R^2 + \omega^2 L^2}{\omega^2 L}$$

Inductor is represented by R_P in series with inductance L_P.

Q at resonance, $$Q_0 = \frac{\omega_0 L}{R}$$

$\omega L >> R$ $\because$ Resistance of the inductor R is small,

$\therefore$ $$R_P = \frac{\omega^2 L^2}{R}$$ neglecting R^2 compared to $\omega^2 L^2$

$L_P = L$ neglecting R^2 compared to $\omega^2 L^2$.

Therefore output circuit is simplified to, as shown in Fig. 8.6.

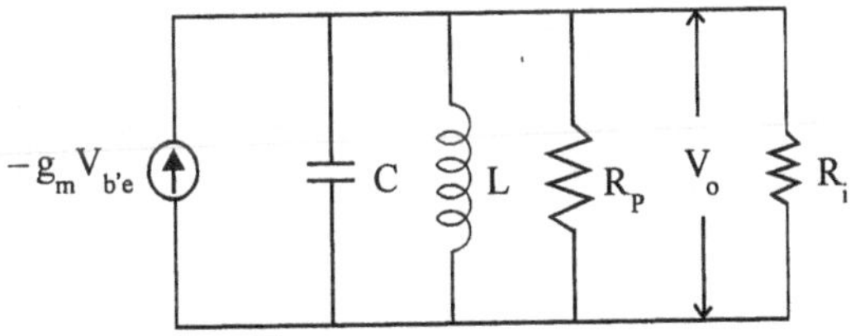

Fig. 8.6 Simplified circuit

$$\frac{1}{R_t} = \frac{1}{R} + \frac{1}{R_P} + \frac{1}{R_i}$$

$$\omega_0 = \frac{1}{\sqrt{LC}}$$

R_i is the input resistance of the next stage

Q_e is defined as, $$Q_e = \frac{\text{Suspectance of L or capacitance of C}}{\text{Conductance of shunt resistance } R_t}$$

R_t = resistance of tuned circuit

$$Q_e - \omega_0 CR_t$$

$$= \frac{R_t}{\omega_0 L} \frac{(1/\omega_0 L)}{1/R_t} = \left(\frac{\omega_0 C}{1/R_t}\right)$$

Let ω_0 be the resonant angular frequency in rad/sec.

$$\omega_0 = \frac{1}{\sqrt{LC}}$$

Output voltage $V_0 = -g_m V_{b'e}.Z$ ($-g_m V_{b'e}$ is the current source).
where Z is the impedance of C, L and R_t in parallel.

Admittance $$Y = \frac{1}{Z} = \frac{1}{R_t} + \frac{1}{j\omega L} + j\omega C$$

Multiplying by R_t throughout and dividing,

or $$Y = \frac{1}{R_t}\left[1 + \frac{R_t}{j\omega L} + j\omega C R_t\right]$$

$$= \frac{1}{R_t}\left[1 + j\frac{\omega_0 \omega C R_t}{\omega_0} + \frac{R_t \omega_0}{j\omega_0 \omega L}\right] \quad \text{(Multiplying and dividing by } \omega_0\text{)}$$

$$Y = \frac{1 + jQ_e\left[\frac{\omega}{\omega_0} - \frac{\omega_0}{\omega}\right]}{R_t} \qquad \left(\because \quad Q_e = \frac{R_t}{w_0 L} = \omega_0 CR_t\right)$$

where $$Q_e = \omega_0 C R_t$$

$$= \frac{R_t}{\omega_0 L} \qquad \because \quad \omega_0 L = \frac{1}{\omega_0 C}$$

$\therefore$ $$Z = \frac{R_t}{1 + jQ_e\left[\frac{\omega}{\omega_0} - \frac{\omega_0}{\omega}\right]}$$

$$Q_e \text{ is defined as } = \frac{\text{Susceptance of L or C}}{\text{Conductance of shunt resistance } R_t}$$

Let δ = Fractional frequency variation

i.e., variation in frequency expressed as a fraction of the resonant frequency

$\therefore$ $$\delta = \frac{\omega - \omega_0}{\omega_0} = \frac{\omega}{\omega_0} - 1$$

$\therefore$ $$\frac{\omega}{\omega_0} = 1 + \delta$$

∴ Rewriting the expression for Z, as

$$Z = \frac{R_t}{1 + jQ_e\left[(1+\delta) - \frac{1}{(1+\delta)}\right]}$$

$$= \frac{X + \delta^2 + 2\delta - X}{1+\delta} = \frac{2\delta\left(1 + \frac{\delta}{2}\right)}{1+\delta}$$

$$Z = \frac{R_t}{1 + j2Q_e\,\delta\left[\frac{1+\delta/2}{1+\delta}\right]}$$

If the frequency ω is close to resonant frequency ω_0, $\delta << 1$.

Therefore Simplified expression for Z is

$$\boxed{Z = \frac{R_t}{1 + j2Q_e\,\delta}}$$

At resonance, $\omega = \omega_0$, $\delta = 0$

At resonance, R_p may also be put as,

$$R_p = Q_0^2\,R = Q_0 = \sqrt{\frac{L}{C}} = \omega_0\,L\,Q_0$$

Expression for $V_{b'e} = V_i \cdot \frac{r_{b'e}}{r_{bb'} + r_{b'e}}$ potential divider network.

Expression for $V_0 = -g_m V_{be} \cdot Z$

Expression for $V_0 = -g_m V_i \cdot \frac{r_{b'e}}{r_{b'e} + r_{b'e}} \cdot Z$

∴ Voltage gain $A = \frac{V_0}{V_i} = -g_m \cdot \frac{r_{b'e}}{r_{b'e} + r_{bb'}} \cdot Z$

$$A = -g_m \cdot \frac{r_{b'e}}{r_{b'e} + r_{bb'}} \cdot \frac{R_t}{1 + j2\delta Q_e}$$

voltage gain at resonance. Since at resonance $\delta = 0$

$$A_{reso} = \frac{-g_m \cdot r_{b'e}}{r_{b'e} + r_{bb'}} \cdot R_t$$

∴ $$\frac{A}{A_{reso}} = \frac{1}{1 + j2\delta Q_e}$$

Magnitude $\left|\frac{A}{A_{reso}}\right| = \frac{1}{\sqrt{1+(2\delta Q_e)^2}}$

Phase angle $\left|\frac{A}{A_{reso}}\right| = -\tan^{-1}(2\,\delta\,Q_e)$

At a frequency ω_1, below the resonant frequency δ has the value,

$$= -\frac{1}{2Q_e};$$

$$\frac{A}{A_{reso}} = \frac{1}{\sqrt{2}} = 0.707$$

ω_1 is the lower 3db frequency.

Similarly ω_2, the upper 3db frequency is

$$\delta = +\frac{1}{2Q_e};\ \frac{A}{A_{reso}} = \frac{1}{\sqrt{2}} = 0.707$$

The 3 db band width $\Delta\omega = (\omega_2 - \omega_1)$

$$= \frac{[(\omega_2 - \omega_0) + (\omega_0 - \omega_1)].\,\omega_0}{\omega_0}$$

$$= [\delta + \delta]\,\omega_0 \quad = 2\,\delta\,\omega_0$$

But $\delta = \frac{1}{2Q_e}$

$$\therefore \quad \Delta\,\omega = \frac{\omega_0}{Q_e} = \frac{\omega_0}{R_t\,\omega_0\,C} = \frac{1}{R_t\,C} \text{ rad/sec.}$$

8.3 TAPPED SINGLE TUNED CAPACITANCE COUPLED AMPLIFIER

The circuit is shown in Fig. 8.7.

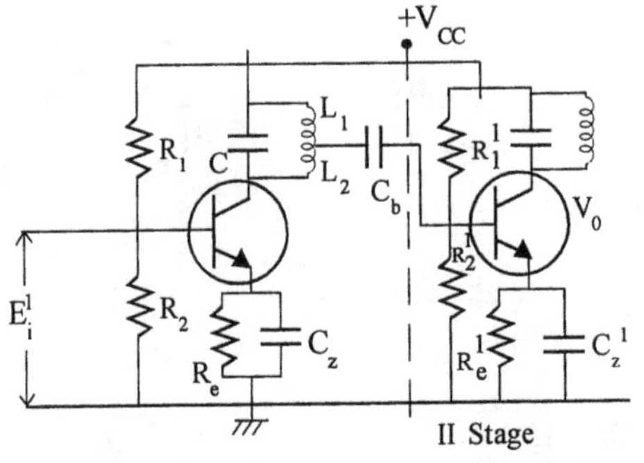

Fig. 8.7 Tapped single tuned capacitive coupled amplifier circuit

8.3.1 Equivalent Circuit on the Output Side of the I Stage

R_i is the input resistance of the II stage.

R_0 is the output resistance of the I stage amplifier. (Fig. 8.8).

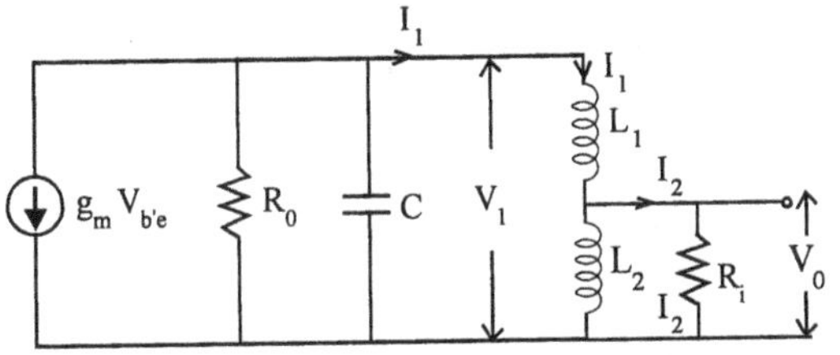

Fig. 8.8 Equivalent circuit

The input |Z| of the common emitter amplifier circuits will be less. So the output impedance of the circuit being coupled to one common emitter amplifier, should also have low |Z| for impedance matching and to get maximum power transfer. So in order to reduce the impedance of the LC resonant circuit, to match the low |Z| of the common emitter circuit, tapping is made in the LC tuned circuit. Tapped single tuned circuits are used in such applications.

8.3.2 Expression for 'Inductance' for Maximum Power Transfer

Let the tapping point divide the impedance into two parts L_1 and L_2.

Let $L_1 = nL$ so that $L_2 = (1 - n)$

Writing Kirchoff's Voltage Law (KVL)

$$V_1 = j\omega L \,.\, I_1 - j\omega\,(L_2 + M)\,I_2 \qquad .....(8.1)$$

$$0 = -j\omega\,(L_1 + M)\,I_1 + (R_i + j\omega L_2)\,I_2 \qquad .....(8.2)$$

Where M is the mutual inductance between L_1 and L_2. Solving equations (8.1(and (8.2),

$$I_1 = \frac{V_1(R_i + j\omega L_2)}{j\omega L(R_i + j\omega L_2) + \omega^2 (L_2 + M)^2} \qquad .....(8.3)$$

Hence the |Z| offered by the coil along with input resistance R_i of the next stage is

$$Z_1 = \frac{V_1}{I_1} = \frac{j\omega L(R_i + j\omega L_2) + \omega^2 (L_2 + M)^2}{(R_i + j\omega L_2)} \qquad .....(8.4)$$

$$= j\omega L + \frac{\omega^2 (L_2 + M)^2}{R_i + j\omega L_2} \qquad .....(8.5)$$

But ωL_2 much less than R_i.

As R_i, the input resistance of transistor circuit II stage is kΩ and much greater than ωL_2

$$Z_1 = j\omega L + \frac{\omega^2 (L_2 + M)^2}{R_i} \qquad .....(8.6)$$

$$M = K\sqrt{L_1 L_2} \qquad M = \text{Mutual Inductance}$$

Where K is the coefficient of coupling. Since $L_1 = nL,\ L_2 = (1-n)L$

$$= K\sqrt{nL(1-n)L} = KL\sqrt{(n-n^2)} \qquad \text{.....(8.7)}$$

Putting K = 1, we get

$$M \simeq L\sqrt{n-n^2} \qquad \text{.....(8.8)}$$

Substituting thus value of M in (8.6),

$$Z_1 \simeq j\omega L \pm \frac{\omega^2\left[(1-n)L + L\sqrt{n-n^2}\right]^2}{R_i} \qquad \text{.....(8.9)}$$

$$\simeq j\omega L + \frac{\omega^2 L^2\left[(1-n) + \sqrt{n-n^2}\right]^2}{R_i} \qquad \text{.....(8.10)}$$

$$(j\omega L + R)$$

The resistance effectively reflected in series with the coil due to the resistance R_i is given by,

$$R_{is} \simeq \frac{\omega^2 L^2\left[(1-n) + \sqrt{n-n^2}\right]^2}{R_i}$$

This is the resistance component; S : series, i : input

This resistance R_{is} in series with the coil L may be equated to a resistance R_{ip} in shunt with the coil where R_{ip} is given by,

$$R_{ip} = (\omega L)^2 / R_{is}$$

So the equivalent circuit is as shown in Fig. 8.9.

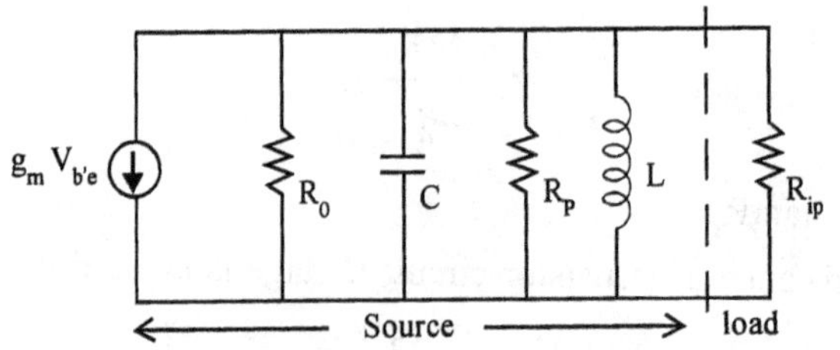

Fig. 8.9 Equivalent circuit

Simplifying, (Fig. 8.10)

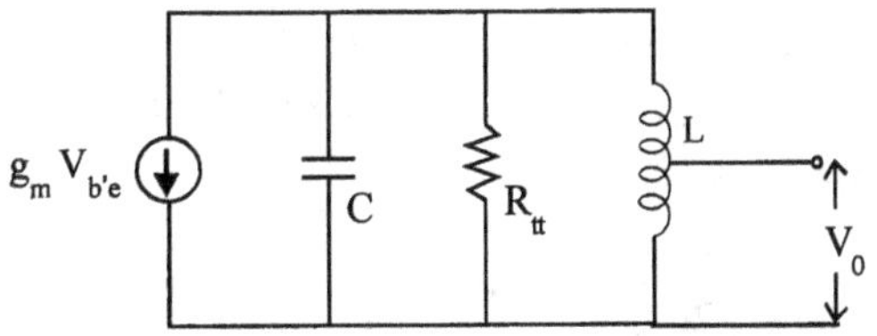

Fig. 8.10 Equivalent circuit after simplification

$$\frac{1}{R_{tt}} = \frac{1}{R_0} + \frac{1}{R_p} + \frac{1}{R_{ip}}$$

$$Q_e = \frac{R_{tt}}{\omega_0 L}$$

tt : tuned tapped circuit.

$$\omega_0 = \frac{1}{\sqrt{LC}}$$

Under the conditions of maximum power transfer theorem, the total resistance appearing in shunt with the coil is = R_{op}

Since it is a resonant circuit, at resonance, the |Z| in purely resistive. For maximum power transfer |Z| = R/2.

∴ $$Q_e = \frac{R_{op}/2}{\omega_0 L}; \quad R_{tt} = R_{op}/2$$

$$R_{op} = 2\, Q_e \cdot \omega_0\, L$$

But $$R_{op} = \frac{R_0 R_p}{R_0 + R_p}$$

∴ $$2\, Q_e\, \omega_0\, L = \frac{R_0\, \omega_0\, Q_0\, L}{R_0 + \omega_0\, Q_0\, L}$$

Solving for L, we get

$$L = \frac{R_0 (Q_0 - 2Q_e)}{2\omega_0\, Q_0\, Q_e}$$

Expression for L for maximum power transfer.

$$\boxed{L = \frac{R_0}{\omega_0}\left[\frac{1}{2Q_e} - \frac{1}{Q_0}\right]}$$

This is the value of L for maximum power transfer.

Expression for voltage gain and Bandwidth are determined in the same way as done for a single tuned circuit. In this circuit we have,

1. R_{tt} instead of R_t (as in single tuned) tapped tuned circuit.
2. Output voltage equals (1 – n) times the voltage developed across the complete coil.

|Z| at any frequency close to w_0 is given by,

$$Z = \frac{R_{tt}}{1 + j\,2\delta\, Q_e} \qquad R_{tt} = \text{resistance of Tapped Tuned Circuit}$$

output voltage $$V_0 = \frac{-g_m V_i\, r_{b'e}}{r_{b'e} + r_{bb'}} \cdot Z\,(1-n)$$

$\therefore$ voltage gain $$A = \frac{V_0}{V_i} = -g_m (1-n) \frac{r_{b'e}}{r_{b'e} + r_{bb'}} \cdot Z$$

$$= -g_m (1-n) \cdot \frac{r_{b'e}}{r_{b'e} + r_{bb'}} \cdot \frac{R_{tt}}{1 + j\,2\delta Q_e}$$

At resonance, voltage gain is

$$A_{reso} = -g_m (1-n) \cdot \frac{r_{b'e}}{r_{b'e} + r_{bb'}} \cdot R_{tt}$$

$\therefore$ $$\boxed{\frac{A}{A_{reso}} = \frac{1}{1 - j\,2\delta Q_e}}$$

8.4 SINGLE TUNED TRANSFORMER COUPLED OR INDUCTIVELY COUPLED AMPLIFIER

The circuit is shown in Fig. 8.11 (a) and (b).

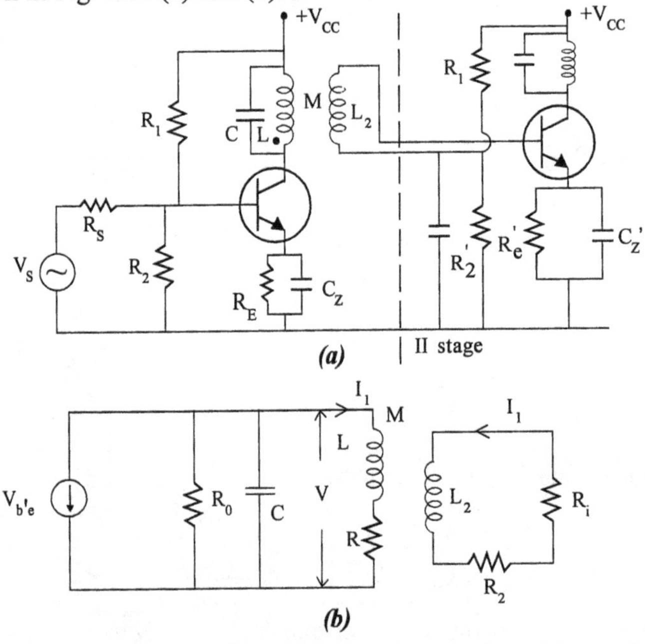

Fig. 8.11 Inductive coupled amplifier circuit (a) and its equivalent (b)

In this circuit, the voltage developed across the tuned circuit is inductively coupled to the next stage. Coil L, of the tuned circuit, and the inductor coupling the voltage to the II stage, L_2 form a transformer with mutual coupling M. This type of circuit is also used, where the input |Z| of the II stage is smaller or different from the tuned circuit. So |Z| matching is done by the transformer depending on its turn ratio. In such requirements, this type of circuit is used.

The resistors R_1, R_2 and R_1' and R_2' are the biasing resistors. The parallel tuned circuit, L and C resonates at the frequency of operation. Fig. 8.11(b) shows output equivalent circuit. Input equivalent circuit will be the same as that of the capacitive coupled circuit.

In the output equivalent circuit, C is the total capacitance, including the stray capacitance, Miller equivalent capacitance $C\left(\frac{A-1}{A}\right)$. L_2 and R_2 are the inductance and resistance of the secondary winding.

8.4.1 Expression for L_2 for Maximum Power Transformer

Writing KVL to the primary and secondary windings,

$$V = I_1 Z_{11} + I_2 Z_{12} \quad .....(8.11)$$

$$0 = I_1 Z_{21} + I_2 Z_{22} \quad .....(8.12)$$

where

$$Z_{11} = R + j\omega L \quad .....(8.13)$$

$$Z_{12} = Z_{21} = j\omega M \quad .....(8.14)$$

$$Z_{22} = R_2 + R_i + j\omega L_2 \quad .....(8.15)$$

Solving eqs. (8.11) and (8.12) for I_1,

$$I_1 = \frac{V.Z_{22}}{Z_{11}Z_{22} - Z_{12}^2} \quad .....(8.16)$$

The impedance seen looking into the primary is,

$$Z_{in} = \frac{V}{I_1} \; \frac{Z_{11}Z_{22} - Z_{12}^2}{Z_{22}}$$

$$= Z_{11} - \frac{Z_{12}^2}{Z_{22}} \quad .....(8.17)$$

Substituting the values of Z_{11}, Z_{22} and Z_{12} in equation (8.17) we get,

$$Z_{in} = (R + j\omega L) + \frac{\omega^2 M^2}{R_2 + R_i + j\omega L_2} \quad .....(8.18)$$

R_i generally much greater than R_2 and ωL_2.

$$\therefore \quad Z_{in} \simeq (R + j\omega L) + \frac{\omega^2 M^2}{R_i} \quad .....(8.19)$$

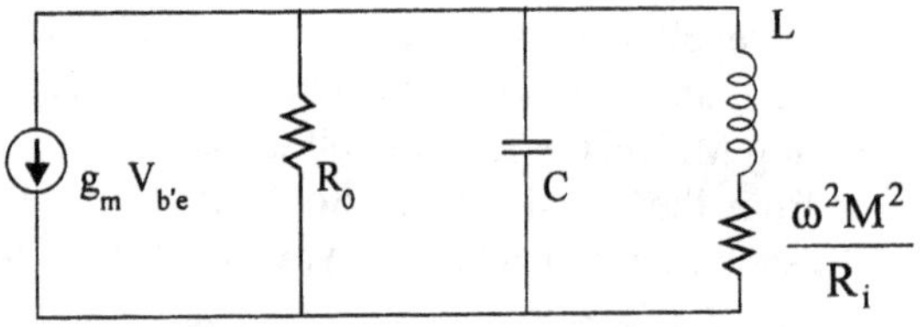

Fig. 8.12 Equivalent circuit

$\frac{\omega^2 M^2}{R_i}$ is the impedance of the secondary side reflected to the primary (Fig. 8.12).

If M is reasonably large, then $R << \frac{\omega^2 M^2}{R_i}$

$$\therefore \quad Z_{in} \simeq j\omega L + \frac{\omega^2 M^2}{R_i} \qquad \text{......(8.20)}$$

$\therefore$ The equivalent circuit may be written as,

Inductance L with series resistance $\frac{\omega^2 M^2}{R_i}$ may be represented as L in shunt with R_{ip} as shown below, where

$$R_{ip} = \frac{(\omega L)^2}{\left(\frac{\omega^2 M^2}{R_i}\right)} = \left(\frac{L}{M}\right)^2 . R_i \qquad \text{.....(8.21)}$$

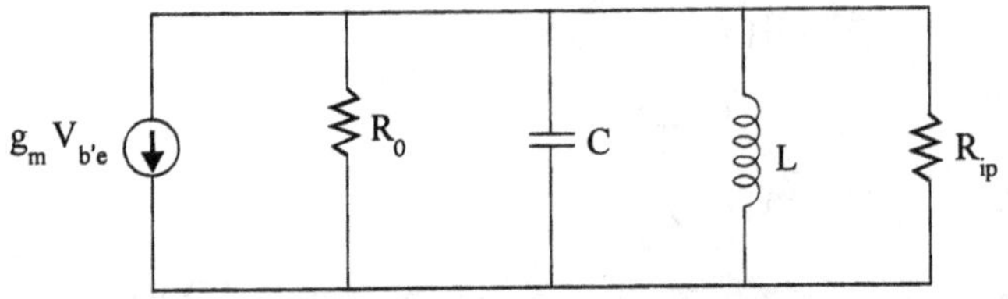

Fig. 8.13 Simplified circuit

For maximum transfer of power at resonance, (Fig. 8.13)

$$R_{ip} = R_0$$

$$\therefore \quad R_0 = \left(\frac{L}{M}\right)^2 . R_i \qquad \text{.....(8.22)}$$

Equation (8.22) gives the value of M for maximum power transfer, (Fig. 8.14)

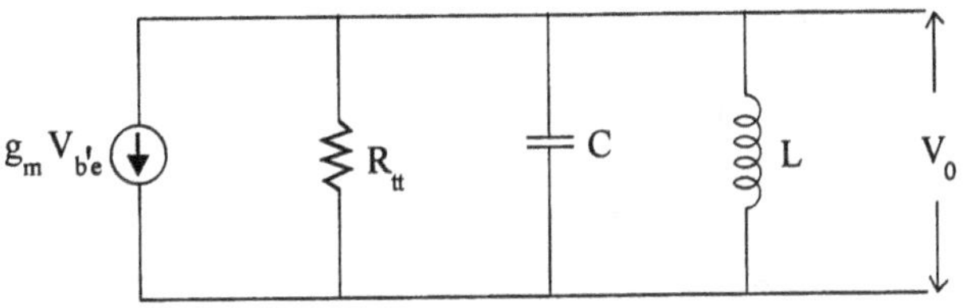

Fig. 8.14 Equivalent circuit

L and L_2 are the primary and secondary windings of inductances.

$$\therefore \qquad M = K\sqrt{LL_2} \qquad .....(8.23)$$

Combining eqs. (8.22) and (8.23) we get

$$R_0 = \left(\frac{L}{K^2 L_2}\right) R_i \qquad .....(8.24)$$

Therefore from equation (8.24), for a given value of R_0 and coefficient of coupling K and R_i, we can determine L_2 for maximum transformer of power.

Shunt resistance R_0 and R_{ip} may be combined to yield the total shunt resistance R_{tt}.

$$\frac{1}{R_{tt}} = \frac{1}{R_0} + \frac{1}{R_{ip}}$$

R_{tt} = Resistance of tapped tuned circuit .....(8.25)

Effective Q of the entire circuit is,

$$Q_e = \frac{R_{tt}}{\omega_0 L} \qquad .....(8.26)$$

where w_0 is the resonant frequency of L and C.

$$\omega_0 = \frac{1}{\sqrt{LC}} \qquad .....(8.27)$$

Under conditions of maximum transfer of power, total resistance appearing in shunt with the coil equals $R_0/2$. Since it is resonant circuit, at resonance, |Z| = resistance only. For maximum power, R = R/2.

$\therefore$ equation 16 becomes

$$Q_e = \frac{R_0/2}{\omega_0 L} \qquad .....(8.28)$$

or

$$R_0 = 2\, Q_e\, \omega_0\, L \qquad .....(8.29)$$

Solving equations (8.11) and (8.12)

$$I_2 = \frac{V.Z_{21}}{Z_{21}.Z_{12} - Z_{11}\,Z_{22}} \quad .....(8.30)$$

$$\therefore \quad V_0 = -I_2 \,.\, R_i \quad .....(8.31)$$

$$= V \,.\, \frac{R_i.Z_{21}}{Z_{11}.Z_{22} - Z_{12}^2} \quad .....(8.32)$$

|Z| of the output circuit at any frequency 'ω' close to ω_0, is given by,

$$\text{Impedance of output circuit } Z = \frac{R_{tt}}{1 + j2\delta Q_e} \quad .....(8.33)$$

$$\therefore \quad V_0 = -g_m \,.\, \frac{V_0\, r_{b'e}}{r_{b'e} + r_{bb'}} \,.\, Z \,.\, \frac{R_i\, Z_{21}}{Z_{11}\, Z_{22} - Z_{12}^2} \quad .....(8.34)$$

$$= -g_m\, V_i \,.\, \frac{r_{b'e}}{r_{b'e} + r_{bb'}} \,.\, \frac{R_i.Z_{21}}{Z_{11}\, Z_{22} - Z_{12}^2} \,.\, \frac{R_{tt}}{1 + j2\delta Q_e} \quad .....(8.35)$$

$\therefore$ Voltage gain A at any frequency ω is,

$$A = \frac{V_0}{V_i} = -g_m \frac{r_{b'e}}{r_{b'e} + r_{bb'}} \,.\, \frac{R_i\, Z_{21}}{Z_{11}\, Z_{22} - Z_{12}^2} \,.\, \frac{R_{tt}}{1 + j2\delta Q_e} \quad .....(8.36)$$

Voltage at resonance,

$$A_{reso} = -g_m \,.\, \frac{r_{b'e}}{r_{b'e} + r_{bb'}} \,.\, \frac{R_i\, Z_{22}}{Z_{11}\, Z_{22} - Z_{12}^2}$$

$$\therefore \quad \boxed{\frac{A}{A_{reso}} = \frac{1}{1 + j2\delta Q_e}} \quad .....(8.37)$$

8.5 EFFECT OF CASCADED SINGLE TUNED AMPLIFIERS ON BANDWIDTH

Several stages of identical tuned amplifiers can be cascaded in order to achieve high gain. However this overall high gain is achieved at the cost of reduced bandwidth. The overall bandwidth achieved by cascading of n-identical tuned amplifiers is narrower than the bandwidth offered by single stage tuned amplifier. And at the same time the overall voltage gain obtained by cascading of n-identical stages of tuned amplifiers is equal to the product of voltage gain of individual stages.

Let us assume n-stages of single tuned direct coupled amplifiers are connected in cascade configuration as shown in Fig.8.15.

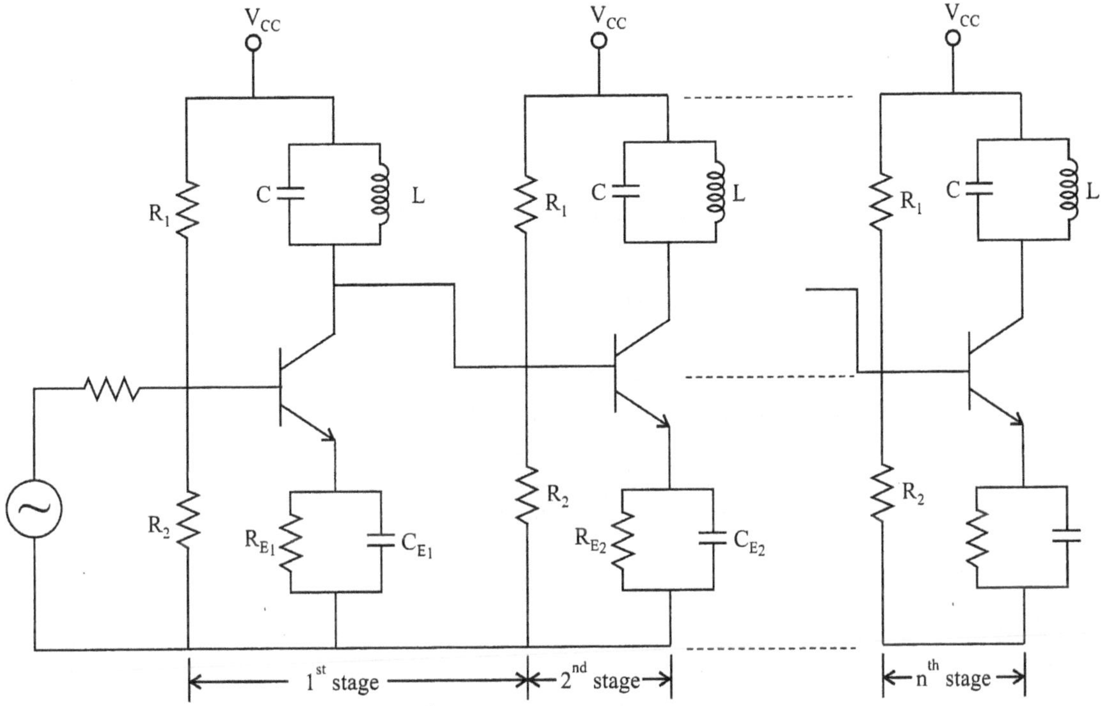

Fig.8.15 Cascading of 'n' identical single tuned amplifiers

The relative gain of single tuned amplifier with respect to gain at resonant frequency (f_0) of single stage tuned amplifier is given by

$$\left|\frac{A}{A_{res}}\right| = \frac{1}{\sqrt{1+(2\,\delta\,\phi_e)^2}} \qquad .....(8.38)$$

where ϕ_e = effective quality factor

δ = fractional frequency variation i.e., variation in frequency expressed in terms of resonant frequency.

Now gain of n-stages cascaded amplifier becomes

$$\left|\frac{A}{A_{res}}\right|^n = \left[\frac{1}{\sqrt{1+(2\,\delta\,\phi_e)^2}}\right]^n$$

$$= \frac{1}{\left[1+(2\,\delta\,\phi_e)^2\right]^{n/2}} \qquad .....(8.39)$$

Now the 3-dB frequencies of n-stage cascaded amplifier can be obtained equating eq. (8.39) to $\frac{1}{\sqrt{2}}$.

$$\left|\frac{A}{A_{res}}\right|^n = \frac{1}{\left[1+(2\,\delta\,\phi_e)^2\right]^{n/2}} = \frac{1}{\sqrt{2}}$$

$$\left[\sqrt{1+(2\,\delta\,\phi_e)^2}\right]^n = \sqrt{2}$$

$$1+(2\,\delta\,\phi_e)^2 = 2^{1/n}$$

$$2\,\delta\,\phi_e = \pm\sqrt{2^{1/n}-1} \qquad \text{.....(8.40)}$$

Substituting for fractional frequency variation (δ) in eq.(8.40)

$$\delta = \frac{\omega-\omega_0}{\omega_0} = \frac{f-f_0}{f_0}$$

$$2\left(\frac{f-f_0}{f_0}\right)\phi_e = \pm\sqrt{2^{1/n}-1}$$

$$2(f-f_0)\,\phi_e = \pm f_0\sqrt{2^{1/n}-1}$$

Now

$$f_2-f_0 = +\frac{f_0}{2\,\phi_e}\sqrt{2^{1/n}-1}$$

$$f_0-f_1 = +\frac{f_0}{2\,\phi_e}\sqrt{2^{1/n}-1}$$

Thus bandwidth of n-stage cascaded single tuned amplifier is

$$B_n = f_2-f_1 = (f_2-f_0)+(f_0-f_1)$$

$$= \frac{f_0}{2\,\phi_e}\sqrt{2^{1/n}-1}+\frac{f_0}{2\,\phi_e}\sqrt{2^{1/n}-1}$$

$$= \frac{f_0}{\phi_e}\sqrt{2^{1/n}-1}$$

$$B_n = B_1\sqrt{2^{1/n}-1} \qquad \text{.....(8.41)}$$

where B_n is bandwidth of n-stage cascaded single tuned amplifier and B_1 is bandwidth of single stage tuned amplifier.Thus bandwidth of n-stage cascaded single tuned amplifier is equal to the bandwidth of single stage tuned amplifier multiplied with a factor of $\sqrt{2^{1/n}-1}$

when $\quad n = 2; \ \sqrt{2^{1/n}-1} = 0.643$

$n = 3; \ \sqrt{2^{1/n}-1} = 0.510$

$n = 4; \ \sqrt{2^{1/n}-1} = 0.434$

Thus the bandwidth of 2 stage single tuned amplifier is reduced by 64.3% and bandwidth of 3 stage amplifier is reduced by 51% and similarly for 4 stages the bandwidth is reduced by 43.4%. This clearly shows that the bandwidth of single stage tuned amplifier is greater than of bandwidth achieved by cascading of n stages.

8.6 CE DOUBLE TUNED AMPLIFIER

In this circuit, voltage developed across the tuned circuit in the collector circuit is inductively coupled to another tuned circuit. Both circuits are tuned to the same frequency, the desired frequency of the input signal.

8.6.1 Advantages

1. It provides larger 3-db band width than the single tuned amplifier. Therefore Gain × Bandwidth product is more. (eventhough gain is same, Bandwidth is more. So gain × Bandwidth product is more).
2. It provides gain - frequency curve having steeper sides and flatter top.

In double tuned amplifier, as Bandwidth is increased the overshoot of gain also increases. So a compromise must be made. The value of the coefficient 'b' ranges from 1.0 to 1.7.

8.6.2 Circuit

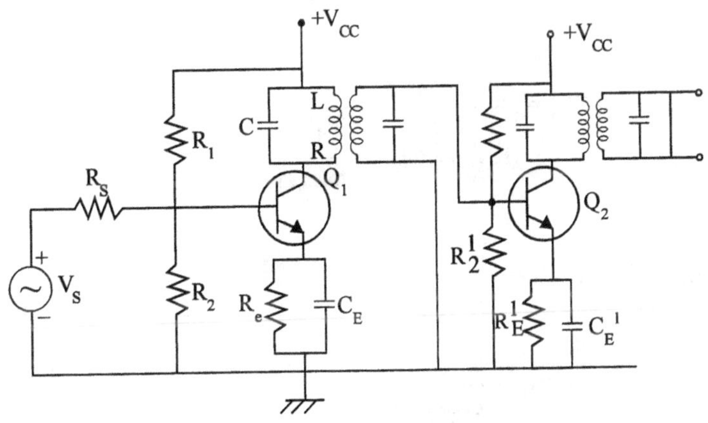

Fig. 8.16 Double tuned amplifier circuit

8.6.3 Equivalent Circuit

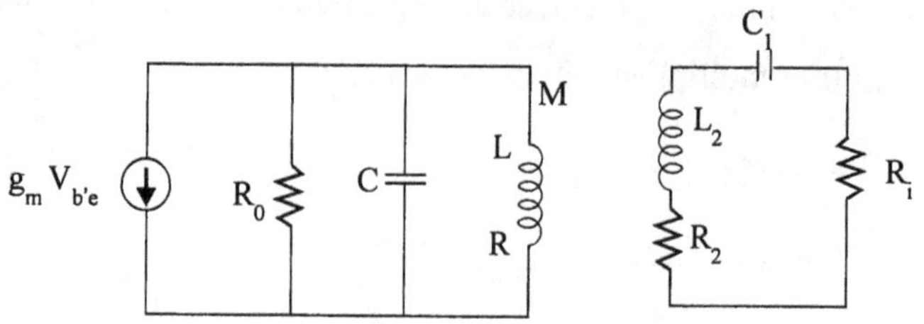

Fig. 8.17 Equivalent circuit

Voltage developed across the tuned circuit in the LCR combination is inductively coupled to another tuned circuit of the same resonant frequency.

R is the resistance associated with L. R_i is the output |Z| of the second circuit. The equivalent circuit can be further simplified as, R_i and R_2 are combined to form R_i'. R_0 in parallel with L-R may be brought in series with inductance and combined with R to from R_i.

The current source $g_m V_{b'e}$ in shunt with 'C' is modified to a voltage source V_1 in series with 'C'.

∴ The modified circuit is as shown in Fig. 8.18.

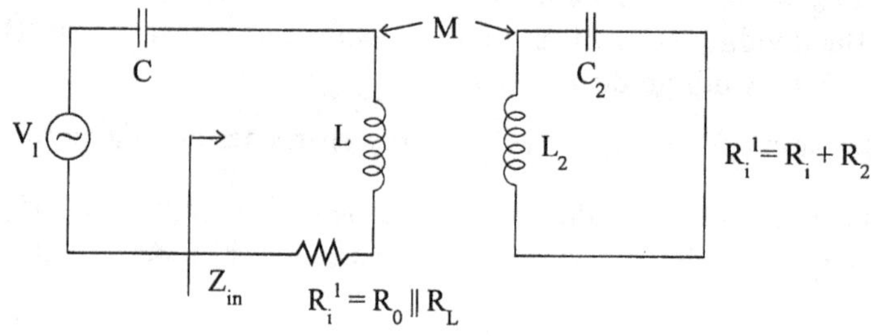

Fig. 8.18 Modified circuit

$$Z_{in} = \frac{\omega^2 M^2}{R_{ik} + \left(\omega L_2 - \frac{1}{\omega L_2}\right)} \quad .....(8.42)$$

At resonance,

$$\omega_0 L = \frac{1}{\omega_0 C} \quad .....(8.43)$$

and

$$\omega_0 L_2 = \frac{1}{\omega_0 C_2} \quad .....(8.44)$$

$\therefore$ At resonance, Z_i is reduced to

$$Z_{in} = \frac{\omega_0^2 M^2}{R_i^1} \qquad .....(8.45)$$

Hence at resonance for maximum transfer of power R_0 must equal Z_{in}. i.e., M is adjusted to critical value M_C, such that

$$R_0 = \frac{\omega_0^2 M_C^2}{R_i} \qquad ......(8.46)$$

or $$\sqrt{R_0' R_i'} = \omega_0 M_0 \qquad .....(8.47)$$

$$= \omega_0 K_C \sqrt{L L_2} \qquad .....(8.48)$$

where *K_C is the critical value of the coefficient of coupling corresponding to the critical value M_C of mutual inductance.*

From (8.48), $$K_C = \frac{\sqrt{R_0' R_i'}}{\omega_0 \sqrt{L L_2}} \qquad .....(8.49)$$

$$= \left(\frac{R_0'}{\omega_0 L}\right)^{1/2} \cdot \left(\frac{R_i'}{\omega_0 L_2}\right)^{1/2} = \frac{1}{\sqrt{Q_1 Q_2}} \qquad ......(8.50)$$

In a general case, $K \neq K_C$

So let $K = b.\ K_C$

Now, $$V_1 = Z_{11} I_1 + Z_{12}.I_2 \qquad .....(8.51)$$

$$0 = Z_{21} I_1 + Z_{22} I_2 \qquad .....(8.52)$$

Solving equation (8.51) and (8.52), for I_2, we get

$$I_2 = \frac{V_1 Z_{21}}{Z_{11} Z_{22} - Z_{12}^2} \qquad .....(8.53)$$

where $$Z_{11} = R_0' + j\,(\omega L - \frac{1}{\omega C}) \qquad ......(8.54)$$

$$Z_{22} = R_i' + j\left(\omega L_2 - \frac{1}{\omega C_2}\right) \qquad .....(8.55)$$

and $$Z_{12} = Z_{21} = j\omega_0 M \qquad .....(8.56)$$

At resonance, $Z_{11} = R_0^{'}$ and $Z_{22} = R_i^{'}$

For maximum transfer of power at resonance,

$$Z_{11} = R_0^{'},\ Z_{22} = R_i^{'},$$

$$Z_{12} = j\omega_0 M_C \qquad .....(8.57)$$

Substituting these values of Z_{11}, Z_{22}, and Z_{12} in equation (8.53), we get the following conditions for maximum transfer of power at resonance.

$$I_{2\,Max} = \frac{-j\,V_1\,\omega_0 M_C}{R_0^1\,R_i^1 + \omega_r^2\,M_C^2} \qquad .....(8.58)$$

M_c is the critical value of Mutual Inductance.

Substituting the value of M_C in equation (8.58), we get

$$I_{2\,Max} = \frac{-j\,V_1\sqrt{R_0^{'}\,R_i^{'}}}{R_0^1\,R_i^1 + R_0^{'}\,R_i^{'}}$$

$$= \frac{-jV_1}{2\sqrt{R_0^{'}\,R_i}} \qquad .....(8.59)$$

Magnitude $$|I_{2\,Max}| = \frac{V_1}{2\sqrt{R_0^{'}\,R_i^{'}}}$$

$$\frac{\omega}{\omega_0} - \frac{\omega_0}{\omega} \pm \sqrt{\left(b^2 - 1 \pm 2b\right)/Q}$$

where
$b = (K/K_C)$
K = Coefficient of coupling
K_C = Critical value of coefficient of coupling corresponding to the critical value of Mutual Inductance Mc.

Tuned amplifiers are used where it is desired to amplify a relatively narrow band of frequencies, centered about some designated mean carrier frequency.

8.7 EFFECT OF CASCADING DOUBLE TUNED AMPLIFIERS ON BANDWIDTH

As with the case of single tuned amplifier, it is also possible to cascade double tuned amplifier at the cost of reduced bandwidth.

The 3-dB bandwidth of n-stage cascaded double tuned amplifier is given by

$$B_n = B\left[2^{1/n} - 1\right]^{1/4}$$

where B_n = Bandwidth of n-stage cascaded double tuned amplifier

B = Bandwidth of single stage double tuned amplifier.

when $$n = 2; \left[\sqrt{2^{1/n} - 1}\right]^{1/4} = 0.896$$

$$n = 3; \left[\sqrt{2^{1/n} - 1}\right]^{1/4} = 0.845$$

$$n = 4; \left[\sqrt{2^{1/n} - 1}\right]^{1/4} = 0.812$$

Thus bandwidth of 2-stage double tuned amplifier is reduced by 89.3% and bandwidth of 3-stage double tuned amplifier is reduced by 84.5% and similarly bandwidth of 4-stage double tuned amplifier is reduced by 81.2%.

KEEN OBSERVATION

From the analysis of effect of bandwidth by cascading single tuned amplifiers and analysis of effect of bandwidth by cascading double tuned amplifier it can be observed that bandwidth is more effected by cascading double tuned amplifier.

8.8 APPLICATIONS OF TUNED AMPLIFIERS

1. Radar
2. Television
3. Communication receivers
4. I.F amplifiers

There are mainly three types tuned voltage amplifiers.

1. Single tuned amplifier
2. Double tuned amplifier
3. Stagger tuned amplifier

8.9 STAGGER TUNING

Tuned amplifiers have large gain, since at resonance, Z is maximum. So A_V is maximum. To get this large A_V over a wide range of frequencies, stagger tuned amplifiers are employed. This is done by taking two single tuned circuits of a certain Bandwidth, and displacing or staggering their resonance peaks by an amount equal to their Bandwidth. The resultant staggered pair will have a Bandwidth, $\sqrt{2}$ times as great as that of each of individual pairs (Fig. 8.19).

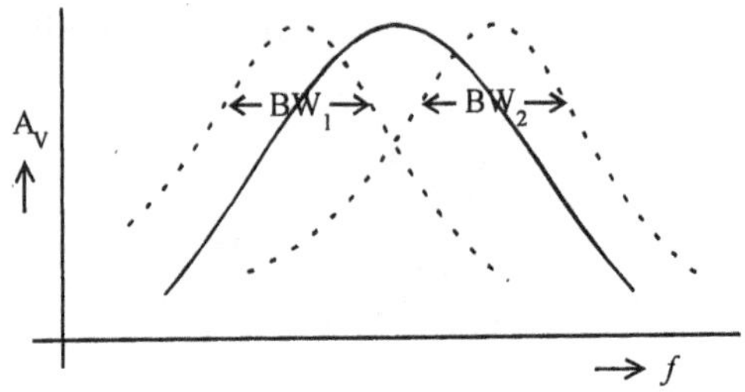

Fig. 8.19 Variation of A_V with f

8.10 SINGLE TUNED TRANSISTOR AMPLIFIER

The circuit is shown in Fig. 8.20.

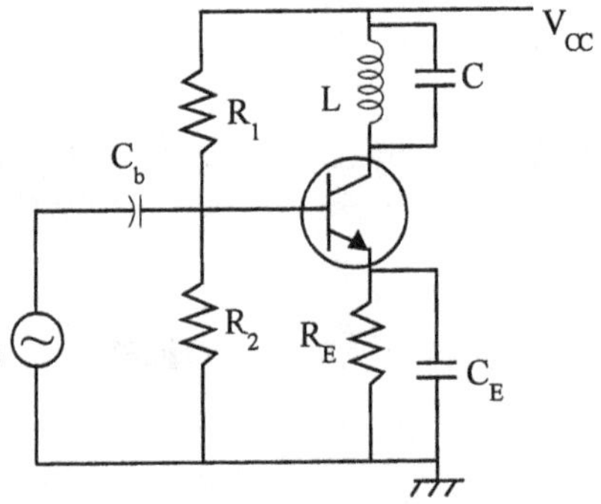

Fig. 8.20 Single tuned amplifier

8.11 STABILITY CONSIDERATIONS

An electronic amplifier circuit or oscillator circuit becomes unstable, i.e., will not perform the desired function, due to various reasons associated with circuit design aspects like Thermal stability, Bias considerations, output circuit, feedback, circuit etc. Let us consider these aspects.

Thermal Effects : Any electronic circuit, when used continously, various components will get heated, due to power dissipation in each of the components. When the active device in the circuit (BJT) also gets heated, due to the dependance of its characteristics on temperature, the operating point changes. So the device will not function as desired and the output from the electronic circuit will not be obtained as per specifications. Heat sinks are to be used, to dissipate the excess thermal energy and to keep the operating temperature of the active device within the limits. Heat sinks are good themal conductors designed suitably, to keep the temperature of the device within limits.

Let

P_J = Power dissipated in watts at the junction of the device

T_A = Ambient Temperature

T_J = Temperature of the junction

θ_T = Themel resistance in °C/W

$T_J = T_A + \theta_T \cdot P_J$

Range of value of θ_J is 0.2 to 2.0 °C/W.

A thin mica spacer is often used to electrically insulate the BJT or semiconductor device from the heat sink. The thermal path for heat dissipation is from semiconductor p-n junction to casing of the device, from the case to heat sink and from heat sink to the ambient.

θ_C = Thermal resistance in °C/W associated with the Casing of the device

θ_H = Thermal resistance in °C/W associated with the Heat sink.

θ_R = Radiation Thermal Resistance

So,

$$T_H = T_J - P_J(\theta_J + \theta_C)$$

$$\text{Total Heat sink resistance} = \theta_H + \frac{\theta_C\,\theta_R}{\theta_C + \theta_R}$$

If P_J, the power dissipated at the junction is independent of T_j then,

$$\frac{dP_J}{dT_J} = \frac{1}{\theta_T}$$

$$\text{At the operating temperature, } \frac{dP_J}{dT_J} > \frac{1}{\theta_T}$$

Bias Considerations : Distortion in Audio amplifiers and other types of circuits depends on :

(i) Input signal level (in mv)

(ii) Source Resistance

(iii) Bias Conditions

(iv) Type of output load and its impedance

(v) Loading effect.

For audio frequency amplifiers, distortion of 1 or 2% is considered maximum for higher fiedility equipment.

A maximum of 5% distorion is set sometimes for radio receivers and public address systems.

Consider the circuit shown in Fig. 8.21.

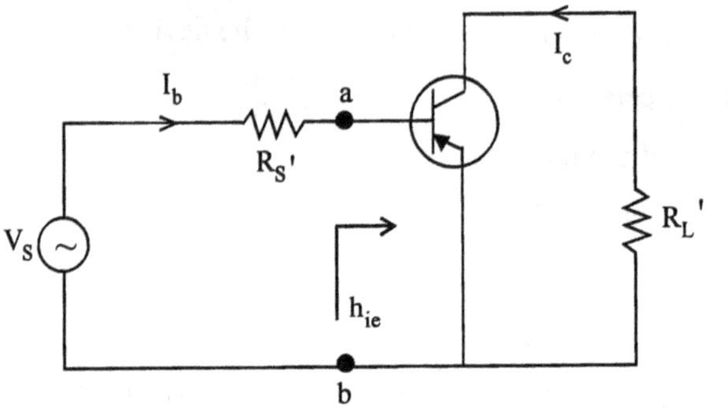

Fig. 8.21

Expression for input power P_S is,

$$P_S = Ib^2 \left(R_S' + h_{ie}\right)$$

Expression for output power P_o is,

$$P_O = I_C^2 . R_L'$$

$$\therefore \text{Power Gain} \qquad A_P = \frac{P_o}{P_s} = \frac{I_C^2 R_L'}{I_b^2\left(R_S' + h_{ie}\right)}$$

$$= \frac{h_{fe}^2 R_L'}{R_S' + h_{ie}}$$

Considering only the BJT, expression for power gain

$$A_{p(BJT)} = \frac{h_{fe}^2 R_L}{h_{ie}} = h_{fe} \cdot g_m \; R_L'$$

Maximum power is transferred to the base-emitter circuit of BJT, when impedance matching is done as,

$$R_S' = h_{ie}$$

$\therefore$ Matched input Power Gain

$$A_P \text{ (Matched input)} = \frac{h_{fe}^2 R_L'}{2h_{ie}}$$

$$A_p = h_{fe} \cdot \frac{g_m . R_L'}{2}$$

To maintain output voltage and power levels stable, the circuit must be designed accordingly.

8.12 TUNED CLASS B AND CLASS C AMPLIFIERS

The efficiency of the output circuit of an amplifier increases as the operation is shifted from class A to B and then to C.

In class C amplifiers, efficiency approaches 100%.

But the difficulty with class 'C' operations is harmonic distortion is more. So the conventional untuned amplifiers use class B pushpull configuration. But, if it is a tuned amplifier, and only one frequency f_0, is to be amplified and power to be handled P_0 is large, then class C operation is preferable, since efficiency is high and harmonic distortion will not be a problem since, only one frequency is to be amplified and the tuned circuit will reject the other frequencies. (Fig. 8.22).

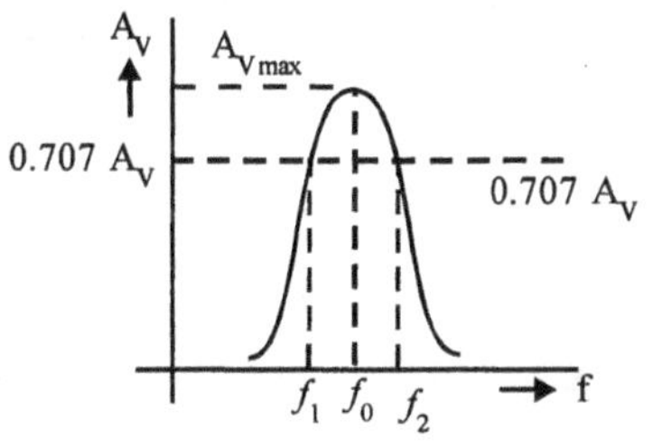

Fig. 8.22 Frequency response

In radio transmitters, the output power often exceeds 50 KW and efficiency is the important criteria. Hence usually class C operation is done. The operation of such amplifier circuits is non linear, since the variation of output current i_L with the input current i_b is non linear, as the signal magnitudes are large. So these amplifiers are also called as *Large signal amplifiers.*

Since they have a tuned circuit and usually operate in the R.F range, they are also called as *R.F power amplifiers.*

Usually these circuits use common emitter configuration, but sometimes common base configuration is also used. A simplified circuit (without emitter stabilization) uses transformer and FET as shown below. Since the circuit is class B and class C configurations, the input signal itself drives the transistor or FET into conduction. So there are no separate biasing resistors.

The circuit may be class B or class C, depending upon the actual conduction angle (180^0 or less).

8.12.1 BIPOLAR JUNCTION TRANSISTOR (BJT) TUNED CLASS B/C AMPLIFIER

The capacitors C_S, C_b and C_C are considered to be short circuits at the operating signal frequencies. (Fig. 8.23).

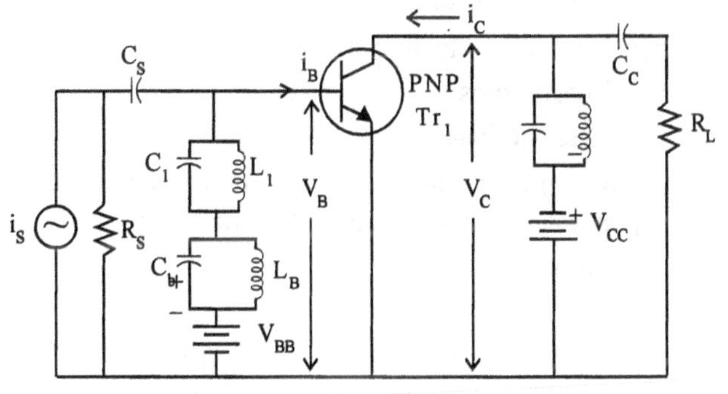

Fig. 8.23 Tuned class B/C amplifier

The transformer circuit L_1, C_1 and R_b, C_b provide high input impedance $|Z|$ to the input signal and so i_b passes to the base of the transformer only and will not get divided across L_1 C_1 or R_b C_b. So the power output will be more.

8.12.2 FET Tuned R.F Amplifier

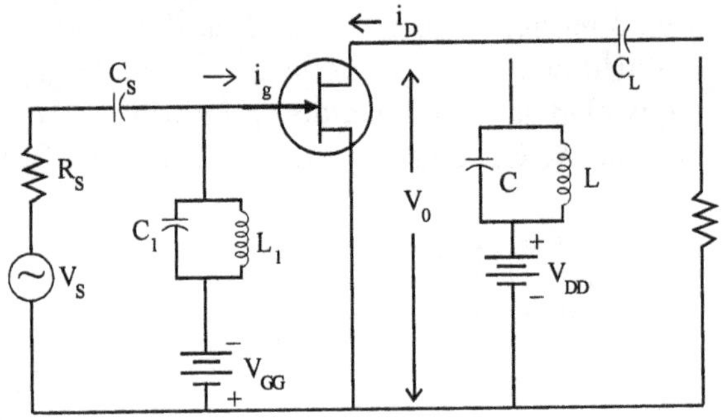

Fig. 8.24 FET tuned RF amplifier

8.12.3 Waveforms

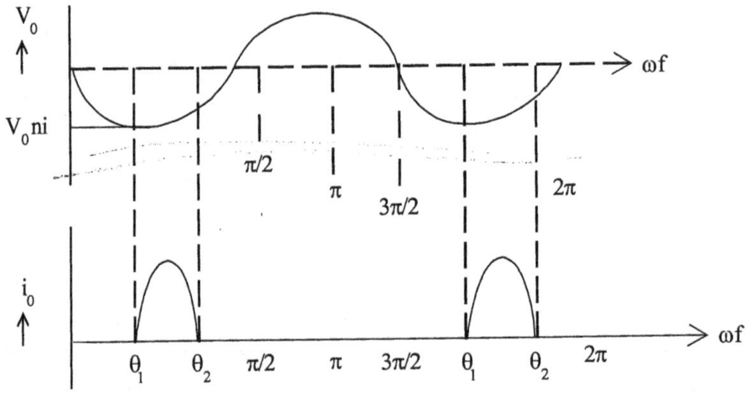

Fig. 8.25 Frequency response

8.12.4 Resonant Circuit

In a tuned amplifier, the functions of the resonant circuit are :

1. To provide correct load impedance to the amplifier.
2. To reject unwanted harmonics
3. To couple the power to the load

The resonant circuits in tuned power amplifiers are sometimes called tank circuits, because the L and C elements store energy like water stored in a tank. Such a circuit is as shown in Fig. 8.26.

Its equivalent is $(R + j\omega L)$ in parallel with R_L.

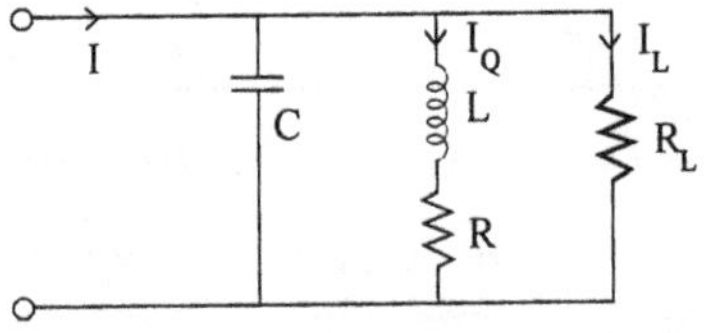

Fig. 8.26 Resonant circuit

$$= \frac{R_L(R + jX_L)}{R_L + R + jX_L} = \frac{R_L R}{R_L + R + jX_L} + j\frac{(X_L R_L)}{R + R + jX_L}$$

Neglecting the losses in the capacitor C, the η of the above circuit is,

Efficiency $= \eta =$ output power across R_L/Input power

$$\eta = \frac{I_L^2 . R_L}{I_L^2 R_L + I_a^2 R}$$

I_L = current through load resistance R_L
I_a = current through inductor inductance with its internal resistance R in series.
At resonance, the |Z| is only, resistive.

R = Series resistance of 'L'. R_L^1 is internal resistance

Let $Q_0 = \frac{\omega_0 L}{R}$

= Q of the coil at resonance

Let $Q_{eff} = \frac{\omega_0 L}{R + R_L}$

= Q of the coil shunted by R_L'. (Taking R_L' into consideration)

$$\therefore \quad \eta = \frac{R_L}{R + R_L}$$

This equation can be written as,

$$\text{Efficiency} = \eta = \frac{\frac{1}{Q_{eff}} - \frac{1}{Q_0}}{1/Q_{eff}} = 1 - \frac{Q_{eff}}{Q_0}$$

Q_0 = Q factor at resonance frequency f_0.

Q_{eff} = Effective value of Q

$Q_0 >> Q_{eff}$, $\eta \simeq 1$ or 100% (or it is very high). If R_L can be varied, Q_{eff} can be made as small as desired. The value of Q_0 is usually large.

The purpose of resonant circuits in tuned circuits is,

1. To provide correct load |Z|
2. To reject unwanted harmonics
3. To couple power to load

These tuned circuits are sometimes called Tank circuits because, like water tank, energy can be stored in inductance and capacitance energy storage elements, if they are ideal.

8.12.5 Tank Circuits

Fig. 8.27 Tank circuits

R_L = equivalent circuits R_L in series with inductance, resistance

Neglecting losses in capacitors,

$$\eta = \frac{I_L^2 R_L}{I_L^2 R_L + I_a^2 R}$$

For the equivalent circuit,

$$R_L{}^1 = \frac{\omega_0^2 L^2}{R_L} \quad (R_L >> \omega_0 L)$$

For the equivalent circuit

$$\eta = \frac{I_b^2 R_L^2}{I_b^2 \left(R + R_L\right)} = \frac{R_L}{R + R_L}$$

Let $$Q_0 = \frac{\omega_0 L}{R} = \text{Q of the coil at resonance.}$$

$$Q_{eff} = \frac{\omega_0 L}{R + R_L^1} = \text{Q of the coil shunted by } R_L.$$

$$\therefore \qquad \eta = \frac{(1/Q_{eff}) - (1/Q_0)}{(1/Q_{eff})}$$

$$= 1 - \frac{Q_{eff}}{Q_0}$$

8.12.6 MUTUAL INDUCTANCE COUPLED OUTPUT RESONANT CIRCUIT

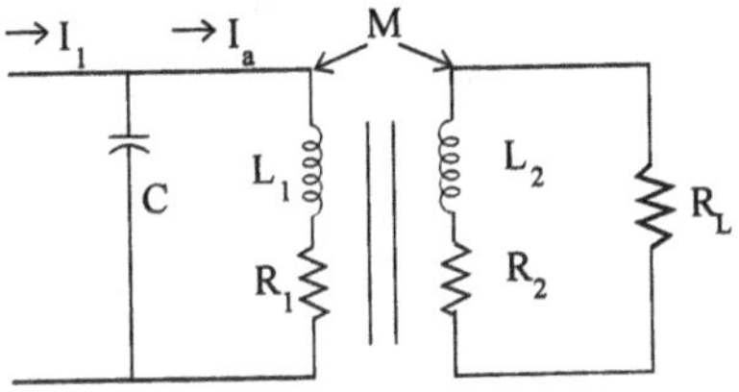

Fig. 8.28 Mutual inductance coupled resonant circuit

8.12.7 EQUIVALENT CIRCUIT

Power dissipated in R_2 and R_L (Fig. 8.29).

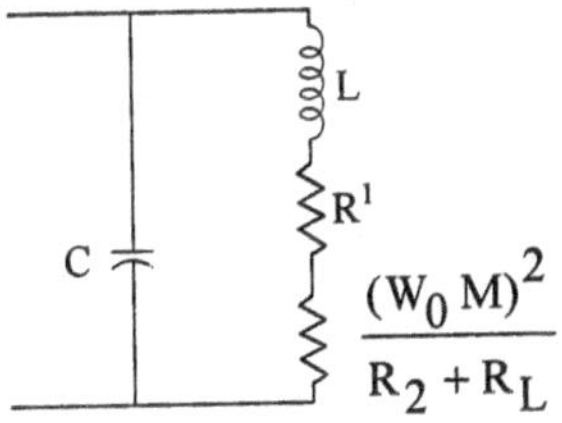

Fig. 8.29 Equivalent circuit

$$P = I_a^2 \frac{\omega_0^2 M^2}{R_2 + R_L}$$

Power dissipated in R_L can be obtained by multiplying the above equation by $\frac{R_L}{R_2 + R_L}$.

8.13 WIDEBAND AMPLIFIERS

The frequency response of a given amplifier can be extended by adding few passive circuit elements to the basic amplifier.

8.13.1 Shunt Compensation

One simple method of shortening the rise time in the response of an amplifier circuit and thus enhancing the high frequency response of the amplifier. (pulse having, sudden change in the input is considered as high frequency variation) is to add an inductor in series with V_{CC} and collector. (Fig. 8.30).

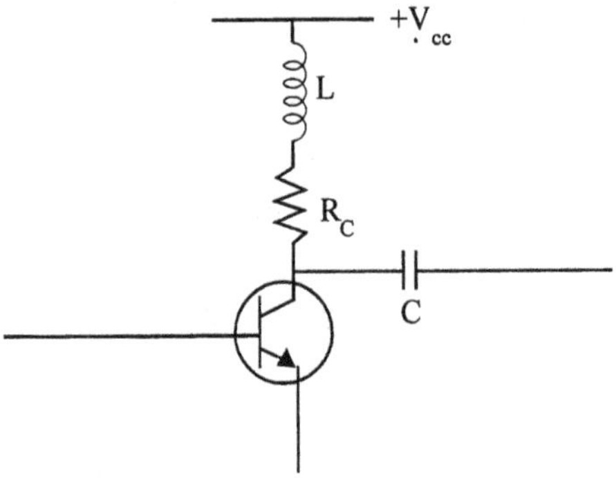

Fig. 8.30 Shunt compensation circuit

In the a.c equivalent circuit, this 'Inductance' will be in parallel with capacitor. Thus 'Inductance Capacitor' combinations changes the output response. Since in shunt in the output stage, this is called *shunt compensated amplifier*. The collector circuit $Z = R_C + j\omega L$.

This increases with frequency. So V_0 increases and gain increases. So if 'Inductance' is not present, the gain will be less. When '*capacitor*' is present in the output circuit, X_C decreases as f increases. Thus L will compensate for capacitor.

So for compensated amplifier, when L is introduced, the damping factor K is defined as,

$$K = R_C \cdot \sqrt{\frac{C}{L}}$$

$$f_2 = \frac{1}{2\pi R_c C}$$

Where f_2 is the upper 3 db point of the uncompensated amplifier. (Fig. 8.31).

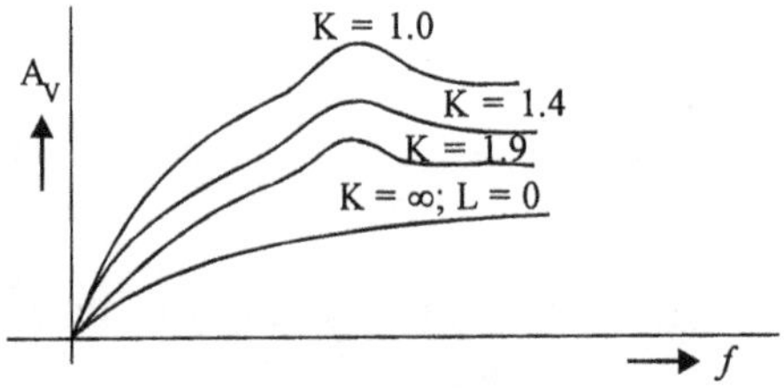

Fig. 8.31 Frequency response

So the frequency response changes, when 'L' is added :

The bandwidth of an amplifier circuit can be increased by decreasing f_1 and increasing f_2. (Figs. 8.32 and 8.33).

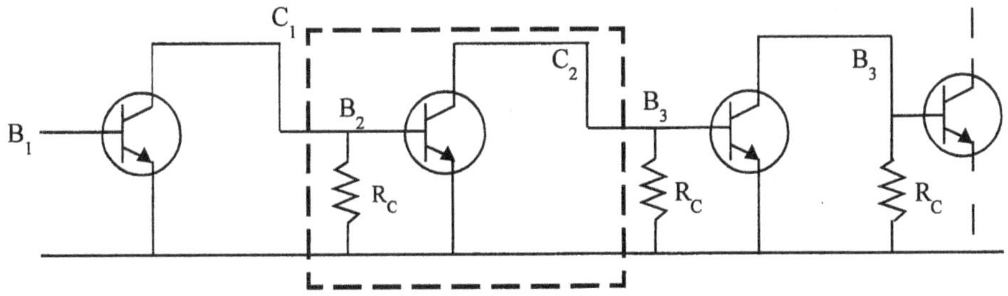

Fig. 8.32 Simplified circuit

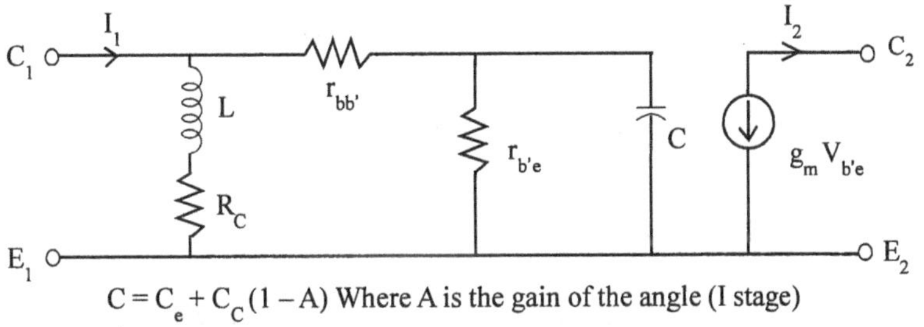

Fig. 8.33 Equivalent circuit

8.13.2 Extension of Low Frequency Range

Low frequency range can be extended by decreasing f_1 (Since the region as seen from the mid frequency range becomes more if f_1 is decreased). For resistance capacitance coupled amplifier,

$$f_1 = \frac{1}{2\pi C_c (R_C + R_L)}$$

Where C_C is the coupling capacitor, f_1 can be decreased by increasing C_C. If the value of C_C were to be large, its cost and size will be more. So there is a limit to which C_C can be increased. Hence, to get lesser value of f_1, R should be increased. This is known as *Compensation.* Instead of changing the value of C, to get decrease in f_1 we are changing the value of R. Thus R is compensating for the effect of C.

Now if R were to be fixed and load Z were to change with frequency, another capacitor C_C is connected in parallel with load resistance R_L so that the net load Z, is Z_L so that Z_L increases as f decrease. So f_1 can be increased. This is known as *Compensation.*

But the value of f_1 is very small (few hundreds MHz) compared to f_2. Therefore jω is increased effectively by increasing f_2.

8.13.3 Circuit for Extending Low Frequency Range

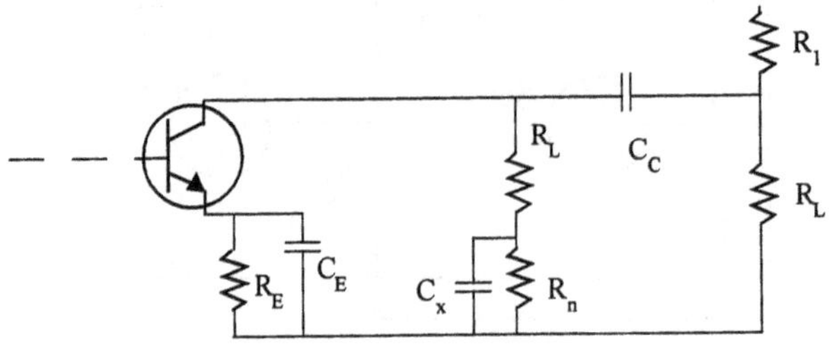

Fig. 8.34 (a) Kinks in frequency response

8.13.4 Low Frequency Equivalent Circuit with Transistor Replaced by a Current Source

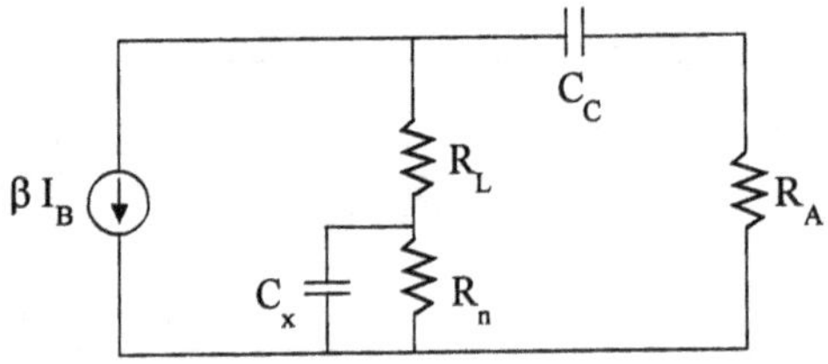

Fig. 8.34 (b) Equivalent circuit with current source

8.13.5 Extension of High Frequency Range

Bandwidth of a given amplifiers can be effectively increased by increasing f_2. (since $f_2 >> f_1$) f_2 can be increased by decreasing C_S. But there is a limit to which C_S value can be reduced f_2 can also be increased by decreasing R_L^1. But if R_L^1 is reduced the mid band gain will reduce. Therefore without changing the value of C_S and R_L^1 and without decreasing the mid band gain, f_2 can be increased by compensating techniques.

There are two different methods :

(i) Series compensation

(ii) Shunt compensation

This classification is depending upon character '*Inductance*' is in series or in shunt with the load resistance in the a.c equivalent circuit.

$$\left|\frac{A_H}{A_M}\right|^2 = \frac{1+m^2\left(\frac{\omega}{\omega_2}\right)^2}{1+m^2\left(\frac{\omega}{\omega_2}\right)^2 + m^2\left(\frac{\omega}{\omega_2}\right)^4}$$

where $m = \omega_2 L/R_1$ (R_1 is the resistance in series or the resistance of the coil itself)

ω_2 is the upper cutoff frequency of the uncompensated amplifier.

For a shunt peaked amplifier, for different values of m, the response will be as shown in Fig. 8.35.

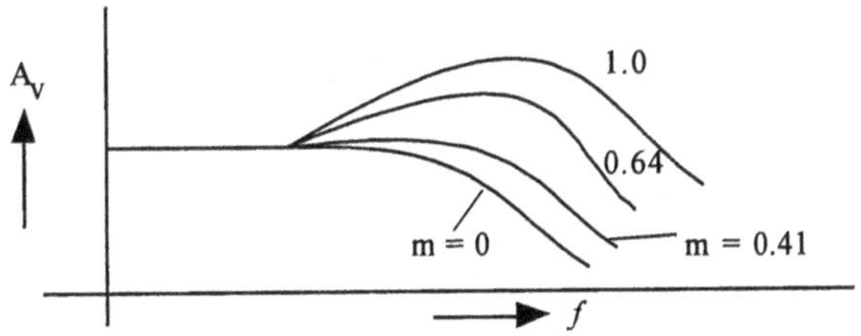

Fig. 8.35 Kinks in frequency response

If m = 0.414, ω_2^1 = 1.72 ω_2 where $\omega_2{}^1$ is the upper cutoff frequency with compensation. For this value of 'm', there will not be peaking of the mid band gain. For higher value of m, there will be peaking. So m = 0.414 value is often used. Because of resonance, the voltage across C_S or the output voltage will be maximum. So overshoot occurs.

8.13.6 Series Peaked Circuit

Consider the circuit shown in Fig. 8.36.

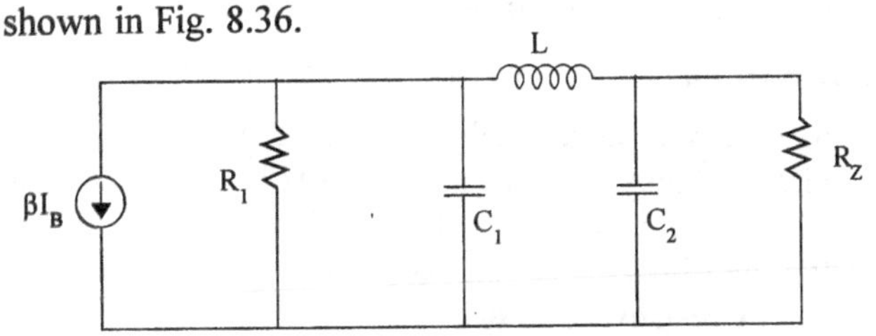

Fig. 8.36 Series peaked circuit

Series peaked circuit is used if $\dfrac{C_1}{C_1 + C_2}$ = a = 0.25.

If any other type of capacitances and shunt capacitances are present, this is not used. Shunt compensated circuit is preferred over series peaked since, for shunt compensated value circuits, the gain falls smoothly beyond f_2. For series compensated circuits, there will be sudden drop in gain beyond f_2.

Because of the presence of both Inductance and Capacitance, in the circuit the f_2 point will be near about the resonant frequency of the circuit. Overshoot or peaking in gain will occur near about the resonant frequency. This will depend upon the Q factor of the circuit. As Q increase peak gain increases.

8.14 IMPEDANCE TRANSFORMATION

When a signal generator is to supply power to a load or amplifier is to supply output power to a load (driving a load), for maximum power transfer the output impedance of the signal generator or amplifier Z_0 must be complex conjugate of load impedance Z_L according to maximum power transfer theorem. In other words, their magnitudes must be the same.

i.e., $$|Z_L| = |Z_0|$$

This is konwn as *Impedance Matching.*

In the case of audio amplifiers, the load is usually a loudspeaker. Its impedance is of the order of 4 Ω, 8 Ω or 16 Ω. But the audio amplifier circuit may not have output impedance equal to these values. So impedance matching condition is not satisfied.

Similarly a radio transmitter will have typically output impedance of 4000 Ω. It is required to supply power to an antenna. Antenna will hae usually impedance of 70 Ω. So impedance matching condition is not satisfied for maximum power transfer, impedance transformation must be used in such coupled circuits.

Some of the methods by which impedance transformation is done in coupoled circuits are :

1. Transformation of impedances with tapped resonant circuits
   (a) by tapping inductors
   (b) by tapping capacitors
2. Reactance section for impedance transformation
3. Image impedances - Reactance matching
4. Reactance T Networks for impedance transformation

8.14.1 Transformation of Impedances with Tapped Resonant Circuits

Consider a parallel LC tuned circuit connected to a generator. At antiresonant frequency, the net impedance of the tuned circuit is only resistive and say it is R_{ar} (resistance at anti resonance). The generator sees a load of R_{ar} at anti resonance. R_{ar} is independent of L and C values of the tuned circuit. This value of R_{ar} may be greater than or less than the generator output resistance for maximum power transfer. The resistive impedance into which the generator supplied power can be reduced by tapping the external generator connections across only a portion of the impedance is shown in Fig. 8.37 (a) and (b).

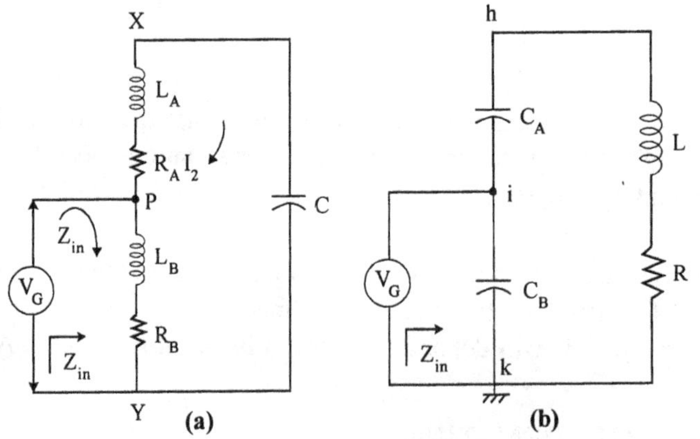

Fig. 8.37 (a) Equivalent circuit with current source
(b) Tapping across only a portion of the impedance

Let the mutual inductance between L_A and L_B is such that the total inductance is,

$$L = L_A + L_B + 2M \qquad .....(8.60)$$

Anti resonance will occur between points P and Y if,

$$\omega\,(L_A + L_B + 2M) = \frac{1}{\omega C} \qquad .....(8.61)$$

Consider the terminals P and Y of Fig. 4.35(a) :
Anti resonance will occur if

$$\omega\,(L_B + M) = \frac{1}{\omega C} - \omega\,(L_A + M) \qquad .....(8.62)$$

Because tapping is done in the inductor, part of the inductor is in series with capacitor C. But still, the net reactance must be capacitive for this branch.

The circuit equations in matrix force for the Fig. 8.37 (a) are :

$$\begin{bmatrix} V_G \\ O \end{bmatrix} = \begin{bmatrix} R_B + jX_{LB} & -(R_B + jX_{LB} + jX_M) \\ -(R_B + jX_{LB} + jX_M) & \left(R_A + R_B + jX_{LA} + jX_{L_B} + j2X_M - jX_C\right) \end{bmatrix} \begin{bmatrix} I_1 \\ I_2 \end{bmatrix} \qquad .....(8.63)$$

Since $\quad Z_{in} = Z_{P,\,Y} = \Delta/\Delta_{11},$

$$Z_{P,\,Y} = \frac{\left(R_B + jX_{LB}\right)\left(R_A + R_B + jX_{LA} + jX_{LB} + j2X_M - jX_C\right) - \left(R_B + jX_{LB} + jX_M\right)^2}{R_A + R_B + jX_{LA} + jX_{LB} + j2X_M - jX_C} \qquad .....(8.64)$$

If $\quad R_A + R_{B'} = R$

$$Z_{P,\,Y} = R_B + jX_{LB} - \frac{\left(R_B + jX_{LB} + jX_M\right)}{R + j\left(X_L - X_C\right)} \qquad .....(8.65)$$

If the Q factor of the circuit is high, and the circuit is in anti resonance, then $X_L = X_C$.

$X_{LB} >> R_B$. So R_B can be neglected.

$$\therefore \qquad Z_{P,Y} = jX_{LB} + \frac{\left(X_{LB} + X_M\right)^2}{R} \qquad .....(8.66)$$

If Q value is reasonable, X_{LB} is small compared to $(X_{LB} + X_M)^2 / R$. So the term X_{LB} can be neglected. Comparing the value of impedance $Z_{2,\,3}$ with resonant impedance across the terminals X and Y, where

$$Z_{X,\,Y} = \frac{R^2 + \omega^2 L^2}{R} = \frac{R^2 + X_L^2}{R} \qquad .....(8.67)$$

Then $$\frac{Z_{P,Y}}{Z_{X,Y}} = \frac{(X_{LB} + X_M)^2}{R^2 + X_L^2} \qquad .....(8.68)$$

Again if Q is high $\left(Q = \frac{\omega L}{R}\right)$, R must be small, or $R \ll X_L$. Then simplifying,

$$\frac{Z_{P,Y}}{Z_{X,Y}} = \frac{(X_{LB} + X_M)^2}{X_L^2} = \frac{(L_B + M)^2}{L^2} \qquad .....(8.69)$$

Considering Fig. 8.37 (b), if the circuit capacitance is split into two capacitors in series, equivalent in capacitance to the single capacitor C, and if the external generator is tapped between the two capacitors,

since $$\omega L = \frac{1}{\omega C},$$

$$Z_{i,k} = \frac{(X_{C2})^2}{R} \qquad .....(8.70)$$

and $$Z_{n,k} = \frac{(X_{C1} + X_{C2})^2}{R} \qquad .....(8.71)$$

The effect of tapping down in the capacitive side of the circuit is :

$$\frac{Z_{i,k}}{Z_{n,k}} = \frac{(X_{C2})^2}{(X_{C1} + X_{C2})^2} = \frac{C_1^2}{(C_1 + C_2)^2} \qquad .....(8.72)$$

This shows that impedance is reduced.

These methods become very convinient at high frequencies.

8.14.2 Reactance L Section for Impedance Transformation

R is the load resistance R_{in} is the net input resistance for the generator. $R < R_{in}$. R is to be matched with R_{in} through impedance transformation. So two reactances of opposite sign i.e., X_L and X_C are to be connected as shown in Fig. 8.38. L and C elements are connected in the shape of L. So it is called as L-section circuit. At a particular frequency, the oppposite reactances X_L and X_C and R can be transformed to match R_{in}.

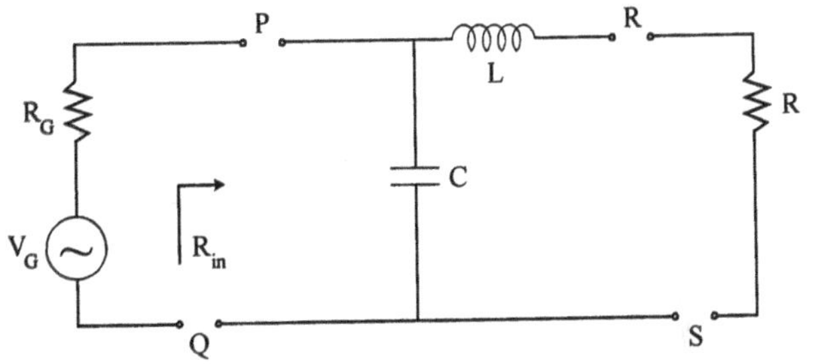

Fig. 8.38 Reactive L section for impedance transformation ($R < R_a$)

The RLC section of the circuit is a parallel resonant circuit. At anti resonance, it appears as a resistance load for the generator. The value of the resistance load depends upon L/C ratio such that R_{in} is matching with R_a for the parallel RLC resonant circuit,

$$\omega = \sqrt{\frac{1}{LC} - \frac{R^2}{L^2}} \qquad .....(8.73)$$

Resistance at anti resonance, $R_{ar} = R_{in} = \dfrac{L}{CR}$

$\therefore \qquad L = R_{in} RC$

Substituing this value of L in eqn. (8.73),

$$\omega^2 = \frac{1}{R_{in} RC^2} - \frac{1}{R_{in}^2 C^2} \qquad .....(8.74)$$

$$\omega C = \sqrt{\frac{R_{in} - R}{R_{in}^2 R}}$$

Value of capacitance C needed for the L-section is,

$$C = \frac{1}{\omega R_{in}} \sqrt{\frac{R_{in}}{R} - 1}$$

Similarly from eqn. (8.73),

$$C = \frac{L}{R_{in} R}$$

Substituting this value in eqn. (8.73) for ω^2,

$$\omega^2 = \frac{R_{in} R}{L^2} - \frac{R^2}{L^2}$$

$$\omega L = \sqrt{R_{in} R - R^2}$$

$$\therefore \quad L = \frac{R}{\omega}\sqrt{\frac{R_{in}}{R} - 1} \qquad .....(8.75)$$

So L is the value of inductance needed for the L section to ensure the desired value of load R_{in} where $R < R_{in}$.

8.14.3 IMAGE IMPEDANCES : REACTANCE MATCHING

The impedances Z_1, Z_2 and Z_3 are arranged in the form of T-Network V_a is the generator having internal impedance Z_{1i}. The load impedance is Z_{2i}. The circuit is shown in Fig. 8.39.

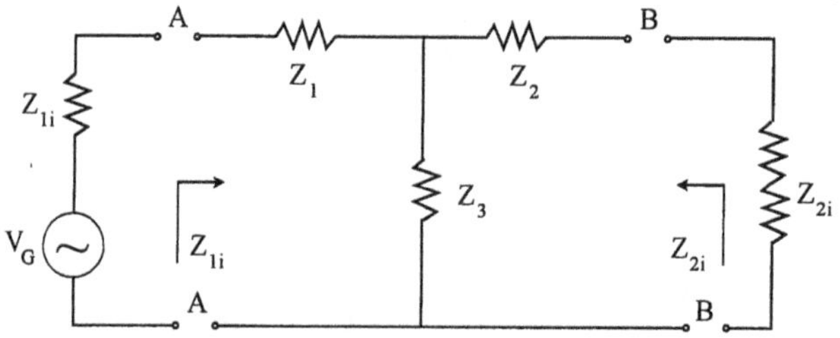

Fig. 8.39 T-Network image impedances

The generator supplies power into the terminals AA. So the impedance betweeen terminals AA must be equal to the generator impedance. The impedance looking into terminals BB must be equal to load Z_{2i}. The impedance at AA looking into one direction is the image impedance of the impedance looking in the other direction. Z_{1i} is called the impedance of the network.

Similarly at B, B terminals the impedance looking is one direction is the same as that looking in the other so that Z_{2i} is also an image impedance at B, B terminals. The network is then said to be matched on an image basis.

The value of image impedance of the T-section is computed as shown below.

The impedance $Z_{1\,in}$ between A, A terminals is,

$$Z_{1\,in} = Z_{1i} = Z_1 + \frac{Z_3(Z_2 + Z_{2i})}{Z_2 + Z_3 + Z_{2i}} \qquad .....(8.76)$$

The impedance looking into B, B terminals is required to be Z_{2i}.

$$Z_{2i} = Z_2 + \frac{Z_3(Z_1 + Z_i)}{Z_1 + Z_3 + Z_{1i}} \qquad .....(8.77)$$

Solving for Z_{1i} and Z_{2i}

$$Z_{1i} = \sqrt{\frac{(Z_1 + Z_3)(Z_1 Z_2 + Z_2 Z_3 + Z_3 Z_1)}{Z_2 + Z_3}} \quad .....(8.78)$$

$$Z_{2i} = \sqrt{\frac{(Z_2 + Z_3)(Z_1 Z_2 + Z_2 Z_3 + Z_3 Z_1)}{Z_1 + Z_3}} \quad .....(8.79)$$

If impedance is measured between terminals AA of the T-section, and terminals B, B are open circuited,

$$Z_{10C} = Z_1 + Z_3 \quad .....(8.80)$$

Similarly if impedance is measured at A, A terminals with B, B terminals short circuited,

$$Z_{1\,SC} = Z_1 + \frac{Z_2 Z_3}{Z_2 + Z_3}$$

$$= \frac{Z_1 + Z_2 + Z_2 Z_3 + Z_3 Z_1}{Z_2 + Z_3} \quad .....(8.81)$$

So the image impedance $Z_{1\,i} = \sqrt{Z_{1OC} Z_{1SC}}$

Similarly for the measurements made at B, B terminals gives

$$Z_{ZC} = \sqrt{Z_{2OC} Z_{2SC}}$$

So a porperly designed T network will have the property of transformation of an impedance to produce matching of a load and a source.

8.14.4 Reactance T Networks for Impedance Transformation

In the T-network shown in Fig. 8.40 the T Network elements are reactances only, either capacitive or inductive. V_a is a generator with internal resistance R_1. It is connected to a lod R_2 through T-Network.

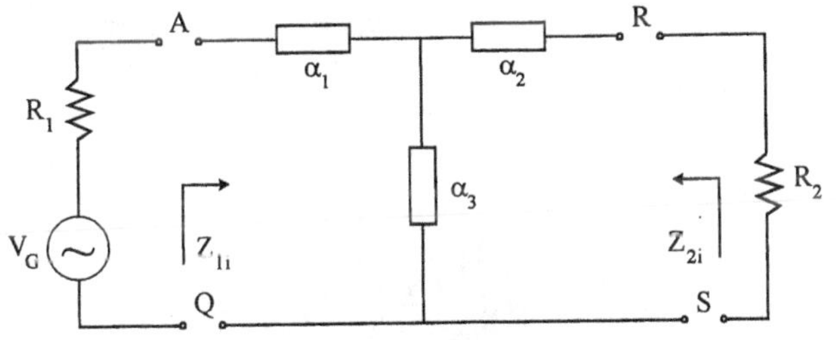

Fig. 8.40 Reactance T-network

If the generator is to transfer maximum power to the laod, it is necessary only that the image impedance Z_{1i} at terminals P,Q must be equal to R_1. In other words, the load impedance R_2 is transformed by the T-network to a value at the P,Q terminals equal to R_1.

When R_2 is connected, the image impedance Z_{1i} must be

$$Z_{1i} = R_1 = jX_1 + \frac{jX_3(R_2 + jX_2)}{R_2 + jX_2 + jX_3} \qquad .....(8.82)$$

X may have either positive sign or negative sign, i.e., it may be either inductive or capacitive in nature. Then, $R_1\ R_2 + j\ R_1\ (X_2 + X_3)$

$$= -(X_1\ X_2 + X_2\ X_3 + X_3\ X_1) + jR_2\ (X_1 + X_3)$$

By equating real terms,

$$R_1\ R_2 = -(X_1\ X_2 + X_2\ X_3\ X_3 + X_1) \qquad .....(8.83)$$

In the above equation, $R_1\ R_2$ term on the LHS is positive. So the term on RHS must also be positive. Since it is negative sign, one or more of the three terms on RHS must be positive. X can be $+jWL$ or $\frac{-j}{WC}$. In order that the product term $X_1\ X_2$ is positive, X_1 can be $+jX_a$ and X_2 can be $-jX_a$ $(-jX_b) = +X_a\ X_b$. This requires that one reactive arm of the T-Network be opposite in sign to the sign of the other two arms. So the T-Network must consist of one capacitance and two inductances or Vice-versa.

By equating imaginary terms,

$$R_1\ (X_2 + X_3) = R_2\ (X_1 + X_3)$$

$$X_2 + X_3 = \frac{R_2}{R1}\ (X_1 + X_3) \qquad .....(8.84)$$

The above equation may be written as,

$$R_1\ R_2 = -\ [(X_1 + X_3)\ (X_2 + X_3) - X_3^{\,2}] \qquad .....(8.85)$$

$$\therefore \qquad R_1\ R_2 = -\left[(X_1 + X_3)^2 \frac{R_2}{R_1} - X_3^2\right]$$

$$X_1 + X_3 = \pm\sqrt{\frac{R_1}{R_2}\left(X_3^2 - R_1 R_2\right)} \qquad .....(8.86)$$

So the value of one of the reactance arms is,

$$X_1 = -X_3 \pm \sqrt{\frac{R_1}{R_2}\left(X_3^2 - R_1 R_2\right)} \qquad .....(8.87)$$

Substituting eqn. (8.87) in (8.85),

$$X_2 + X_3 = \pm \frac{R_2}{R_1}\sqrt{\frac{R_1}{R_2}\left(X_3^2 - R_1 R_2\right)}$$

or

$$X_2 + X_3 = \pm \sqrt{\frac{R_1}{R_2}\left(X_3^2 - R_1 R_2\right)} \qquad .....(8.88)$$

The above equations are the design equations for the T-Network, in terms of the values of X_3.

SUMMARY

- Tuned amplifiers are also classified as (i) small signal and (ii) large signal amplifiers. Small signal tuned amplifieres are classified as (a) single tuned (b) Double tuned and (c) Staggered tuned amplifiers.
- Large signal tuned amplifiers are classified similar to power amplifiers.
- The circuits, their equivalent circuits and expressions for resonant frequency and band widht are derived.
- Compensation circuits, extension of low frequency range and high frequency range are explained.
- Tuned amplifiers are extensively used in receivers, audio, T.V., radio and communication receivers.

OBJECTIVE TYPE QUESTIONS

1. Radio Frequency Range is ________________.
2. VHF Frequency Range is ________________.
3. Parallel Tuned Circuit is also known as ________________.
4. Generally Tuned Amplifiers are used in ________________.
5. In Tuned Amplifiers, Harmonic Distortion is ________________.
6. Small signal Tuned Amplifiers are operated in ________________ mode.
7. Large signal Tuned Amplifiers are operated in ________________ modes.
8. Small signal Tuned Amplifiers are classified as
   (a) ________________ (b) ________________ (c) ________________
9. Single Tuned Amplifiers are classified as ________________.
10. In Tuned Amplifiers equivalent circuits, the model used for Transistors (BJT) is ________________.
11. Expression for $\frac{A}{A_{reso}}$ for tapped single tuned capacitance coupled amplifier with usual notation is ________________.
12. Double Tuned Amplifier provides ________________ Bandwidth than single tuned amplifier.
13. Common Emitter (C.E) Double Tuned amplifier provides Gain - Frequency curve with ________________ sides ________________ top.
14. In Radars and Television ________________ type of amplifiers are used.
15. The purpose of resonant circuits in tuned circuits is ________________.
16. Give the classification of tuned amplifiers.
17. What are the applications of tuned amplifiers ?
18. Give the frequency response of a tuned amplifier.
19. What is the expression for Rp in terms of ω, L, R ? What is Rp ?
20. What is δ called ? What is the expression in terms of ω, ω_o ?
21. What is A/Ao ?
22. Why do we go for tapped single tuned amplifier ?
23. What is maxium power transfer theorem ?
24. What is a double tuned amplifier ?
25. What is staggered tuning ?
26. At what frequencies are the tuned amplifiers operate ?
27. What is a tank circuit ?
28. What is the expression for harmonic distortion in tuned amplifiers ?
29. How do you build a class B tuned amplifier ?

ESSAY TYPE QUESTIONS

1. (a) What are the different types of Tuned Amplifiers ?
   (b) How are Tuned amplifiers classified ?
   (c) What are the applications of Tuned amplifiers ?
2. Draw the circuit for single tuned capacitive coupled amplifier and explain its working. Draw its equivalent circuit and derive the expression for (A/A_{reso})
3. Draw the circuit for tapped single tuned capacitance coupled amplifier and explain its working. Draw the frequency response. Derive the expression for L for maximum power transfer.
4. Draw the circuit for single tuned inductively coupled amplifier. Draw its equivalent circuit and derive the expression for $\left(\frac{A}{A_{reso}}\right)$.
5. Draw the circuit for Double Tuned Amplifier. Explain its working. What are the advantages of this amplifier ? Derive the expression for I_2 Max.
6. Draw the circuit for BJT tuned class B/C amplifier. Explain its working.
7. Draw the circuit for JFET tuned R.F. amplifier and explain its working.
8. Explain the principle and working of wide band amplifiers.

ANSWERS TO OBJECTIVE TYPE QUESTIONS

1. > 20 KHzs
2. 30 MHzs - 300 MHzs
3. Anti Resonant Circuit
4. R. F amplifiers, communication Receivers
5. Less
6. Class A
7. Class B, class C or class AB
8. (a) Single Tunded (b) Double Tuned (c) Staggered Tuned
9. (a) Capacitive coupled (b) Transformer coupled or Inductive coupled
10. Hybrid - π equivalent circuit
11. $\dfrac{A}{A_{reso}} = \dfrac{1}{1+j2\delta Q_e}$ where $\delta = \left(\dfrac{\omega-\omega_0}{\omega_0}\right)$ $Q_e = \omega_o C R_t = \dfrac{R_t}{\omega_0 L}$.
12. Larger
13. Steaper sides and flatter top
14. Tuned amplifiers
15. (a) To provide properly matching load impedance
    (b) To reset unwanted harmonics
    (c) To couple power to load.
16. Single tuned, double tuned, stagger tuned amplifiers.
17. RF amplifiers, communication receivers.
18.

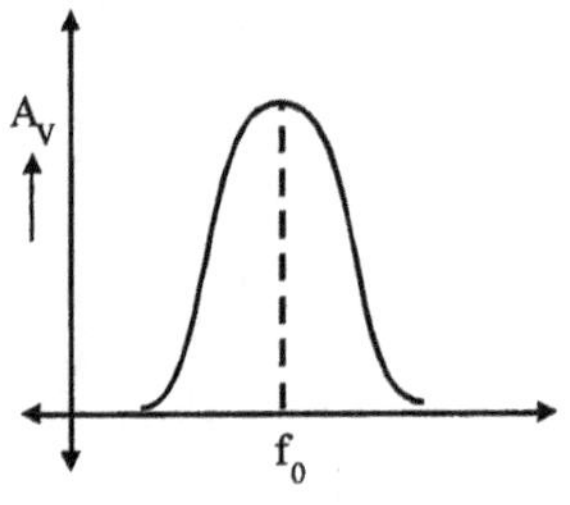

19. $R_p = \frac{\omega^2 L^2}{R}$, R_p is the series internal resistance of inductor represented as a shunt element.

20. δ - fractional frequency variation. $\delta = \frac{\omega - \omega_o}{\omega_o}$

21. $\frac{A}{A_o} = \frac{1}{1 + j2\delta Q_e}$

22 For impedance matching.

23. For a linear circuit, for the maximum power to he delivered to the load the output or load impedance should be the complex conjugate of the effective or equivalent input impedance of the circuit.

24. It has two resonant circuits both tuned to the same frequency.

25. Two resonant circuits tuned to different frequencies.

26. High or radio frequencies.

27. Parallel LC circuit is called a tank circuit.

28. There is very negligible harmonic distortion in tuned amplifiers as they are narrow band amplifiers.

29. By removing the bias resistors from class A amplifier, we can build a class B amplifier.

Appendices

APPENDIX - 1

Colour Codes for Electronic Components

RESISTOR COLOUR CODE :

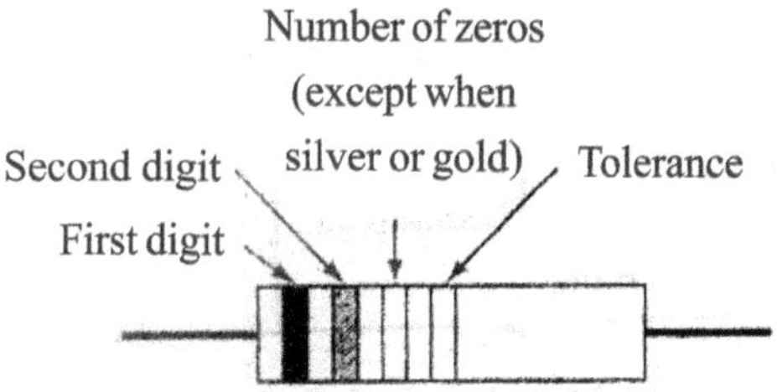

First Three Bands				Fourth Band
Black	– 0	Blue	– 6	Gold + 5%
Brown	– 1	Violet	– 7	Silver + 10%
Red	– 2	Grey	– 8	None + 20%
Orange	– 3	White	– 9	
Yellow	– 4	Silver	0.01	
Green	– 5	Gold	0.1	

CAPACITOR COLOUR CODE :

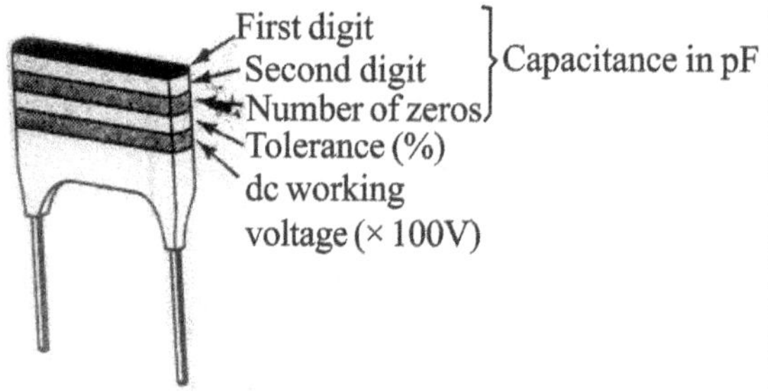

Colour	Figure Significant	Tolerance (%)
Black	0	20
Brown	1	1
Red	2	2
Orange	3	3
Yellow	4	4
Green	5	5
Blue	6	6
Violet	7	7
Grey	8	8
White	9	9
Silver	0.01	10
Gold	0.1	5
No Band		20

INDUCTOR COLOUR CODE :

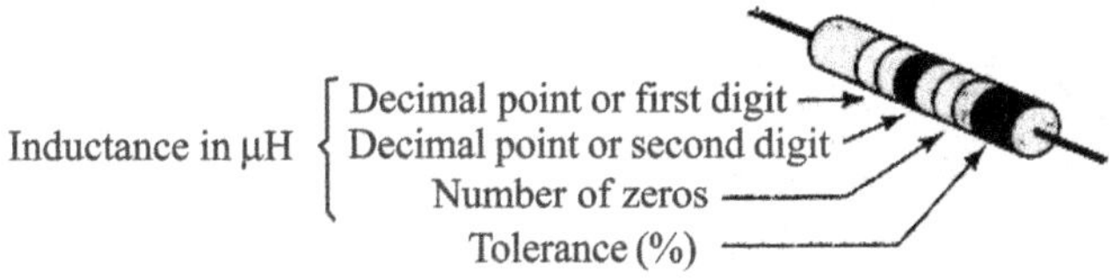

Color	Significant Figure	Tolerance (%)
Black	0	
Brown	1	
Red	2	
Orange	3	
Yellow	4	
Green	5	
Blue	6	
Violet	7	
Grey	8	
White	9	
Silver		10
Gold	Decimal point	5
No Band		20

COLOUR CODE MEMORY AID : W_G VIBGYOR BB (W_G Vibgyor BB)

Memory aid	Color	Number
Black	Black	0
Bruins	Brown	1
Relish	Red	2
Ornery	Orange	3
Young	Yellow	4
Greenhorns	Green	5
Blue	Blue	6
Violets	Violet	7
Growing	Grey	8
Wild	White	9
Smell	Silver	0.01 10%
Good	Gold	0.1 5%

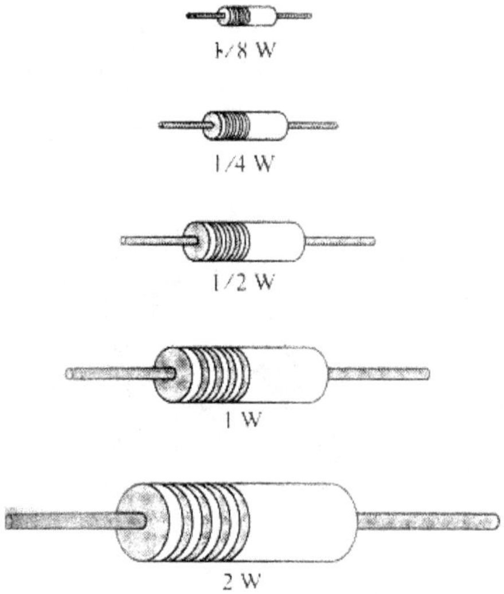

Fig. A-1.1 Relative size of carbon composition resistors with various power ratings

Specifications of Power Transistors

Device	Type	P_D(W)	I_C(A)	V_{CEO}(V)	V_{CBO}(V)	h_{FE} MIN	Max	f_T M.HZs
2N 6688	NPN	200	20	200	300	20	80	20
2N 3442	NPN	117	10	140	160	20	70	0.08
BUX 39	NPN	120	30	90	120	15	45	8
ECP 149	PNP	30	4	40	50	30	–	2.5

Darlington Pair

2N 6052	PNP	150	12	100	100	750	–	4
2N 6059	NPN	150	12	100	100	750	–	4

APPENDIX - 2

Resistor and Capacitor Values

Typical Standard Resistor Values (+ 10% Tolerance)							
Ω	Ω	Ω	kΩ	kΩ	kΩ	MΩ	MΩ
–	10	100	1	10	100	1	10
–	12	120	1.2	12	120	1.2	–
–	15	150	1.5	15	150	1.5	15
–	18	180	1.8	18	180	1.8	–
–	22	220	2.2	22	220	2.2	22
2.7	27	270	2.7	27	270	2.7	–
3.3	33	330	3.3	33	330	3.3	–
3.9	39	390	3.9	39	390	3.9	–
4.7	47	470	4.7	47	470	4.7	–
5.6	56	560	5.6	56	560	5.6	–
6.8	68	680	6.8	68	680	6.8	–
–	82	820	8.2	82	820	–	–

Typical Standard Resistor Values (+ 10% Tolerance)										
pF	pF	pF	pF	µF	µF	µF	µF	µF	µF	µF
5	50	500	5000		0.05	0.5	5	50	50	5000
–	51	510	5100		–	–	–	–	–	–
–	56	560	5600		0.056	0.56	5.6	56	–	5600
–	–	–	6000		0.06	–	6	–	–	6000
–	62	620	6200		–	–	–	–	–	–
–	68	680	6800		0.068	0.68	6.8	–	–	–
–	75	750	7500		–	–	–	75	–	–
–	–	–	8000		–	–	8	80	–	–
–	82	820	8200		0.082	0.82	8.2	82	–	–
–	91	910	9100		–	–	–	–	–	–
10	100	1000		0.01	0.1	1	10	100	1000	10,000
–	110	1100		–	–	–	–	–	–	
12	120	1200		0.012	0.12	1.2	–	–	–	
–	130	1300		–	–	–	–	–	–	
15	150	1500		0.015	0.15	1.5	15	150	1500	
–	160	1600		–	–	–	–	–	–	
18	180	1800		0.018	0.18	1.8	18	180	–	
20	200	2000		0.02	0.2	2	20	200	2000	
24	240	2400		–	–	–	–	240	–	
–	250	2500		–	0.25	–	25	250	2500	
27	270	2700		0.027	0.27	2.7	27	270	–	
30	300	3000		0.03	0.3	3	30	300	3000	
33	330	3300		0.033	0.33	3.3	33	330	3300	
36	360	3600		–	–	–	–	–	–	
39	390	3900		0.039	0.39	3.9	39	–	–	
–	–	4000		0.04	–	4	–	400	–	
43	430	4300		–	–	–	–	–	–	
47	470	4700		0.047	0.47	4.7	47	–	–	

Physical constants

Charge of an electron	: e :	1.60×10^{-19} coulombs
Mass of an electron	: m :	9.09×10^{-31} Kg
e/m ratio of an electron	: e/m :	1.759×10^{11} C/Kg
Plank's constant	: h :	6.626×10^{-34} J-sec
Boltzman's constant	: $\overline{K}$:	1.381×10^{-23} J/°K
	: K :	8.62×10^{-5} ev/°K
Avogadro's number	: N_A :	6.023×10^{23} molecules/mole
Velocity of light	: c :	3×10^{8} m/sec
Permeability of free space	: m_o :	1.257×10^{-6} H/m
Permittivity of free space	: $\hat{I}_o$:	8.85×10^{-12} F/m
Intrinsic concentration in silicon at 300 °K	: n_i	$= 1.5 \times 10^{10}$ /cm^3
Intrinsic resistivity in silicon at 300 °K	: r_i	= 230,000 W–cm
Mobility of electronics in silicon	: m_n	= 1300 cm^2/ V–sec
Mobility of holes in silicon	: m_p	= 500 cm^2/ V–sec
Energy gap at in silicon at 300 °K	:	= 1.1 ev.

APPENDIX - 3

Capacitors

Capacitance

The farad (F) *is the SI unit of capacitance.*

The farad is the capacitance of a capacitor that contains a charge of 1 coulomb when the potential difference between its terminals is 1 volt.

Leakage Current

Despite the fact that the dielectric is an insulator, small leakage currents flow between the plates of a capacitor. The actual level of leakage current depends on the insulation resistance of the dielectric. Plastic film capacitors, for example, may have insulation resistances higher than 100,000 MΩ. At the other extreme, an electrolytic capacitor may have a *microampere* (or more) of leakage current, with only 10 V applied to its terminals.

Polarization

Electrolytic capacitors normally have one terminal identified as the most positive connection. Thus, they are said to be polarized. This usually limits their application to situations where the polarity of the applied voltage will not change. This is further discussed for electrolytic capacitors.

Capacitor Equivalent Circuit

An ideal capacitor has a dielectric that has an infinite resistance and plates that have zero resistance. However, an ideal capacitor does not exist, as all dielectrics have some leakage current and all capacitor plates have some resistance. The complete equivalent circuit for a capacitor [shown in Fig. A-3.1(a)] consists of an ideal capacitor C in series with a resistance R_D representing the resistance of the plates, and in parallel with a resistance R_L representing the leakage resistance of the dielectric. Usually, the plate resistance can be completely neglected, and the equivalent circuit becomes that shown in Fig. A-3.1(b). With capacitors that have a very high leakage resistance (e.g., mica and plastic film capacitors), the parallel resistor is frequently omitted in the equivalent circuit, and the capacitor is then treated as an ideal capacitor. This cannot normally be done for electrolytic capacitors, for example, which have relatively low leakage resistances. The parallel R_C circuit in,

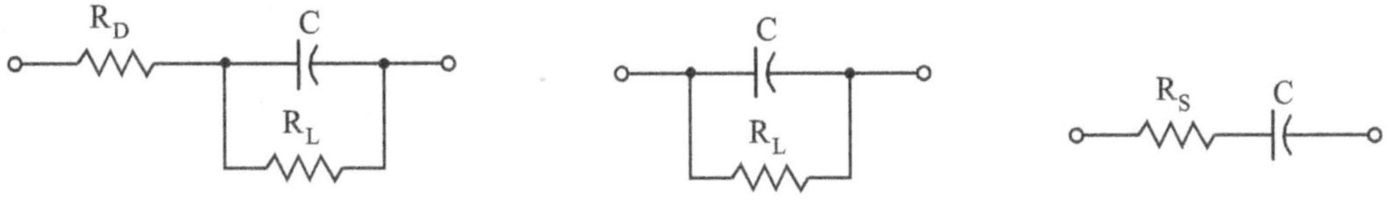

(a) Complete equivalent circuit *(b) Parallel equivalent circuit* *(c) Series equivalent circuit*

Fig. A. 3.1

A capacitor equivalent circuit consists of the capacitance C, the leakage resistance R_L *in parallel with* C, *and the plate resistance* R_D *in series with C and* R_L.

Fig. A. 3.1 (b) can be shown to have an equivalent series *RC* circuit, as in Fig. A. 3.1(c). This is treated in Section 20-6.

A variable air capacitor is made up of a set of movable plates and a set of fixed plates separated by air.

Because a capacitor's dielectric is largely responsible for determining its most important characteristics, capacitors are usually identified by the type of dielectric used.

Air Capacitors

A typical capacitor using air as a dielectric is illustrated in Fig. A.3.2. The capacitance is variable, as is the case with virtually all air capacitors. There are two sets of metal plates, one set fixed and one movable. The movable plates can be adjusted into or out of the spaces between the fixed plates by means of the rotatable shaft. Thus, the area of the plates opposite each other is increased or decreased, and the capacitance value if altered.

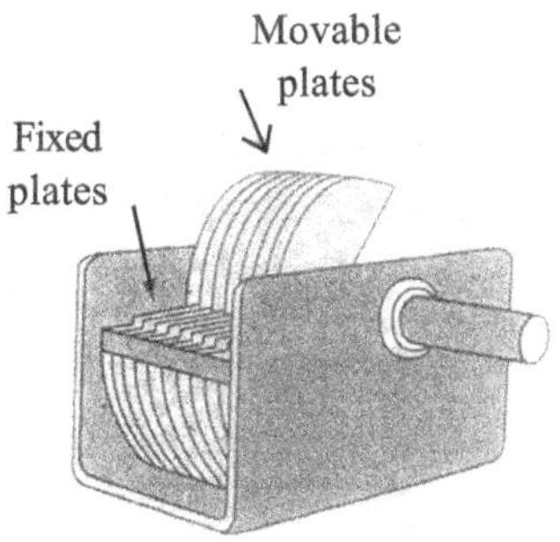

Fig. A. 3.2 A variable air capacitor is made up of a set of movable plates and a set of fixed plates separated by air

Paper Capacitors

In its simplest form, a paper capacitor consists of a layer of paper between two layers of metal foil. The metal foil and paper are rolled up, as illustrated in Fig. A.3.3 (a); external connections are brought out from the foil layers, and the complete assembly is dipped in wax or plastic. A variation of this is the metalized paper construction, in which the foil is replaced by thin films of metal deposited on the surface of the paper. One end of the capacitor sometimes has a band around it [see Fig. A.3.3 (b)]. This does not mean that the device is polarized but simply identifies the terminal that connects to the outside metal film, so that it can be grounded to avoid pickup of unwanted signals.

Paper capacitors are available in values ranging from about 500 pF to 50μF, and in dc working voltages up to about 600 V. They are among the lower-cost capacitors for a given capacitance value but are physically larger than several other types having the same capacitance value.

Plastic Film Capacitors

The construction of plastic film capacitors is similar to that of paper capacitors, except that the paper is replaced by a thin film that is typically polystyrene or Mylar. This type of dielectric gives insulation resistances greater than 100 000 MΩ. Working voltages are as high as 600 V, with the capacitor surviving 1500 V surges for a brief period. Capacitance tolerances of + 2.5% are typical, as are temperature coefficients of 60 to 150 ppm/°C.

Plastic film capacitors are physically smaller but more expensive than paper capacitors. They are typically available in values ranging from 5 pF to 0.47 μF.

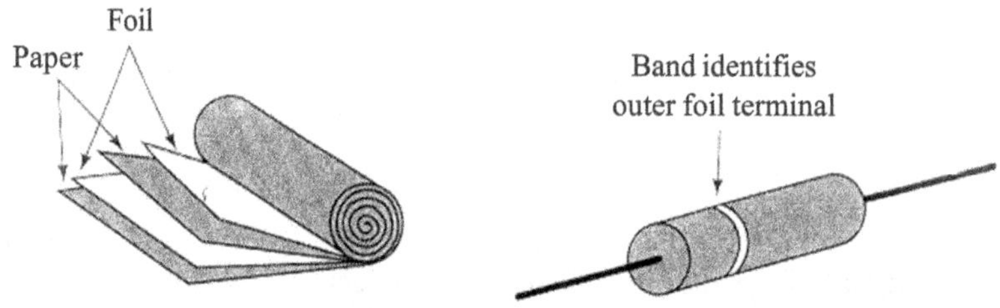

(a) Construction of a paper capacitor *(b) Appearance of a paper capacitor*

Fig. A. 3.3 In a paper capacitor, two sheets of metal foil separated by a sheet of paper are rolled up together. External connections are made to the foil sheets.

Mica Capacitors

As illustrated in Fig. A. 3.4(a), mica capacitors consist of layers of mica alternated with layers of metal foil. Connections are made to the metal foil for capacitor leads, and the entire assembly is dipped in plastic or encapsulated in a molded plastic jacket. Typical capacitance values range from 1pF to 0.1μF, and voltage ratings as high as 35 000 V are possible. Precise capacitance values and wide operating temperatures are obtainable with mica capacitors. In a variation of the process, silvered mica capacitors use films of silver deposited on the mica layers instead of metal foil.

Ceramic Capacitors

The construction of a typical ceramic capacitor is illustrated in Fig. A. 3.4(b). Films of metal are deposited on each side of a thin ceramic disc, and copper wire terminals are connected to the metal. The entire units is then encapsulated in a protective coating of plastic. Two different types of ceramic are used, one of which has extremely high relative permitivity. This gives capacitors that are much smaller than paper or mica capacitors having the same capacitance value. One disadvantage of this particular ceramic dielectric is that its leakage resistance is not as high as with other types. Another type of ceramic gives leakage resistances on the order of 7500 MΩ. Because of its lower permitivity, this ceramic produces capacitors that are relatively large for a given value of capacitance.

The range of capacitance values available with ceramic capacitors is typically 1 pF to 0.1 μF, with dc working voltages up to 1000 V.

(a) Construction of mica capacitor

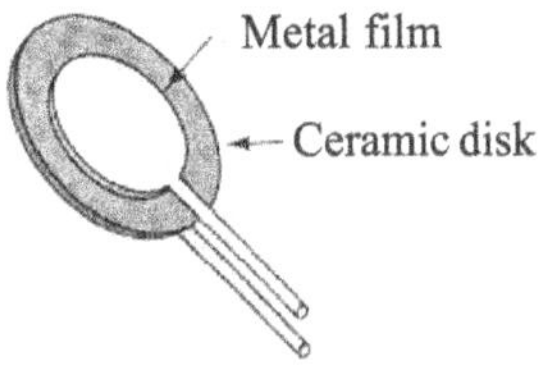

(b) Ceramic capacitor

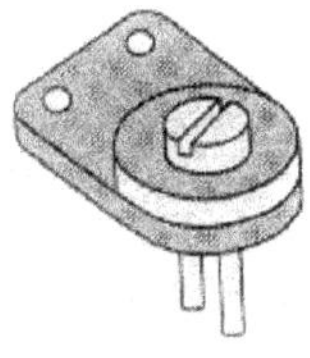

(c) Ceramic trimmer

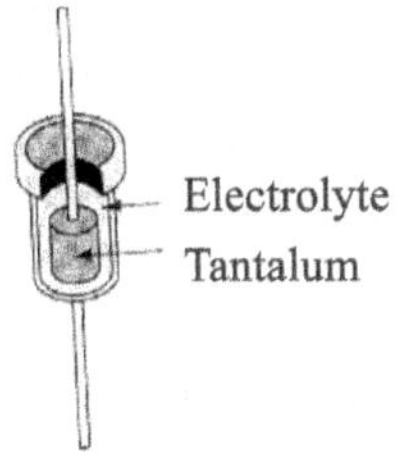

(d) Construction of tantalum capacitor

Fig. A. 3.4 Mica capacitors consist of sheets of mica interleaved with foil. A ceramic disc silvered on each side makes a ceramic capacitor; in a ceramic trimmer, the plates area is screwdriver adjustable. A tantalum capacitor has a relatively large capacitance in a small volume.

Fig. A. 3.4(c) shows a variable ceramic capacitor known as a *trimmer*. By means of a screwdriver, the area of plate on each side of a dielectric can be adjusted to alter the capacitance value. Typical ranges of adjustment available are 1.5 pF to 3 pF and 7 pF to 45 pF.

Electrolytic Capacitors

The most important feature of electrolytic capacitors is that they can have a very large capacitance in a physically small container. For example, a capacitance of 5000 μF can be obtained in a cylindrical package approximately 5 cm long by 2 cm in diameter. In this case the dc working voltage is only voltage is only 10V. Similarly, a 1 F capacitor is available in a 22 cm by 7.5 cm cylinder, with a working voltage of only 3 V. Typical values for electrolytic capacitors range from 1 μF through 100 000 μF.

The construction of an electrolytic capacitor is similar to that of a paper capacitor (Fig. A.3.5(a)). Two sheets of aluminium foil separated by a fine gauze soaked in electrolyte are rolled up and encased in an aluminium cylinder for protection. When assembled, a direct voltage is applied to the capacitor terminals, and this causes a thin layer of aluminium oxide to form on the surface of the positive plate next to the electrolyte (Fig. A.3.5(b)). The aluminium oxide is the dielectric, and the electrolyte and positive sheet of foil are the capacitor plates. The extremely thin oxide dielectric gives the very large value of capacitance.

It is very important that electrolytic capacitors be connected with the correct polarity. When incorrectly connected, gas forms within the electrolyte and the capacitor may **explode**! Such an explosion blows the capacitor apart and spreads its contents around. This could have **tragic consequences** for the eyes of an experimenter who happens to be closely examining the circuit when the explosion occurs. The terminal designated as positive must be connected to the most positive of

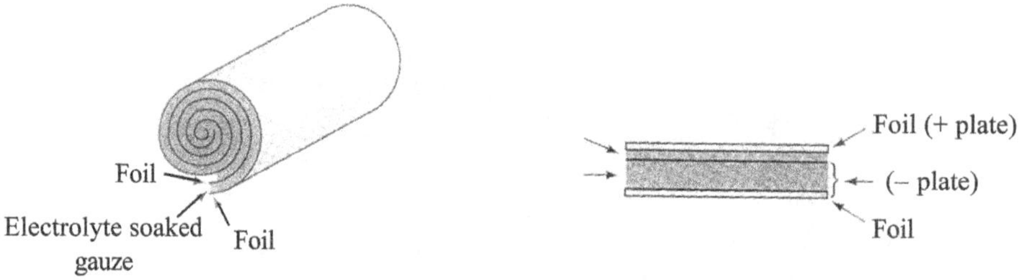

(a) Rolled-up foil sheets and electrolyte-soaked gauze

(b) The dielectric is a thin layer of aluminium oxide

Fig. A. 3.5 An electrolyte capacitor is constructed of rolled-up foil sheets separated by electrolyte-soaked gauze, the dielectric is a layer of aluminium oxide at the positive plate

the two points in the circuit where the capacitor is to be installed. Fig. A. 3.6 illustrates some circuit situations where the capacitor must be correctly connected. Nonpolarized electrolytic capacitors can be obtained. They consist essentially of two capacitors in one package connected *back to back*, so one of the oxide films is always correctly biased.

Electrolytic capacitors are available with dc working voltages greater than 400 V, but in this case capacitance values do not exceed 100 mF. In addition to their low working voltage and polarized operation. Another disadvantage of electrolytic capacitors is their relatively high leakage current.

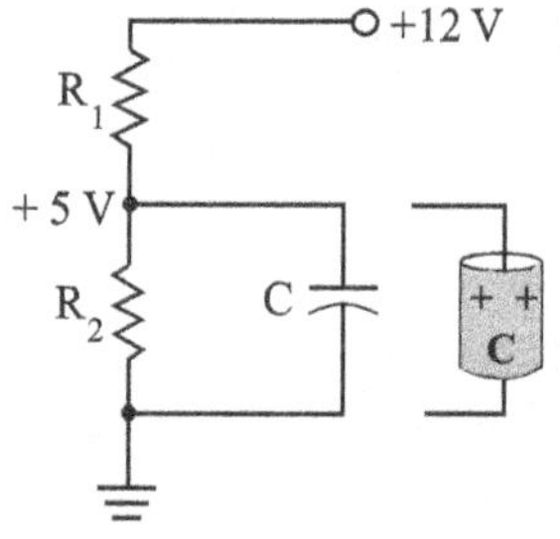

(a) Capacitor connected between +5 V and ground

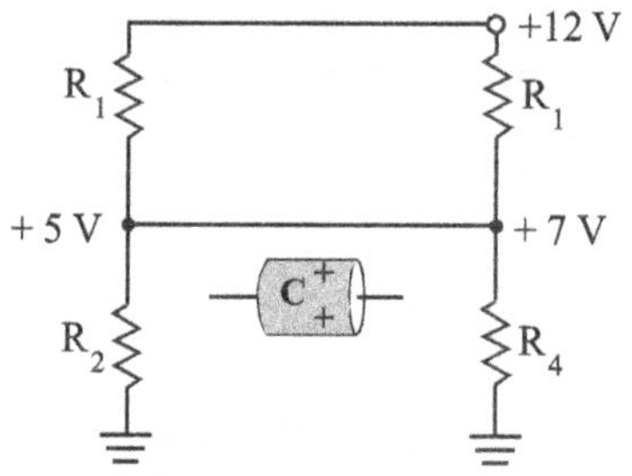

(b) Connected between +7 V and +5 V

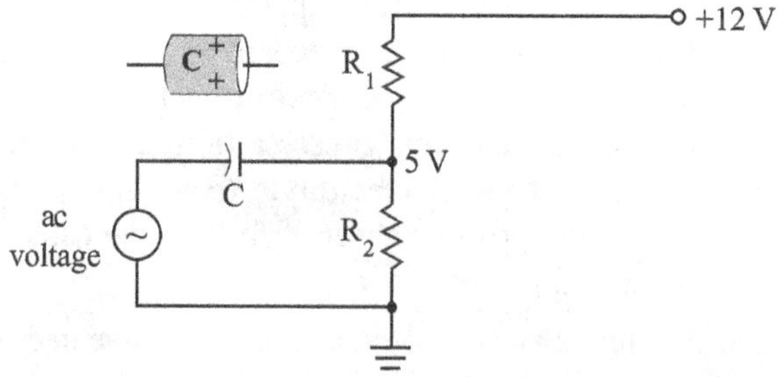

(c) Connected between + 5.7 V and a grounded ac voltage source

Fig. A. 3.6 It is very important that polarized capacitors be correctly connected. The capacitor positive terminal voltage must be more positive than the voltage at the negative terminal.

Tantalum Capacitors

This is another type of electrolytic capacitor. Powdered tantalum is sintered (or baked), typically into a cylindrical shape. The resulting solid is quite porous, so that when immersed in a container of electrolyte, the electrolyte is absorbed into the tantalum. The tantalum then has a large surface area in contact with the electrolyte (Fig. A. 3.5). When a dc *forming voltage* is applied, a thin oxide film is formed throughout the electrolyte-tantalum contact area. The result, again, is a large capacitance value in a small volume.

Capacitor Color Codes

Physically large capacitors usually have their capacitance value, tolerance and dc working voltage printed on the side of the case. Small capacitors (like small resistors) use a code of colored bands (or sometimes colored dots) to indicate the component parameters.

There are several capacitor color codes in current use. Here is one of the most common.

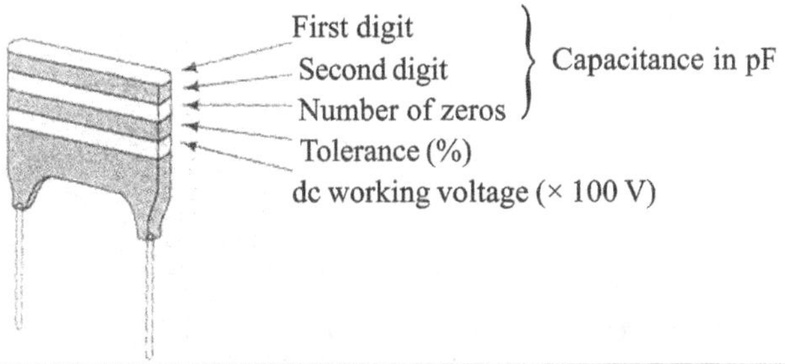

Color	Significant Figure	Tolerance (%)
Black	0	20
Brown	1	1
Red	2	2
Orange	3	3
Yellow	4	4
Green	5	5
Blue	6	6
Violet	7	7
Grey	8	8
White	9	9
Silver	0.01	10
Gold	0.1	5
No band		20

A typical tantalum capacitor in a cylindrical shape 2 cm by 1 cm might have a capacitance of 100 mF and a dc working voltage of 20 V. Other types are available with a working voltage up to 630 V, but with capacitance values on the order of 3.5 mF. Like aluminium-foil electrolytic capacitors, tantalum capacitors must be connected with the correct polarity size of the inductor, the maximum current can be anything from about 50 mA to 1 A. The core in such an inductor may be made adjustable so that it can be screwed into or partially out of the coil. Thus, the coil inductance is variable. Note the graphic symbol for an inductor with an adjustable core [Fig. A. 3.6(b)].

APPENDIX - 4

Inductors

Magnetic Flux and Flux Density

*The weber** (Wb) *is the SI unit of magnetic flux.*

The weber is defined as the magnetic flux which, linking a single-turn coil, produces an emf of 1 V when the flux is reduced to zero at a constant rate in 1 s.

*The tesla**** (T) *is the SI unit of magnetic flux density.*

The tesla is the flux density in a magnetic field when 1 Wb of flux occurs in a plane of 1 m^2*; that is, the tesla can be described as 1* Wb/m^2.

Inductance

The SI unit of inductance is the henry (H).

The inductance of a circuit is 1 henry (1 H) when an emf of 1 V is induced by the current changing at the rate of 1 A/s.

Molded Inductors

A small molded inductor is shown in Fig. A. 4.2(c). Typical available values for this type range from 1.2 μH to 10 mH, maximum currents of about 70 mA. The values of molded inductors are identified by a color code, similar to molded resistors. Fig. A. 4.2(d) shows a tiny-film inductor used in certain types of electronic circuits. In this case the inductor is simply a thin metal film deposited in the form of a spiral on ceramic base.

Laboratory Inductors

Laboratory-type variable inductors can be constructed in decade box format, in which precision inductors are switched into or out of a circuit by means of rotary switches. Alternatively, two coupled coils can be employed as a variable inductor. The coils may be connected in series or in parallel, and the total inductance is controlled by adjusting the position of one coil relative to the other.

Color Code For Small Inductors

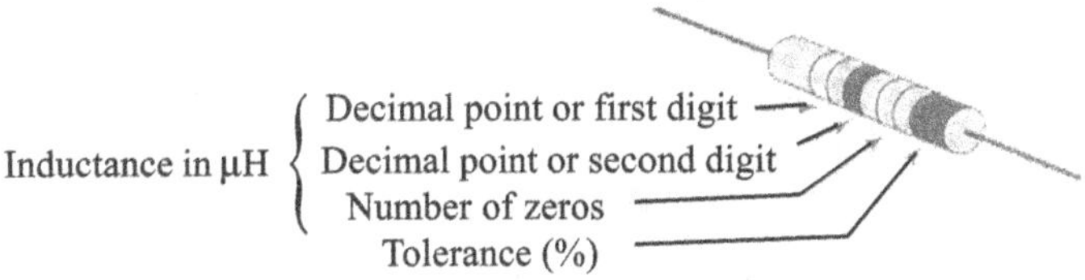

Color	Significant Figure	Tolerance (%)
Black	0	
Brown	1	
Red	2	
Orange	3	
Yellow	4	
Green	5	
Blue	6	
Violet	7	
Grey	8	
White	9	
Silver		10
Gold	decimal point	5
No band		20

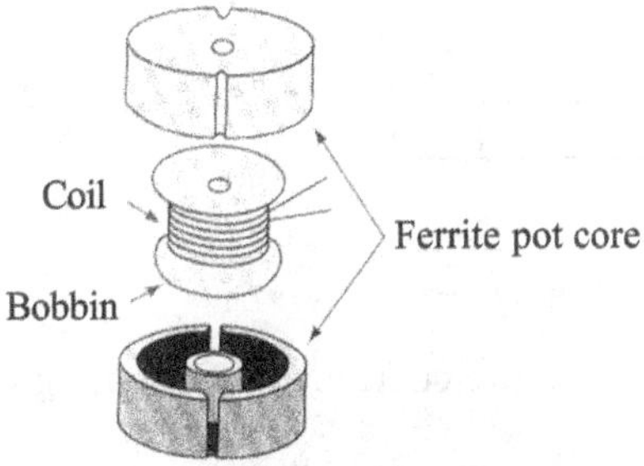

***Fig. A. 4.1** Some low-current, high-frequency inductors are wound on bobbins contained in a ferrite pot core. The ferrite core increases the winding inductance and screens the inductor*

Coil to protect adjacent components against flux leakage and to protect the coil from external magnetic fields. The coil is wound on a bobbin, so its number of turns is easily modified.

Three different types of low-current inductors are illustrated in Fig. A. 4.2. Fig. A 4.2(a) shows a type that is available either as an air-cored inductor or with a ferromagnetic core. With an air core, the inductance values up to about 10 mH can be obtained. Depending on the thickness of wire used and the physical.

(a) Inductor with air core or ferromagnetic core

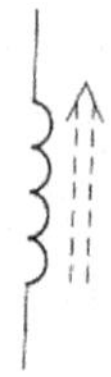

(b) Circuit symbol for an inductor with an adjustable ferromagnetic core

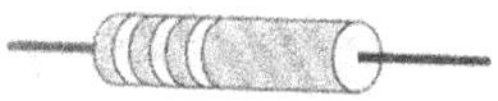

(c) Molded inductor

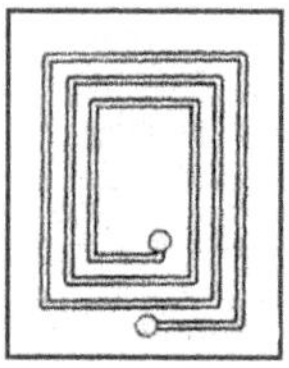

(d) Thin-film inductor

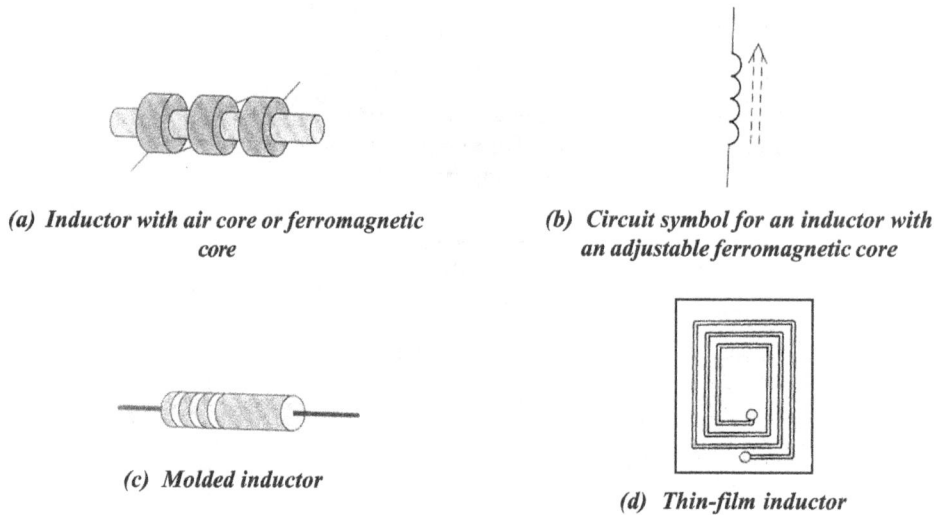

Fig. A. 4.2 Small inductors may be wound on an insulating tube with an adjustable ferrite core, molded like small resistors, or deposited as a conducting film on an insulating material

If the mutual inductance between two adjacent coils is not known, it can be determined by measuring the total inductance of the coils in series-aiding and series-opposing connections. Then,

$$L_a = L_1 + L_2 + 2M \quad \text{for series-aiding}$$

and

$$L_b = L_1 + L_2 - 2M \quad \text{for series-opposing}$$

Subtracting,

$$L_a - L_b = 4M$$

Therefore,

$$\boxed{M = \frac{L_a - L_b}{4}}$$

$$M = k\sqrt{L_1 L_2}$$

From these two equations, the coefficient of coupling of the two coils can be determined.

Stray Inductance

Inductance is (change in flux linkages) / (change in current). So every current-carrying conductor has some self-inductance, and every pair of conductors has inductance. These *stray inductance* are usually unwanted, although they are sometimes used as components in a circuit design. In dc applications, stray inductance is normally unimportant, but in radio frequency ac circuits it can be considerable nuisance. Stray inductance is normally minimized by keeping connecting wires as short as possible.

Summary of Formulae

Induced emf : $e_L = \dfrac{\Delta\Phi}{\Delta t}$

Induced emf : $e_L = \dfrac{\Delta\Phi}{\Delta t}$

Inductance : $L = \dfrac{e_L}{\Delta i / \Delta t}$

Inductance : $L = \dfrac{\Delta\Phi N}{\Delta i}$

Flux change : $\Delta\Phi = \mu, \mu_0 \, \Delta i \, N \, \dfrac{A}{t}$

Self - inductance : $L = \mu, \mu_0 \, \Delta i \, N^2 \, \dfrac{A}{t}$

Mutual inductance : $M = \dfrac{e_L}{\Delta i / \Delta t}$

Induced emf : $e_L = \dfrac{\Delta\Phi N_s}{\Delta t}$

Mutual inductance : $M = \dfrac{\Delta\Phi N_s}{\Delta i}$

Mutual inductance : $M = k \, \dfrac{\Delta\Phi N_s}{\Delta i}$

Mutual inductance : $M = k\sqrt{L_1 L_2}$

Energy stored : $W = \dfrac{1}{2} LI^2$

Energy stored : $W = \dfrac{B^2 Al}{2\mu_0}$

Inductances in series : $L_s = L_1 + L_2 + L_3 + \ldots..$

Inductances in parallel : $\dfrac{1}{L_P} = \dfrac{1}{L_1} + \dfrac{1}{L_2} + \dfrac{1}{L_3} + \ldots..$

Total inductance (series-aiding) : $L = L_1 + L_2 + 2M$

Total inductance (series-opposing) : $L = L_1 + L_2 - 2M$

Mutual inductance : $M = \dfrac{L_0 - L_b}{4}$

APPENDIX - 5

Miscellaneous

Ionic Bonding

In some insulating materials, notably rubber and plastics, the bending process is also covalent. The valence electrons in these bonds are very strongly attached to their atoms, so the possibility of current flow is virtually zero. In other types of insulating materials, some atoms have parted with outer-shell electrons, but these have been accepted into the orbit of other atoms. Thus, the atoms are *ionized*; those which gave up electrons have become *positive ions*, and those which accepted the electrons become negative ions. This creates an electrostatic bonding force between the atoms, termed ionic bonding. Ionic bonding is found in such materials as glass and porcelain. Because there are virtually no free electrons, no current can flow, and the material is an insulator.

Insulators

Fig. A. 5.1 shows some typical arrangements of conductors and insulators. Electric cable usually consists of conducting copper wire surrounded by an insulating sheath of rubber or plastic. Sometimes there is more than one conductor, and these are, of course, individually insulated.

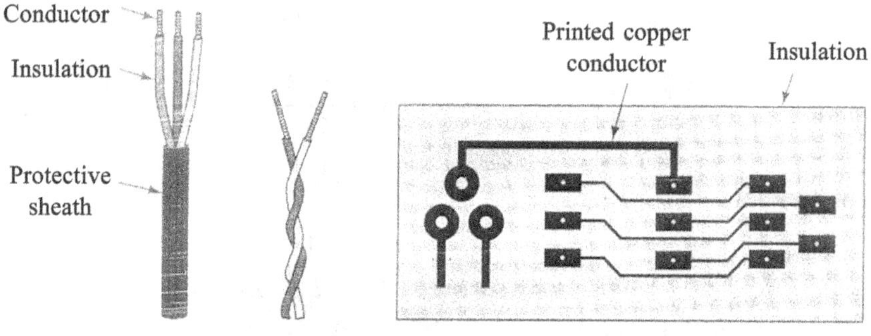

Fig. A. 5.1 Conductors employed for industrial and domestic purposes normally have stranded copper wires with rubber or plastic insulation. In electronics equipment, flat cables of fine wires and thin printed circuit conductors are widely used.

Conductors

The function of a conductor is to conduct current form one point to another in an electric circuit. As discussed, electric cables usually consist of copper conductors sheathed with rubber or plastic

Screened
coaxial
cable   Conductor   Screen

Circular
multiconductor
cable

Flat
multiconductor
cable

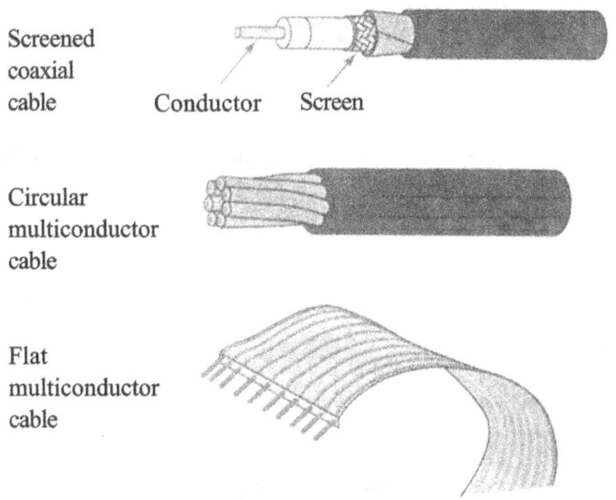

Fig. A. 5.2 Many different types of cables are used with electronics equipment.

insulating material. Cables that have to carry large currents must have relatively thick conductors. Where very small currents are involved, the conductor may be a thin strip of copper or even an aluminium film. Between these two extremes, a wide range of conductors exist for various applications. Three different types of cables used in electronics equipment are illustrated in Fig. A. 5.2 conductor and a circular plaited conducting screen, as well as an outer insulating sheath. The other two are multiconductor cables, one circular, and one flat.

Because each conductor has a finite resistance, a current passing through it causes a voltage drop from one end of the conductor to the other (Fig. A. 5.3). When conductors are long and/or carry large currents, the conductor voltage drop may cause unsatisfactory performance of the equipment supplied. Power ($I^2 R$) is also dissipated in every current-carrying conductor, and this is, ofcourse, wasted power.

$$+ \xleftarrow{\hspace{3cm}} E \xrightarrow{\hspace{3cm}} -$$

$$I \rightarrow$$

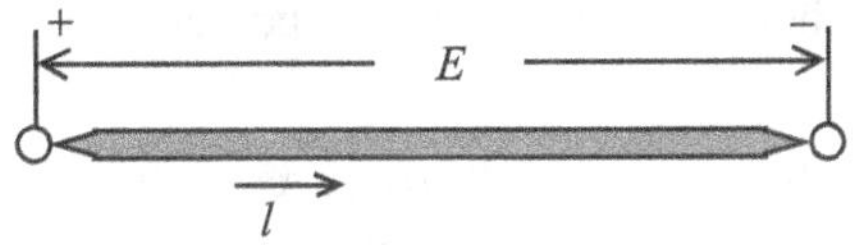

(a) Current flow through a conductor produces a voltage drop along the conductor

$$\xleftarrow{+} E \xrightarrow{-}$$

$$I \rightarrow \quad R$$

Conductor resistance

Cable resistance

$$R = \frac{E}{I}$$

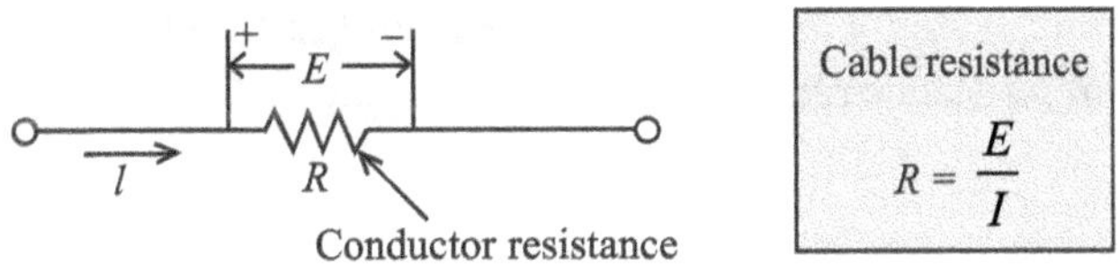

(b) Conductor resistance causes voltage drop when a current flows

Fig. A. 5.3 Conductor resistance (R) is determined by applying the voltage drop and current level to Ohm's law. The resistance per unit length (R/l) is then used to select a suitable wire gauge.

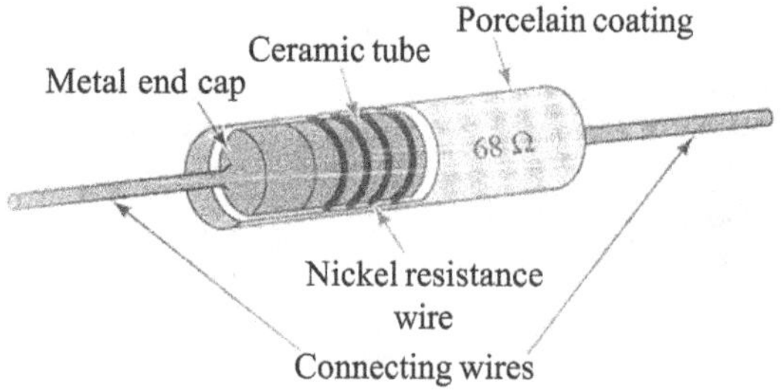

(a) Wire-wound resistor

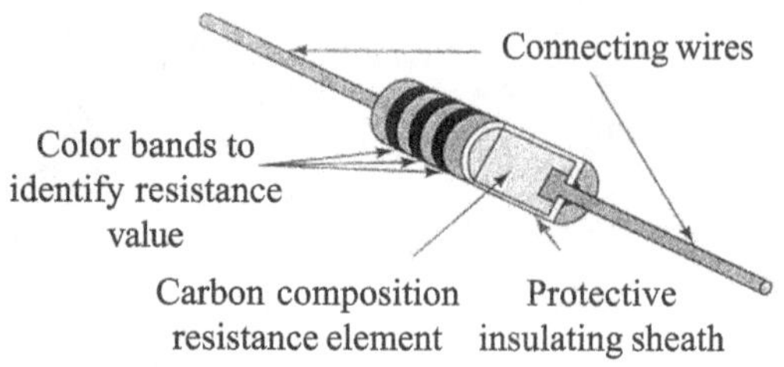

(b) Carbon composition resistor

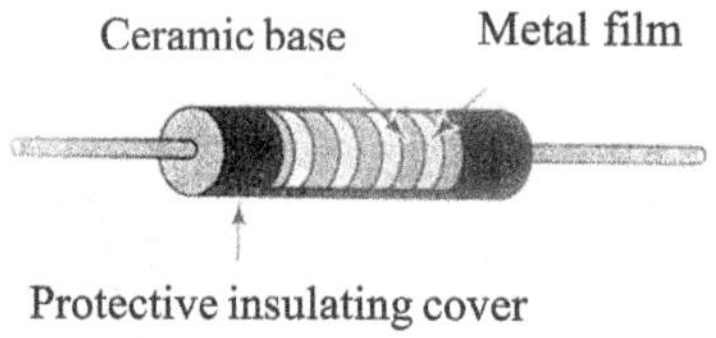

(c) Metal film resistor

Fig. A. 5.4 Individual resistors are typically wire-wound or carbon composition construction. Wirewound resistors are used where high power dissipation is required. Carbon composition type is the least expensive. Metal film resistance values can be more accurate than carbon composition type.

The illustration in Fig. A. 5.5(a) shows a coil of closely wound insulated resistance wire formed into partial circle. The coil has a low-resistance terminal at each end, and a third terminal is connected to a movable contact with a shaft adjustment facility. The movable contact can be set to any point on a connecting track that extends over one (unisulated) edge of the coil. Using the adjustable contact, the resistance from either end terminal to the center terminal may be adjusted from zero to the maximum coil resistance.

Another type of variable resistor, known as a decade resistance box, is shown in Fig. A. 5.5(c). This is a laboratory component that contains precise values of switched series-connected resistors. As illustrated, the first switch (from the right) controls resistance values in 1Ω steps from 0Ω to 9Ω and the second switches values of 10Ω, 20Ω, 30Ω, and so on. The decade box shown can be set to within + 1Ω of any value from 0Ω to 9999Ω. Other decade boxes are available with different resistance ranges.

Resistor Tolerance

Standard (fixed-value) resistors normally range from 2.7Ω to 22MΩ. The resistance tolerances on these standard values are typically + 20%, + 10%, + 5% or + 1%. A tolerance of + 10% on a 100Ω resistor means that the actual resistance may be as high as 100Ω + 10% (i.e., 110Ω) or as low as 100Ω – 10% (i.e., 90Ω). Obviously, the resistors with the smallest tolerance are the most accurate and the most expensive.

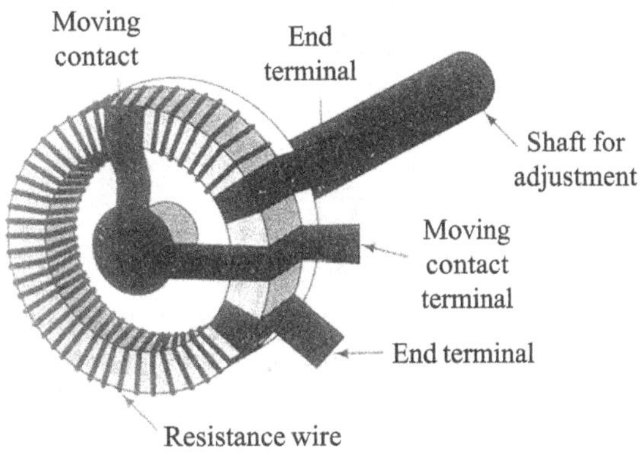

(a) Typical construction of a resistor variable resistor (and potentiometer)

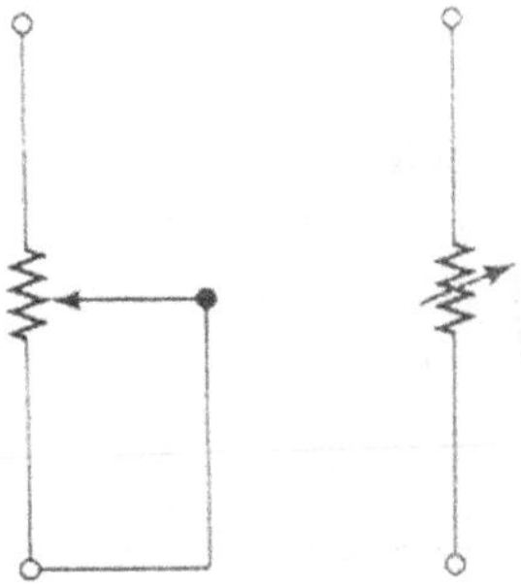

(b) Circuit symbols for a variable resistor

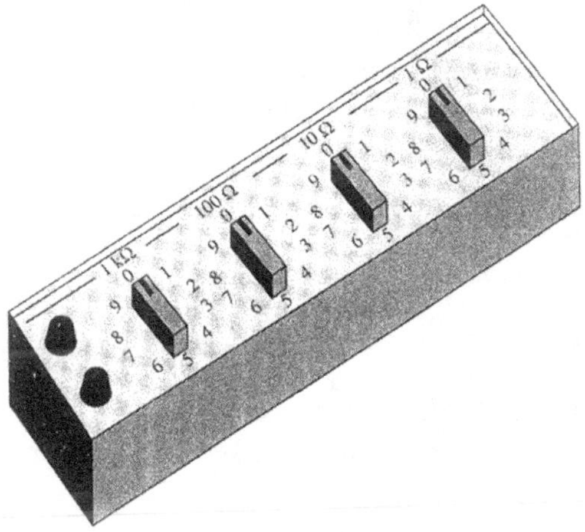

(c) Decade resistance box

Fig. A. 5.5 *Small variable resistors are used in electronic circuit construction. Large decade resistance boxes are employed in electronics laboratories.*

More Resistors

14 pin dual-in-line package

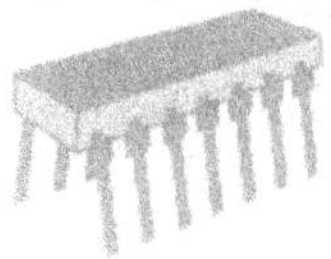

Internal resistor arrangement

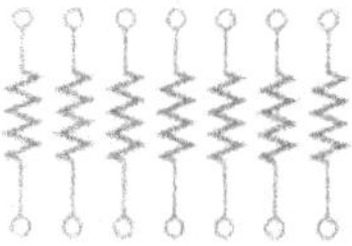

Resistor Networks

Resistor networks are available in *integrated circuits* type *dual-in-lin* package. One construction method uses a *thick film* technique in which conducting solutions are deposited in the required form.

Photoconductive Cell

This is simply a resistor constructed of photoconductive material (cadmium selenide or cadmium sulfide). When dark, the cell resistance is very high. When illuminated, the resistance decreases in proportion to the level of illumination.

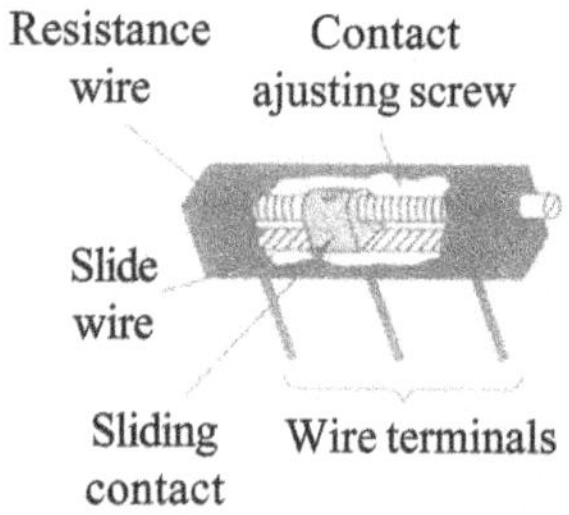

Low Power Variable Resistor

A small variable resistor suitable for mounting directly on a circuit board. A threaded shaft, which is adjustable by a screwdriver, sets the position of the moving contact on a resistance wire.

High Power Resistor

High power resistors are usually wire-wound on the surface of a ceramic tube. Air flow through the tube helps to keep the resistor from overheating.

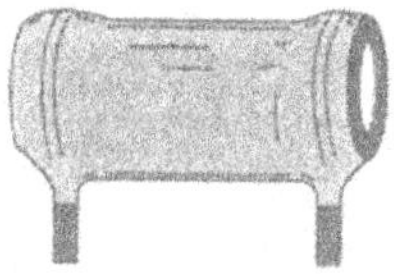

Two memory aids for determining the direction of the magnetic flux around a current-carrying conductor are shown in Fig. A. 5.6. The right-hand-screw rule as illustrated in Fig. A. 5.6(a) shows a wood screw being turned clockwise and progressing into a piece of wood. The horizontal direction of the screw is analogous to the direction of current in a conductor, and the circular motion of the screw shows the direction of magnetic flux around the conductor. In the right-hand rule, illustrated in Fig. A. 5.6(b), a right hand is closed around a conductor with the thumb ointing in the (conventional) direction of current flow. The fingers point in the direction of the magnetic lines of force around the conductor.

Because a current-carrying conductor has a magnetic field around it, when two current-carrying conductors are brought close together there will be interaction between the fields. Fig. A. 5.7(a) shows the effect on the fields when two conductors carrying in opposite direction are adjacent. The directions of the magnetic passes through the center of the coil. Therefore, the one-turn coil acts like a little magnet and has a magnetic field with an identifiable N pole and S pole. Instead of a single turn, the coil may have many turns, as illustrated in Fig. A. 5.7(c). In this case the flux generated by each of the individual current-carrying turns tends to link up and pass out of one end of the coil and back into the other end. This type of coil, known as a solenoid, obviously has a magnetic field pattern very similar to that of a bar magnet.

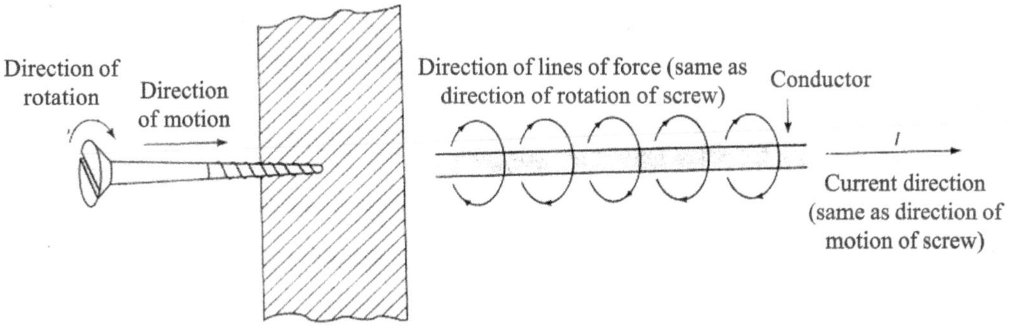

(a) Right-hand screw rule

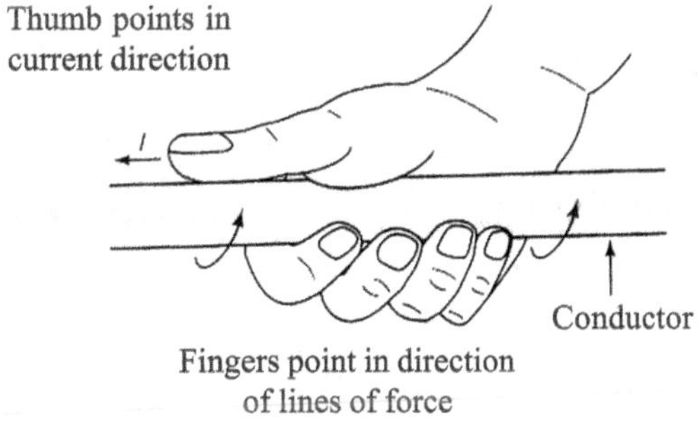

(b) Right-hand rule

Fig. A. 5.6 The right-hand-screw rule and the right-hand rule can be used for determining the direction of the magnetic lines of force around a current-carrying conductor.

The right-hand rule for determining the direction of flux from a solenoid is illustrated in Fig. A. 5.7(d). When the solenoid is gripped with the right hand so that the fingers are pointing in the direction of current flow in the coils, the thumb points in the direction of the flux (i.e., toward the N-pole end of the solenoid).

Electromagnetic Induction

It has been demonstrated that a magnetic flux is generated by an electric current flowing in a conductor. The converse is also possible; that is, a magnetic flux can produce a current flow in a conductor.

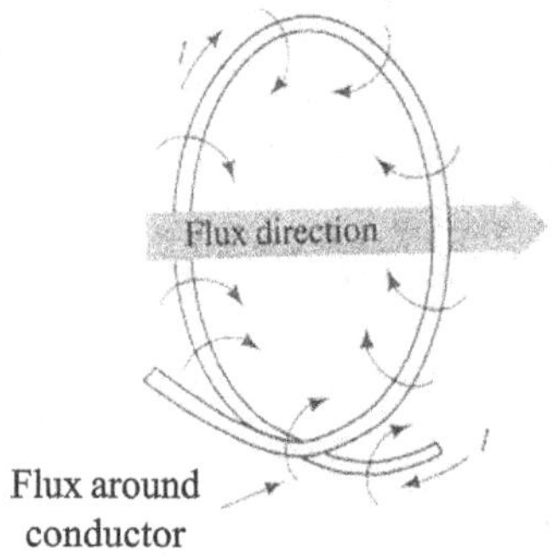

(a) All of the flux passes through the center of the coil turn

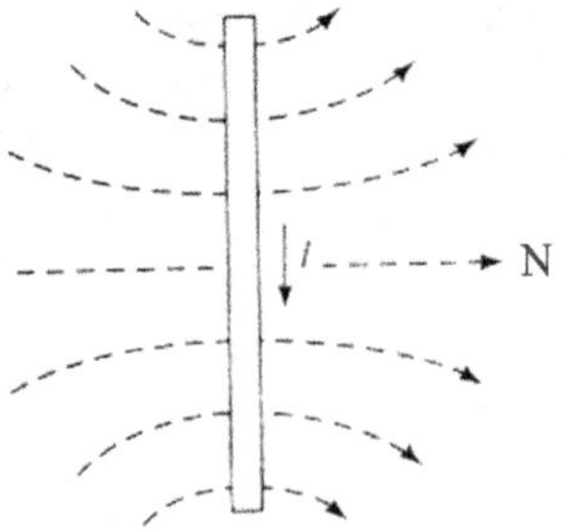

(b) Side view of coil turn and flux

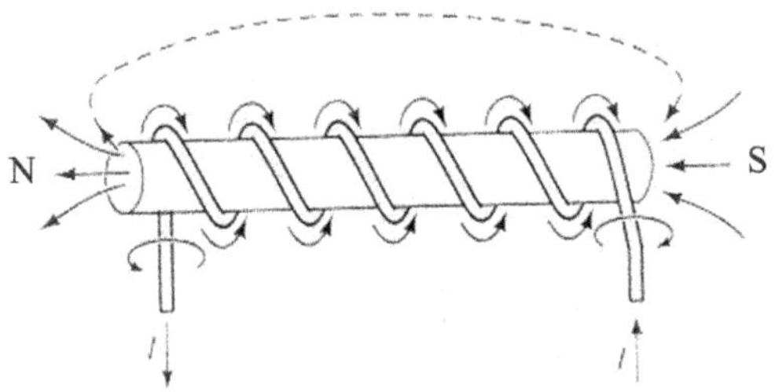

(c) A solenoid sets up a flux like a bar magna

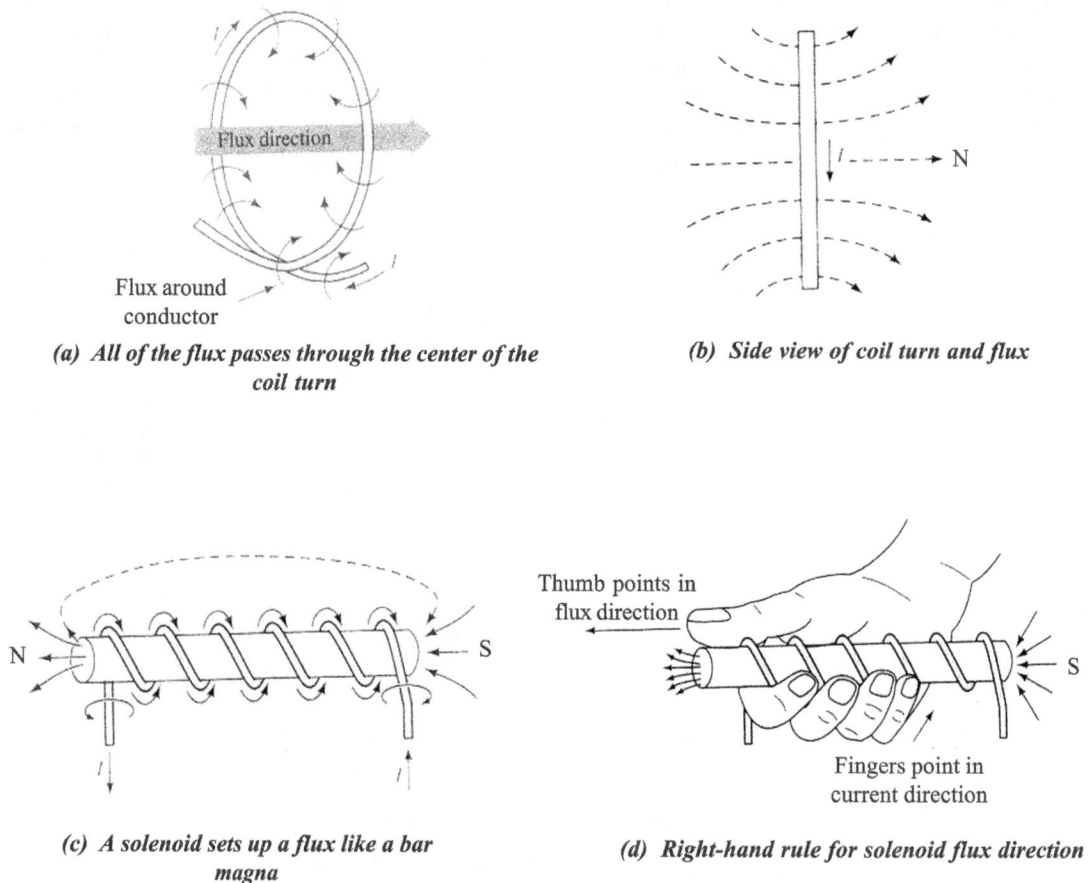

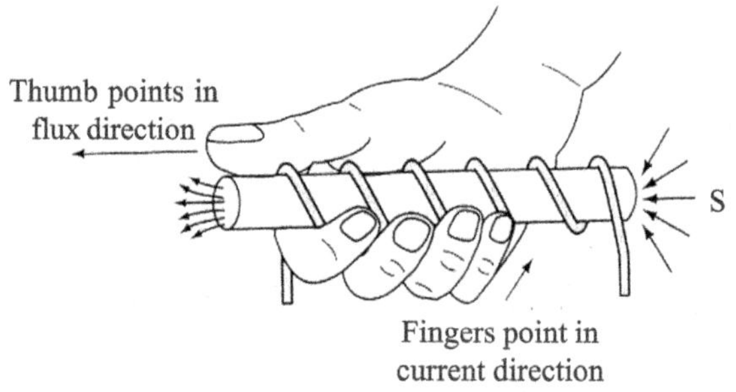

(d) Right-hand rule for solenoid flux direction

Fig. A. 5.7 In current-carrying coils, the magnetic lines of force around the conductors all pass through the center of the coil.

Consider Fig. A. 5.8(a), in which a handled bar magnet is shown being brought close to a coil of wire. As the bar magnet approaches the coil, the flux from the magnet *brushes across* the coil conductors or cuts the conductors. This produces a current flow in the conductors proportional to the total flux that cuts the coil. If the coil circuit is closed by a resistor (as shown broken in the figure), a current flows. Whether or not the circuit is closed, an *electromotive force* (emf) can be measured at the coil terminals. This effect is known as *electromagnetic* induction.

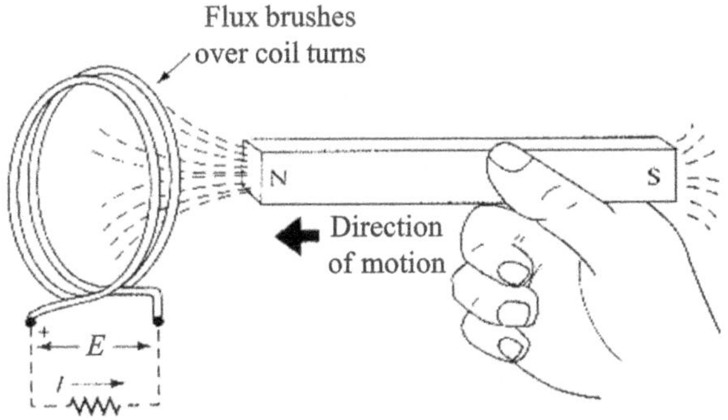

(a) emf induced in a coil by the motion of the flux from the bar magnet

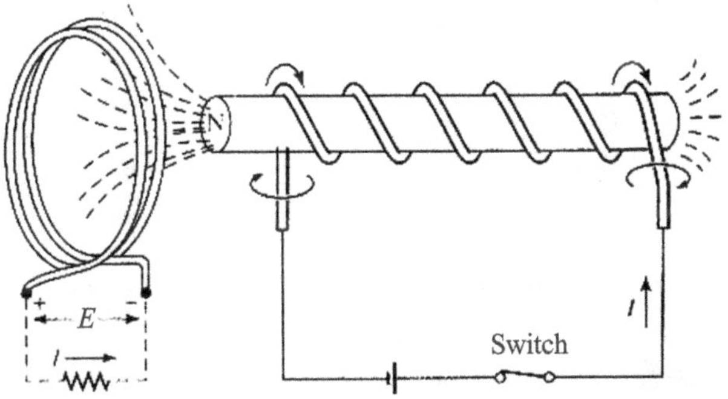

(b) emf induced in a coil by the motion of the flux from the solenoid when the current is switched on or off

Fig. A. 5.8 An electromotive force (emf) is induced in a coil when the coil is brushed by a magnetic field. The magnetic field may be from a bar magnet or from a current-carrying coil.

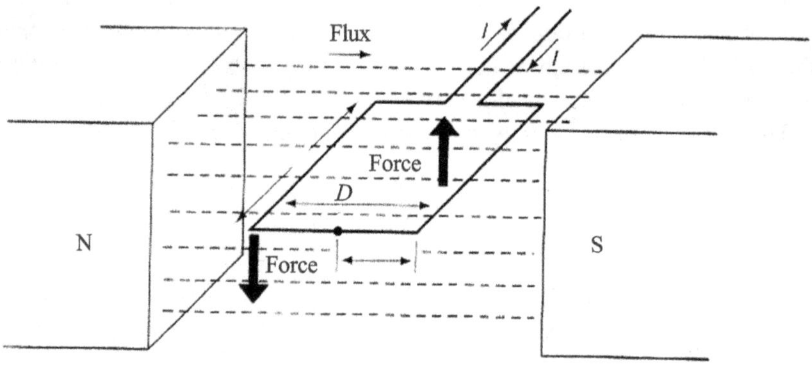

(a) Single-turn coil pivoted in a magnetic field

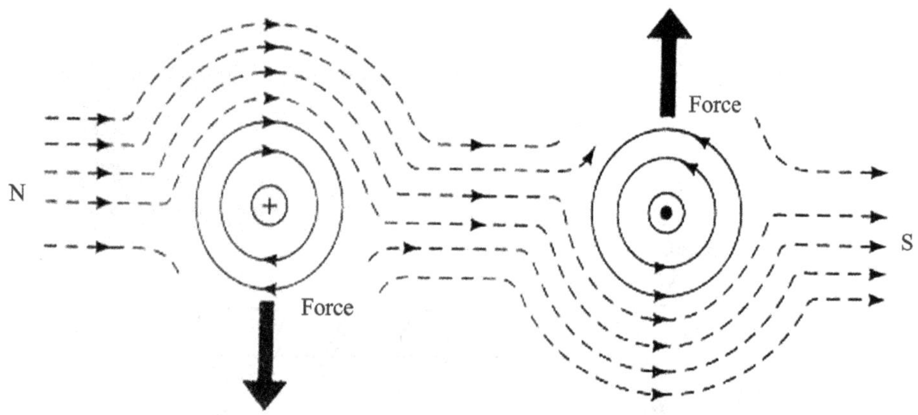

(b) Showing the force on each side of a single-turn pivoted in a magnetic field

***Fig. A. 5.9** A force is exerted on each side of a current-carrying coil pivoted in a magnetic field. This force tends to cause the coil to rotate*

Fibre Optic Cables :

Light and Other Rays

Light rays or waves are a type of energy called electromagnetic energy. They shine out, or radiate, from their source, so they are called electromagnetic radiation. Light rays are just a tiny part of a huge range of rays and waves called the electromagnetic spectrum (EMS)

The parts of the electromagnetic spectrum have different wavelengths and different names.

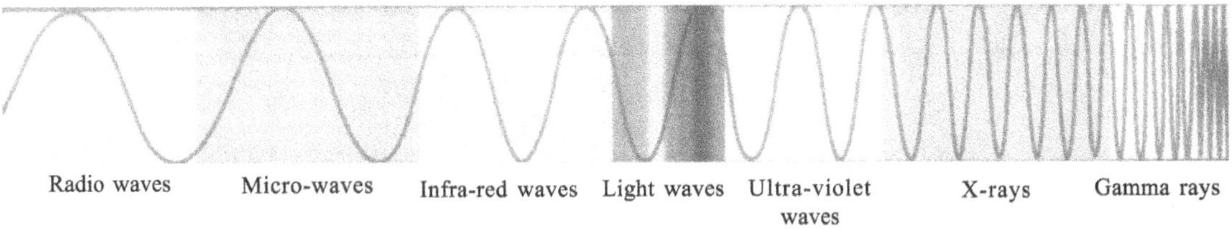

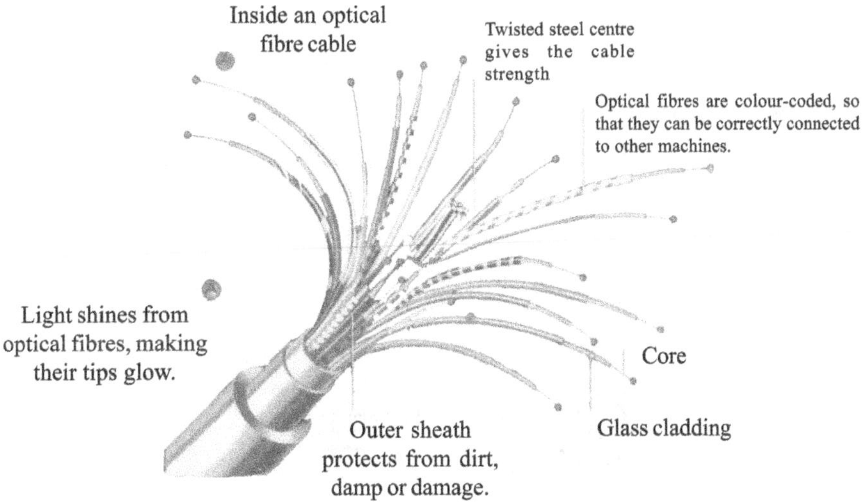

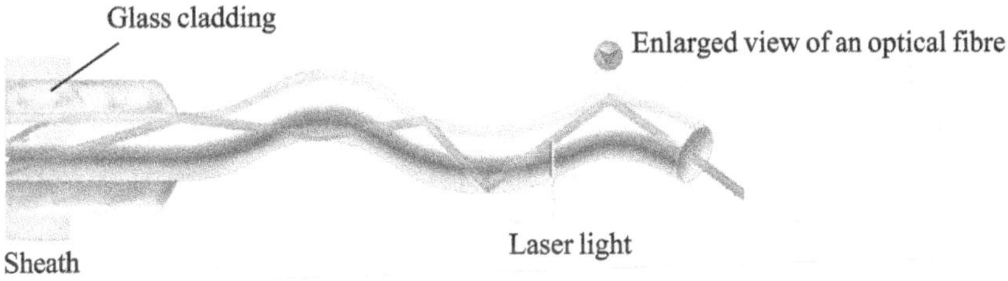

APPENDIX - 6

Circuit Symbols

dc voltage source or single-cell battery

Multicell battery

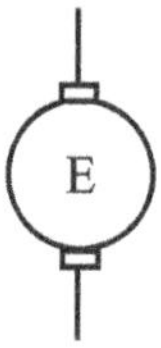

Generator

Current source

ac voltage source

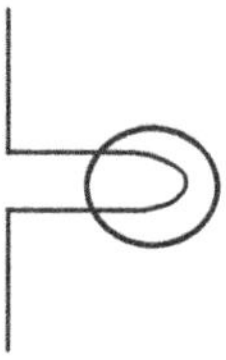

Lamp

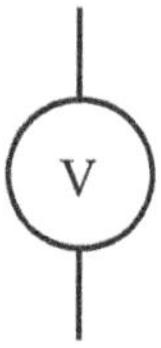

Voltmeter

Ammeter

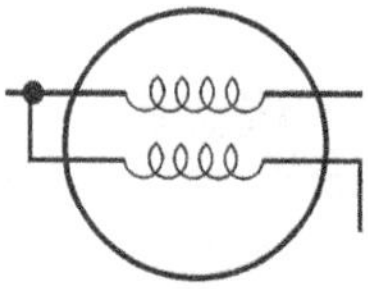

Wattmeter

Chassis

Ground

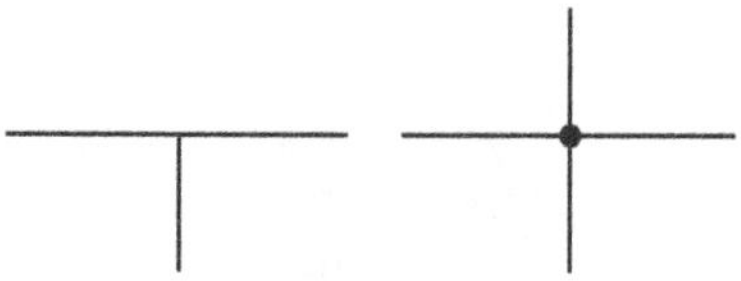

Conductor connection or junction

Unconnected crossing conductors

Fuse

dc voltage source or single-cell battery

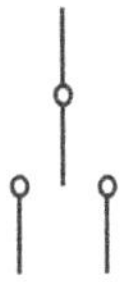
Two-way switch

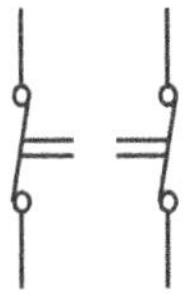
Double-pole switch

Resistor

Variable resistor

Potentiometer

Capacitor

Variable capacitor

Air-cored inductor

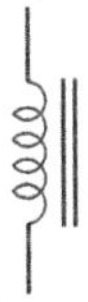
Iron-cored inductor

Variable inductor

APPENDIX - 7

Unit Conversion Factors

The following factors may be used for conversion between non-SI units and SI units.

To Convert	To	Multiply By
Area Units		
acres	square meters (m^2)	4047
acres	hectares (ha)	0.4047
circular mils	square meters (m^2)	5.067×10^{-10}
square feet	square meters (m^2)	0.0909
square inches	square centimeters (cm^2)	6.452
square miles	hectares (ha)	259
square miles	square kilometers (km^2)	2.59
square yards	square meters (m^2)	0.8361
Electric and Magnetic Units		
amperes/inch	amperes/meter (A/m)	39.37
gauses	teslas (T)	10^{-4}
gilberts	ampere (turns) (A)	0.7958
lines/sq. inch	teslas (T)	1.55×10^{-5}
Maxwells	webers (Wb)	10^{-8}
mhos	Siemens (S)	1
Oersteds	amperes/meter	79.577
Energy and Work Units		
Btu	joules (J)	1054.8
Btu	kilowatt-hours (kWh)	2.928×10^{-4}
ergs	joules (J)	10^{-7}
ergs	kilowatt-hours (kWh)	0.2778×10^{-13}
foot-pounds	joules (J)	1.356
foot-pounds	kilogram meters (kgm)	0.1383

To Convert	To	Multiply By
Force Units		
dynes	grams (g)	1.02×10^{-3}
dynes	newtons (N)	10^{-5}
pounds	newtons (N)	4.448
poundals	newtons (N)	0.1383
grams	newtons (N)	9.807×10^{-3}
Illumination Units		
foot-candles	lumens/cm^2	10.764

To Convert	To	Multiply By
Linear Units		
angstroms	meters (m)	1×10^{-10}
feet	meters (m)	0.3048
fathoms	meters (m)	1.8288
inches	centimeters (cm)	2.54
microns	meters (m)	10^{-6}
miles (nautical)	kilometers (km)	1.853
miles (statute)	kilometers (km)	1.609
mils	centimeters (cm)	2.54×10^{-3}
yards	meters (m)	0.9144
Power Units		
horsepower	watts (W)	745.7
Pressure Units		
atmospheres	kilograms/sq. meter (kg/m^2)	10 332
atmospheres	kilopascals (kPa)	101.325
bars	kilopascals (kPa)	100
bars	kilograms/sq.meter (kg/m^2)	1.02 × 10–4
pounds/sq. foot	kilograms/sq.meter (kg/m^2)	4.882
pounds/sq. inch	kilograms/sq.meter (kg/m^2)	703
Temperature Units		
degrees Fahrenheit (°F)	degrees celsius (°C)	(°F – 32)/1.8
degrees Fahrenheit (°F)	degrees kelvin (K)	273.15 + (°F – 32)/1.8

Velocity Units		
miles/hour (mph)	kilometers/hour (km/h)	1.609
knots	kilometers/hour (km/h)	1.853
Volume Units		
bushels	cubic meters (m^3)	0.035 24
cubic feet	cubic meters (m^3)	0.028 32
cubic inches	cubic centimeters (cm^3)	16.387
cubic inches	liters (1)	0.016 39
cubic yards	cubic meters (m^3)	0.7646
gallons (U.S.)	cubic meters (m^3)	3.7853×10^{-3}
gallons (imperial)	cubic meters (m^3)	4.546×10^{-3}
gallons (U.S.)	liters (1)	3.7853
gallons (imperial)	liters (1)	4.546
gills	liters (1)	0.1183
pints (U.S.)	liters (1)	0.4732
pints (imperial)	liters (1)	0.5683

Gauge	Diameter (mm)	Copper Wire Resistance (Ω/km)	Diameter (mil)	Copper Wire Resistance (Ω/km)
36	0.127	1360	5	415
37	0.113	1715	4.5	523
38	0.101	2147	4	655
39	0.090	2704	3.5	832
40	0.080	3422	3.1	1044

To Convert	To	Multiply By
quarts (U.S.)	liters (1)	0.9463
quarts (imperial)	liters (1)	1.137
Weight Units		
ounces	grams (g)	28.35
pounds	kilograms (kg)	0.453 59
tons (long)	kilograms (kg)	1016
tons (short)	kilograms (kg)	907.18

The siemens* is the unit of conductance.

$$\text{conductance} = \frac{1}{\text{resis tan ce}}$$

APPENDIX - 8

American Wire Gauge Sizes and Metric Equivalents

Gauge	Diameter (mm)	Copper wire Resistance (Ω/km)	Diameter (mill)	Copper wire Resistance (Ω/1000 ft)
0000	11.68	0.160	460	0.049
000	10.40	0.203	409.6	0.062
00	9.266	0.255	364.8	0.078
0	8.252	0.316	324.9	0.098
1	7.348	0.406	289.3	0.124
2	6.543	0.511	257.6	0.156
3	5.827	0.645	229.4	0.197
4	5.189	0.813	204.3	0.248
5	4.620	1.026	181.9	0.313
6	4.115	1.29	162	0.395
7	3.665	1.63	144.3	0.498
8	3.264	2.06	128.5	0.628
9	2.906	2.59	114.4	0.792
10	2.588	3.27	101.9	0.999
11	2.30	4.10	90.7	1.26
12	2.05	5.20	80.8	1.59
13	1.83	6.55	72	2
14	1.63	8.26	64.1	2.52
15	1.45	10.4	57.1	3.18
16	1.29	13.1	50.8	4.02
17	1.15	16.6	45.3	5.06
18	1.02	21.0	40.3	6.39

Contd...

19	0.912	26.3	35.9	8.05
20	0.813	33.2	32	10.1
21	0.723	41.9	28.5	12.8
22	0.644	52.8	25.3	16.1
23	0.573	66.7	22.6	20.3
24	0.511	83.9	20.1	25.7
25	0.455	106	17.9	32.4
26	0.405	134	15.9	41
27	0.361	168	14.2	51.4
28	0.321	213	12.6	64.9
29	0.286	267	11.3	81.4
30	0.255	337	10	103
31	0.227	425	8.9	130
32	0.202	537	8	164
33	0.180	676	7.1	206
34	0.160	855	6.3	261
35	0.143	1071	5.6	329

MCQs from Gate Examination, Yearwise from 1993 to 2014

One Mark Questions

1. In a voltage –voltage feedback as shown, below, which one of the following statement is true if the gain K is increased (gate-13).

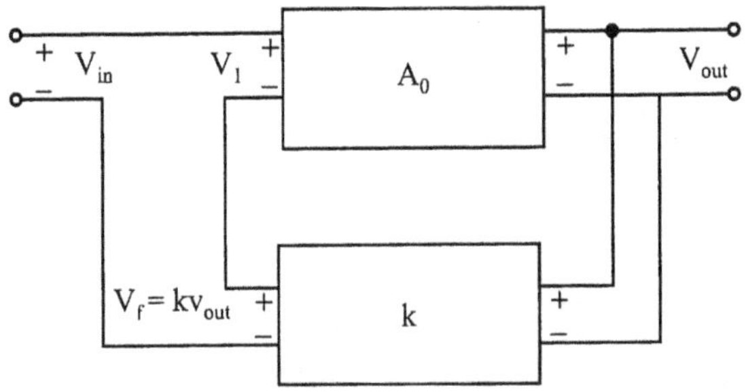

   (a) The input impedance increase and output impedance decreases
   (b) The input impedance increases and output impedances also increases
   (c) The input impedance decreases and output impedance decreases
   (d) The input impedance decreases and output impedance increases

2. To obtain very high input and output impedance decreases in a feedback amplifier, the mostly used is (gate-95)

   (a) Voltage series (b) Current series
   (c) Voltage shunt (d) Current shunt

3. In a shunt-shunt negative feedback amplifier as compared to the basic amplifier (gate-98)
   (a) Both input and output impedance decreases
   (b) Input impedance decreases but output impedance increases
   (c) Input impedance increases but output impedance decreases
   (d) Both input and output impedance increases

4. In a transistor push-pull amplifier (gate-93)
   (a) There is no dc present in the output
   (b) There is no distortion in the output
   (c) There is no even harmonics in the output
   (d) There is no odd harmonics in the output

5. A class-A transformer coupled, transistor power amplifier is required to delivered a power output of 10 Watts. The maximum power rating of the transistor should not be less than (gate-94)
   (a) 5 W (b) 10 W (c) 20 W (d) 40 W

6. Crossover distortion behavior is characteristics of (gate-99)
   (a) Class A output stage
   (b) Class B output stage
   (c) Class AB output stage
   (d) Common-base output stage

7. A cascade connection of two voltage amplifiers A_1 and A_2 is shown in the figure. The open loop gain A_{V0}, input resistance R_{in}, and output resistance R_o for A_1 and A_2 are as follows:
   A_1: A = 10, R = 10 kΩ, R = 1 kΩ
   A_2: A = 5, R = 5 kΩ, R = 200 Ω
   The approximate overall voltage gain V_{out}/V_{in} is __________. (gate-14)

8. A good current buffer has (gate-14)
   (a) Low input impedance and low output impedance
   (b) Low input impedance and high output impedance
   (c) High input impedance and low output impedance
   (d) High input impedance and high output impedance

9. In a multi-stage R-C coupled amplifier the coupling capacitor (gate-93)
   (a) Limits the low frequency response
   (b) Limits the high frequency response
   (c) Does not affect the frequency response
   (d) Blocks the d.c. component without affecting the frequency response

10. A good current buffer has (gate-93)
   (a) Low input impedance and low output impedance
   (b) Low input impedance and high output impedance
   (c) High input impedance and low output impedance
   (d) High input impedance and high output impedance

11. The desirable characteristics of a transconductance amplifier are (gate-94)
   (a) High input resistance and low output resistance
   (b) High input resistance and low output resistance
   (c) Low input resistance and high output resistance
   (d) Low input resistance and low output resistance

ANSWERS

(1). a	(2). b	(3). a	(4). a or c	(5). c	(6). b
(7). 34.722	(8). b	(9). a	(10). b	(11). a	

Two Marks Questions

1. The most commonly used amplifier in sample & hold circuit is(gate-00)
   (a) A unity gain of inverting amplifier
   (b) A unity gain of non- inverting amplifier
   (c) An inverting amplifier with a gain of 10
   (d) An inverting amplifier with a gain of 100

2. In a negative feedback amplifier using voltage-service (i.e., voltage sampling, service mixing) feedback (gate-02).
   (a) R_i decreases & and R_0 decreases
   (b) R_i decreases & and R_0 increases
   (c) R_i increases & and R_0 decreases
   (d) R_i increases & and R_0 increases

3. An amplifier without feedback has a voltage gain of 50, input resistance of 1 kΩ of output resistance of 2.5 Ω. The input resistance of current-shunt negative feedback amplifier using the above amplifier with a feedback factor of 0.2 is (gate-03).
   (a) $\frac{1}{11}$ kΩ (b) $\frac{1}{5}$ kΩ (c) 5 kΩ (d) 11 kΩ

4. Voltage series feedback (also called series shunt feedback) result in (gate -04).
   (a) Increases in both input & output impedance
   (b) Decreases in both input & output impedance
   (c) Increases in both input impedance & decreases output impedance
   (d) Decreases in both input impedance & increases output impedance

5. The effect of current shunt feedback in an amplifier to (gate-05).
   (a) Increases the input resistances & decreases the output resistance
   (b) Increases both input & output resistance
   (c) Decreases both input & output resistance
   (d) Decreases the input resistances & increases the output resistance

6. Consider the common-collector amplifier in the figure (bias circuitry ensures that the transistor operates in forward active region, but has been omitted for simplicity). Let I_c be the collector current, V_{BE} be the base-emitter voltage and V_T be the thermal voltage. Also, g_m and r_o are the small-signal transconductance and output resistance of the transistor, respectively.
   Which one of the following conditions ensures a nearly constant small signal voltage gain for a wide range of values of R_E? (gate-14)

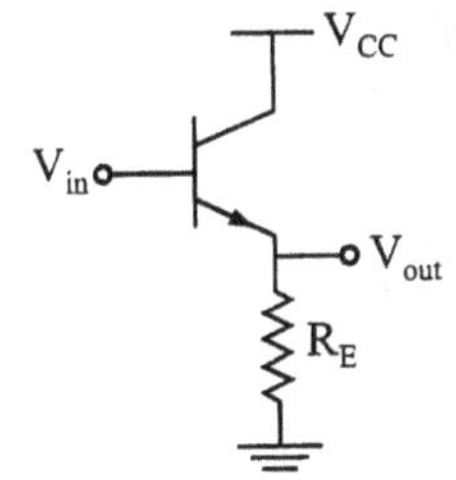

(a) $g_m R_E << 1$ (b) $I_c R_E >> V_T$ (c) $g_m r_o >> 1$ (d) $V_{BE} >> V_T$

7. For the common collector amplifier shown in the figure, the BJT has high β, negligible V_{CE} (sat), and V_{BE} = 0.7 V. The maximum undistorted peak-to-peak output voltage V_0 (in Volts) is _______. (gate-14)

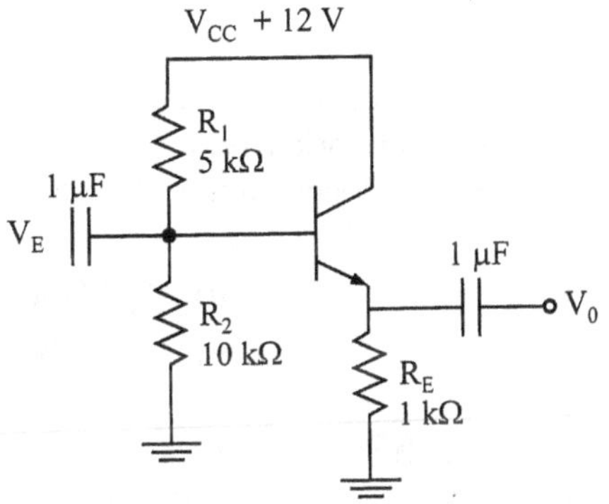

8. In case of class A amplifiers the ratio (efficiency of transformer coupled amplifier)/(efficiency of transformer less amplifier) is (gate-87)

(a) 2.9 (b) 1.36 (c) 1.0 (d) 0.5

A regulated power supply, shown in figure below, has an unregulated input (UR) of 15 Volts and generates a regulated output V_{out}.

Use the component values shown in the figure.

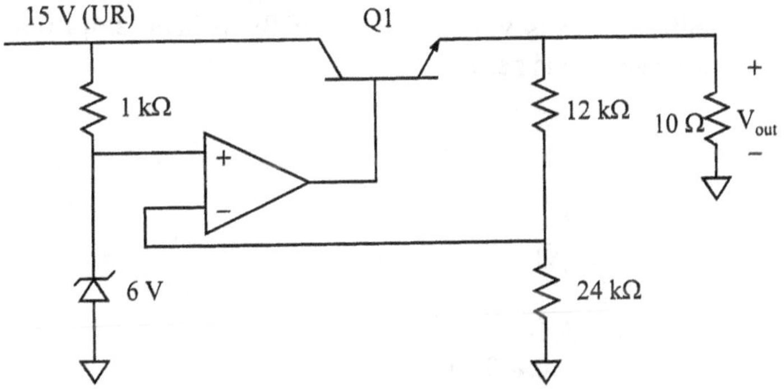

In the figure above, the ground has been shown by the symbol ∇

9. The power dissipation across the transistor Q1 shown in the figure is: (gate-06)
   (a) 4.8 Watts (b) 5.0 Watts (c) 5.4 Watts (d) 6.0 Watts

10. If the unregulated voltage increases by 20%, the power dissipation across the transistor Q1 (gate-06)
    (a) Increases by 20% (b) Increases by 50%
    (c) Remains unchanged (d) Decreases by 20%

11. Negative feedback in (gate-97)
    1. Voltage series configuration
    2. Current shunt configuration
    (a) Increases input impedence (b) Decreases input impedence
    (c) Increases closed loop gain (d) Leads to oscillation

12. An amplifier has an open-loop gain of 100, an input impedance of 1 kΩ, and an output impedance of 100 Ω. A feedback network with a feedback factor of 0.99 is connected to the amplifier in a voltage series feedback mode. The new input and output impedance respectively are (gate-99)
    (a) 10 kΩ and 1 Ω (b) 10 Ω and 10 Ω
    (c) 100 kΩ and 1 Ω (d) 100 kΩ and 1 kΩ

ANSWERS

(1). b	(2). c	(3). a	(4). c	(5). d	(6). b
(7). 9.39 to 9.41	(8). b	(9) b	(10) d	(11) 1.a 2.b	(12) c

Answers with Explanation

1. Ans (b)
   The sample and hold circuit is a non-inverting amplifier with unity gain.

2. Ans(c)
   Voltage-series feedback means voltage amplifier. In voltage amplifier input resistance is infinite and output resistance is zero.

3. Ans(a)
   Given $A_v = 50$

$$R_i = 1\ k\Omega$$
$$R_o = 2.5\ k\Omega$$
$$\beta = 0.2$$
$$D = 1 + A\beta$$
$$= 1 + (50 \times 0.2)$$
$$= 11$$

In current shunt (or) current amplifier: R_i decreases and R_o is increases

$$R_i = 1/11\ k\Omega$$

4. Ans(c)
   Voltage series feedback is a voltage amplifier. Input impedance increases and output impedance decreases.

5. Ans(d)
   Current shunt feedback is also called as current amplifier. So Input resistance decreases and output resistance increases.

6. Ans(a)

$$V_o = (1 + h_{fe})\, I_B R_E$$

$$V_{in} = h_{ie} I_B + (1 + h_{fe})\, I_B R_E$$

$$A_v = \frac{V_o}{V_{in}} = 1$$

For wide range of R_E if A_v has to be constant the condition $(1+h_{fe})R_E >> h_{ie}$

$$(1 + h_{fe})\, R_E >> h_{ie}$$

$$(1 + \beta)\, R_E >> \frac{\beta}{g_m}$$

$$g_m = \frac{I_c}{V_t}$$

$$R_E >> 1/g_m$$

$$R_E >> V_t / I_c$$

$$I_c R_E >> V_t$$

7. Ans (9.4 V)

$$V_B = \frac{12\ k}{10\ k + 5\ k}(10\ k) = 8\ V$$

$$V_{o(DC)} = V_B - 0.7 = 8 - 0.7 = 7.3\ V$$

But the transfer characteristics is restricted between ground and V_{cc} is between 0 and 12 V
The peak to peak maximum voltage with symmetrical swing is 2(4.7) = 9.4 V

8. Ans (b)

$$\text{Class-A amplifier ratio} = \frac{\text{efficiency of transformer coupled amplifier}}{\text{efficiency of transformer less amplifier}}$$

$$= \frac{50\%}{25\%} = 2$$

9. Ans (d)

The voltage across 24 k = 6 V

Due to virtual ground concept

So voltage across 12 k = 3 V

$$V_{out} = 6 + 3 = 9\text{ V}$$

$$V_{CE} = 15 - V_{out}$$

$$= 15 - 9 = 6\text{V}$$

$$I_c = \frac{V_{out}}{12} + \frac{V_{out}}{(24+12)}$$

$$= \frac{9}{12} + \frac{9}{36} = 1\text{ A}$$

$$\text{Power (P)} = V_{CE} \times I_c$$

$$= 6\text{ V} \times 1\text{ A}$$

$$P = 6\text{ W}$$

10. Ans (b)

Unregulated voltage increases by 20%

i.e., New regulated voltage = 18 V

$$V_{CE} = 18 - 9 = 9\text{ V}$$

$$I_c = 1\text{ A}$$

$$P = 9 \times 1 = 9\text{ V}$$

$$\%\text{ increase} = (9 - 6/6) \times 100$$

$$= 50\%$$

11. Ans 1.(a), 2.(b)

In voltage series configuration, input impedance increases and output impedance decreases. Current shunt configuration, input impedance decreases and output impedance increases.

12. Ans (c)

Given open loop gain A = 100

Feedback factor β = 0.99

Input impedance = 1 kΩ

Output impedance = 100 kΩ

For voltage series feedback is

$$Z_{if} = Z_i(1 + A\beta)$$

$$= 1\text{ k}(1 + 100 \times 0.99)$$

$$Z_{if} = 100\text{ k}\Omega$$

$$Z_{of} = \frac{Z_o}{1+A\beta} = \frac{100}{1+99} = 1\ \Omega$$

Index

A

B

C

Q

R

S

T

V

W

Z

Bibliography

1. Electronic Devices and Circuits

   - J. Millman, C.C Halkias, McGraw Hill

2. Micro-electronics

   - J. Millman, C.C Halkias, McGraw Hill

3. Inegrated Electronics

   - J. Millman, C.C Halkias, 2008 TMH

4. Microelectric Circuits

   Sedra and Samith 5th Ed. 2009 Oxford University Press

www.ingramcontent.com/pod-product-compliance
Lightning Source LLC
Chambersburg PA
CBHW081458190326
41458CB00015B/5276